Social Solutions

for Climate Change

*How to inspire action
through social media*

Sherry Nouraini, PhD

Social Solutions for Climate Change: How to inspire action through social media

Published by Science Outreach Press

Vista, California, USA

Cover design by Suzanne Santillan

Book design by Andrea Leljak

Editing and book production by eFrog Press

First edition

Publisher's Cataloging-in-Publication Data
provided by Five Rainbows Cataloging Services

Names: Nouraini, Sherry.
Title: Social solutions for climate change : how to inspire action through social media / Sherry
    Nouraini, PhD.
Description: Vista, CA : Science Outreach Press, 2016. | Includes bibliographical references.
Identifiers: LCCN 2016913098 | ISBN 978-0-9979193-0-1 (pbk.) | ISBN 978-0-9979193-1-8 (ebook)
Subjects: LCSH: Communication in climatology. | Communication in the environmental sciences--
    Social aspects. | Climatic changes--Social aspects. | Environmental policy--Public opinion. |
    BISAC: SCIENCE / Global Warming & Climate Change. | SCIENCE / Philosophy & Social Aspects.
Classification: LCC QC902.9 .N68 2016 (print) | LCC QC902.9 (ebook) | DDC 363.738/74--dc23.

for Earth

Contents

Preface

The arrival of the World Wide Web, blogs, and social networking sites has transformed the way we communicate. We catch up with our friends and family and share our life experiences on Facebook; we consume the latest news, organize, and have intellectual debates on Twitter; snap selfies on Instagram; and make professional connections on LinkedIn. Using the tools of social media for communicating science in general—and climate science as well as global warming in particular—has also been on the rise. Many scientists and environmental activists have taken to blogging and using the tools of social media to raise awareness about global warming, or to debunk myths about climate change and side effects of vaccinations. Finding success with these efforts has been challenging as the public continues to be disengaged with climate change issues and doubt benefits of vaccinations or other advances brought about by modern science. This resistance to accepting scientific facts stems partly from organized and well-funded climate change denial and antivaccine efforts. But climate activists and life scientists must share some of the blame for public resistance to accepting facts as they are using the new tools of communication but are not well versed in the art of communicating with the public. In addition, their communication efforts seem to be ad hoc and not informed by a sound strategy. These are the issues this book addresses.

Challenges with climate change and science communication have not escaped the attention of the science and climate change community. Science and climate change communication has become an active field of research in the hope of understanding why people don't accept and act on facts that are right in front of them. There exists a body of knowledge about effective strategies for communicating climate change with the public, policy makers, and business leaders. This body of knowledge has been informed by a fifty-year climate change communication effort, and research in science and climate change communication. Leveraging this resource, the Scripps Institution of Oceanography has taken a proactive approach to climate change mitigation by creating a Master of Advanced Studies in Climate Science and Policy. The aim of this program is to prepare educators, science policy analysts and advocates, managers in cleantech and related industries, communication specialists, journalists, and practicing scientists to engage stakeholders and decision-makers in climate action (https://scripps.ucsd.edu/masters/mas/climate-science-and-policy).

I had the privilege of developing a curriculum for—and teaching—the social media component of this program. While doing research to develop this curriculum, I realized that information necessary to create a strategic plan for communicating climate change and inspiring action was fragmented. In addition, despite a plethora of "how-to" books on social media, none were comprehensive and relevant enough to serve as a template for strategic, focused, and effective use of social media and blogging. I also reviewed how scientists and activists use social media for outreach, and saw little sign of effectiveness in their efforts. Research in conversations around climate change in the blogosphere and social media confirmed my suspicions, as you will see later. *Social Solutions for Climate Change* is my contribution to further the cause of climate change communication and mitigation.

In the pages of this book the reader will be introduced to three fictional sample profiles: Linda Goldberg, Paul Berg, and Alex Donovan. The students in the inaugural Social Media for Climate Science and Policy class created these three profiles in order to define a clear audience for their outreach efforts. These sample profiles are used throughout this book so the reader can see how to apply what he/she learns to his/her own communication efforts.

Content in *Social Solutions for Climate Change* is divided into three sections. Section 1 has been structured to first review the current state of climate change communication and what activists have learned through their efforts. Next, the reader learns how to lay a solid foundation and develop a strategy for science and climate change communication, whether through the tools of social media, or any other medium. Finally, findings from science communication, moral and social psychology, and behavioral economics research in understanding humans' decision-making will be presented, and the relevance of these findings to inspiring climate action will be discussed.

Section 2 of this book covers the technical aspects of blogging and social media. Armed with the foundation built in Section 1, the reader will learn how to write blog articles relevant to their target audience, how to search and find communities of this audience, and how to measure the effectiveness of their activities. There is a plethora of social networking platforms, and among these, I have chosen to include Twitter, Facebook, Pinterest, LinkedIn, and Instagram. Note that it is not necessary to learn and be active in all of these networks, and the choice will depend on a reader's target audience in Section 1.

In Section 3, the reader will learn how to compile everything they learned in chapters 1 through 9 into a plan of action. Most of the instructions in this book are targeted to individuals for small-scale interactions with their personal communities.

Although this book has been designed to enhance climate change communication, the methods and strategies used are applicable to any discipline that seeks to motivate a specific audience to notice, hear, understand, and take action.

Science and climate change communication strategies, as well as blogging and social media tools, are in constant flux. To be successful, there is an undisputed need for a support group, where one can get updated about new findings and technology changes, and ask questions when they arise, so I have created a Facebook group for the readers of this book. I invite you to visit the link below and request to join this group in order to network with other readers, stay up to date with new information, and get support. Visit us at http://facebook.com/groups/social4climate.

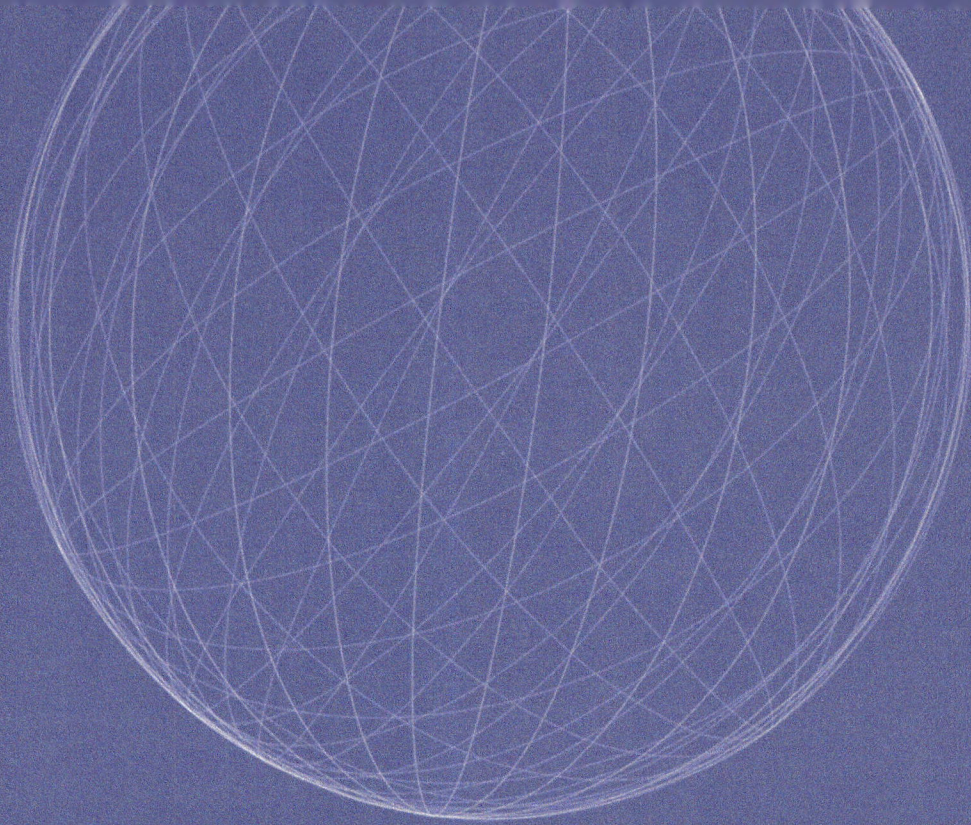

Section 1

Laying
the Foundation

Chapter 1

State of Climate Change Communication

There is near unanimous consensus among climate scientists that human activities are affecting the climate in unprecedented ways and that we need to act in order to reduce the harmful effects of global warming (1.1–1.4). Earth's global average temperature has risen by about 1.4 degrees Fahrenheit since the late-nineteenth century, mainly because of the unprecedented rapid rise in the level of CO_2 in the atmosphere. If we continue the current rate of burning fossil fuels and the resulting rapid rise in atmospheric CO_2 levels, global temperature is projected to increase another 4 to 8 degrees Fahrenheit before we reach year 2100 (1.5)!

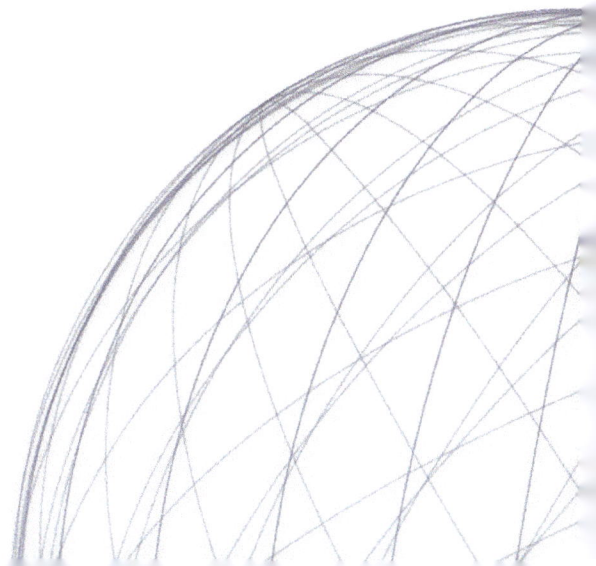

Rapid rise in temperatures has devastating consequences for our planet, and we are already seeing the effects. Arctic sea ice has been shrinking dramatically, and the rate of loss is accelerating (1.6). Glaciers have been melting rapidly, leading to sea level rise, as well as jeopardizing water supplies for billions of people worldwide (1.6). Oceans are becoming rapidly acidified as they act as a sink that absorbs the excess CO_2 we put in the atmosphere. Ocean acidification has documented deleterious effects on shelled organisms such as oysters, and will continue to harm oceanic life. (1.7). The rapid rate of global warming is changing ecological landscapes. As the planet warms, some organisms are moving up mountains, and marine species are moving to deeper depths in search for a cooler climate. (1.8–1.10). Accelerated rate of sea level rise has made storm surges higher, leading to increased coastal flooding and pushing salt water into aquifers, which are a major source of fresh water for coastal communities (1.6, 1.11). These and other consequences of global warming have dire implications for economic impact, as well as health and well-being of humans across the globe (1.12). There is urgent need for action. Unfortunately, the majority of the American public and the global population don't seem to care!

State of public opinion on climate change

Excellent work by the Yale Project on Climate Change Communication (YPCCC), and other organizations, has shed some light on the roots of the skepticism by the public. In summary, here is what we know about the public's mindset about climate change:

- **Climate change is viewed as a debate rather than a consensus** (1.13).
  A majority of the American public believes that the science around climate change is not settled. At the global level, only 4 out of 10 people are even aware of climate change, a large number of which either live in areas that are most affected by climate change, or where a great amount of CO_2 originates (1.14).

- **Trust in sources of information about climate change is scattered and not necessarily on climate scientists**. This fact may be one of the underlying reasons why climate change is still viewed as a debate, as some nonexpert opinion leaders are still publicly doubtful about occurrence of climate change itself or its underlying causes! Surveys by YPCCC (1.13) has shown that, other than climate scientists, the American public places a similar level of trust in the opinions of their family and friends, other kinds of scientists, television reporters, and their primary care doctors. The Pope also has the trust of many Americans.

With his declaration of global warming as a moral issue, more Americans in general—and the Catholic population in particular—have become aware of the impact of climate change (1.15). However, this also shows that knowledge of scientific evidence for global warming may not be the deciding factor for the public to act on climate issues. There are other forces at play, which need to be taken into consideration when communicating climate change.

- **Effects of global warming are viewed as a distant concern**. A majority of Americans feel global warming is not affecting them, their communities, or their families, and they view it as something that will harm only future generations (1.13).

- **The majority of Americans express pessimism and apathy about mitigating climate change** (1.13). The American public also correctly places the responsibility for creating change on corporations, industry, U.S. congress, the president, and citizens themselves (1.13). Interestingly, the source of this sentiment is not a lack of understanding about a need for taking action, or what needs to be done to implement change. It is just that the public has no faith that any of these groups will take the responsibility seriously.

What we know from fifty years of communicating climate change

Convincing elected officials and the public about a need for action toward climate change mitigation has been a fifty-year battle (1.16). Communication efforts by climate scientists and NGOs have taught us valuable lessons about factors that influence the public's tendency to engage in conversations about climate change and to take action toward mitigation of its consequences. These lessons have been clearly summarized in a number of freely available publications:

1. The public's attitudes about climate change are greatly influenced by individual ideology, identity, and worldview (1.17).
2. Group settings are more effective for public engagement in—and taking action on—climate change (1.17,1.18).
3. When it comes to communicating climate change, how you frame your message is more important than the content of your message (1.17).

These findings, combined with the knowledge that the public tends to place trust in their friends and families for information about climate change, position the tools of social media as one of the most, if not only, powerful avenues to engage the public and encourage action. In addition, members of the public join online forums and groups based on their own ideologies, identities, and worldviews. The online tools of social media are the modern gathering places for exchange of ideas, creating connections, and organizing collective action. Social networking sites are also where the public keeps in touch and connects with family and friends. In fact, surveys have shown that the American public cites connecting with family and friends as the primary reason they use the tools of social media (1.19). If we want to have a productive conversation with the public about climate change and how it affects the things they care about, we should be where they are, engaging them and those who influence their opinions. This awareness has not escaped climate change activists, nongovernmental organizations (NGOs), and climate scientists, leading them to take to social media to spread their messages.

Using social media and blogs for climate change communication

There is no shortage of online chatter about climate change, and the quality and effectiveness of these sound bites on influencing action is an active area of research. An excellent literature review on this topic by Mike S. Shafer (1.20) reveals a number of interesting findings on how stakeholders use online tools for climate communication:

1. NGOs and climate change deniers dominate the online climate change discussion, whereas scientists, scientific institutions, and politicians play a limited role, although the upcoming (as I write this) 2016 primary and federal elections have turned that tide recently for politicians.

2. Quality of scientific discussions around climate change tend to be rather poor, deviating from evidence-based, logical deliberations, to the extent that the blogosphere has been coined as "rantosphere" and comes across as rather uncivil by some investigators.

3. Although research shows that consumption of online content about climate change tends to correlate with the intent to have individuals modify their lifestyles, there is no evidence that it leads to people actually taking action by the public consuming this content.

Data-driven investigations of chatter about climate change on Twitter has revealed echo chambers of climate change activists on one side and deniers on the other (1.21). Similar research on sharing of content about science news versus conspiracy theories on Facebook has also revealed echo chambers of like-minded clusters (1.22). In other words, climate change and science activists may be sharing information on social media networks in an effort to raise awareness, but only those who already agree with their point of view are listening, while those who disagree are ignoring them. The underlying cause of such selective content consumption may be due to the fact that the public tends to pay attention to information that confirms and is consistent with their preexisting ideology, identity, and worldview (1.17).

Evidently, there is a need for more research on the effects of online climate change communication in driving behavior change. There is also a need for change of behavior on the part of climate scientists and activists. They need to be more active online, to raise the quality of climate change discussions, as well as modify their communication strategies in engaging the public. Instead of becoming billboards and talking *at* the public about scientific data and taking action, climate scientists and activists should find a common ground and talk with the public in small groups of niche audiences based on an overlap of values and interests (1.23). The tools of social media, if used strategically, provide a perfect opportunity to find and communicate with these niche audiences. What is needed is a sound strategy.

The barrier to entry for using social media for communication is low; all you have to do is create a profile and start posting your thoughts on these platforms. However, without taking the time to listen to online conversations, learn from them, and create a content strategy, communicating through social media will not be effective and can quickly become overwhelming. It is the goal of this book to teach climate scientists and the scientific community in general how to develop a social media and content strategy for communicating climate change and global warming with the public.

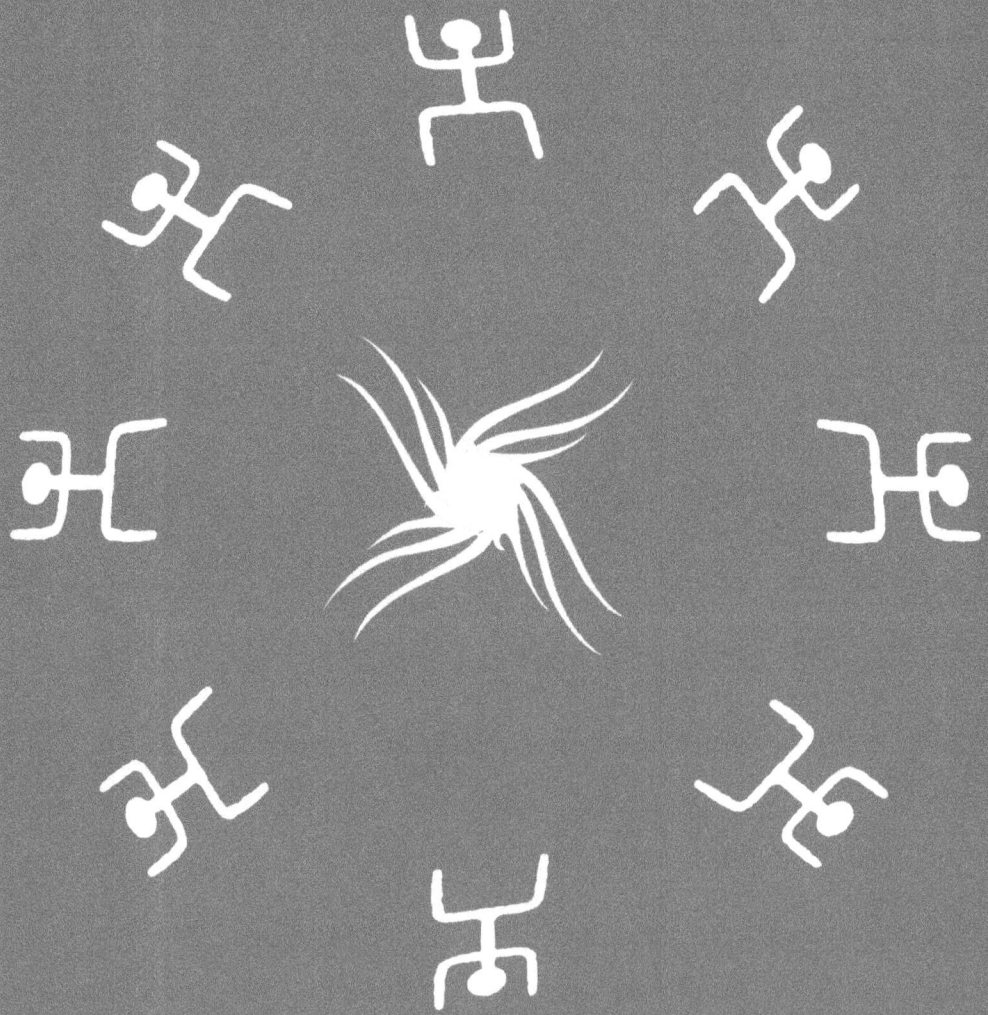

Chapter 2

Defining You, Your Tribe, and Your Purpose

Motivating the public toward reducing carbon pollution and taking action against climate change requires more than just sharing facts. As discussed in Chapter 1, the public's attitude toward climate change and willingness to listen to conversations around this topic is highly influenced by ideology, identity, and worldview (1.17). The implications of this finding are that a successful one-size-fits-all climate change communication strategy is likely to fail. Rather, success requires understanding and focusing on niche audiences and approaching them through their trusted communities. However, infiltrating communities as an outsider may do nothing to encourage a change of behavior, and in fact it may backfire. There should be a common ground and an alignment of values between the communicator and the audience. Instead of publicly broadcasting information and raising alarm about climate change in the hope that *someone* will listen, it would be better for communicators to approach

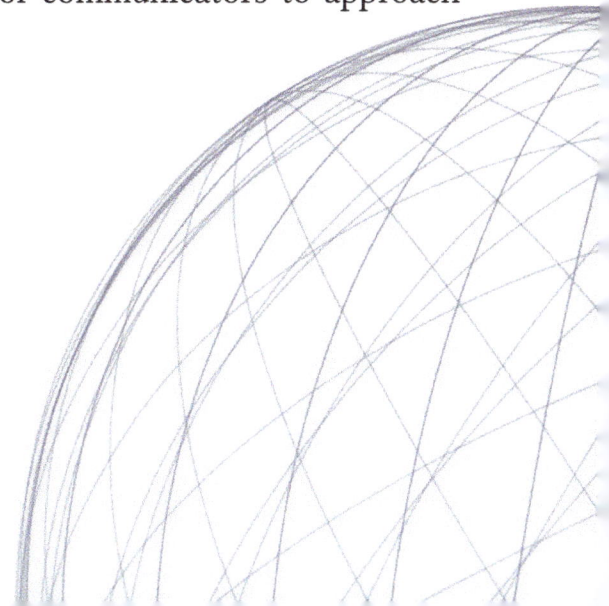

those who share their interests and values. Alternatively, communicators can partner with influencers and opinion leaders outside of their own realm of influence to deliver their message. Organizations such as Moms Clean Air Force is an example of the former approach. Founded by a mom, this organization focuses on the mom community based on a shared concern for the future of their children. Reaching out to Pope Francis to garner support for climate action among the Catholic community illustrates the power of the latter approach. Therefore, before embarking on your communication journey, whether it is to raise awareness or ignite climate action, you need to do the following:

1. Understand who you are.
2. Understand with whom you wish to communicate.
3. Find communities formed around shared values with your audience.
4. Focus and customize your message to that particular community.
5. Alternatively, identify influencers in these communities and partner with them to get your message heard.

This chapter is aimed at helping you reach that understanding so that you can build your communication efforts on a solid foundation.

Getting to know yourself

Have you ever been asked to describe who you are in terms of your interests, values, moral foundations, and political views? What guides your decisions when voting on issues, the groups with which you associate, or hobbies you enjoy? Answering these and other self-directed questions will undoubtedly be transformational—not only for you but also for your communication efforts. This exercise will be the first step toward discovering the niche audience on whom your climate change communication efforts should focus.

Get to know yourself

If you had to describe yourself to someone, how would you answer the following:

Occupation

Education

Religious affiliation

Interests

Hobbies

Political affiliation

Membership and group affiliations

The answers to these questions will help you determine groups, affiliations, and persons with whom you share values and whom you may want to educate about climate change or motivate toward action.

Understanding your audience

Now that you have a better understanding of your own values and group affiliations, you are in a position to choose a niche audience to influence toward climate action. Your audience may be members of your book club or your family, or any other group whose members know you and are willing to listen to what you have to say from a position of trust. You could also create your niche audience based on people who are not directly associated with you right now, but who share your interests. This alternative approach will require you to find and join these communities, take the time to build trust first, and then begin a conversation about climate change. If you want to focus your efforts on communities that are out of your reach or that don't share your values, finding and building relationships with those who do influence them will be key. Whatever path you choose, begin with clearly defining your target audience.

Creating an audience avatar

In the preface section, I mentioned that part of this book is based on a curriculum I developed for a Social Media for Climate Science and Policy course. I also mentioned that you would be introduced to three sample audience profiles, which were created by the students as part of an assignment for this course. It is now time to meet them. Specific information for these profiles is presented in Tables 2.1, 2.5, and 2.6, in order to illustrate how to create an audience profile.

Once you decide on your niche audience, use Table 2.1 as a guide to create a detailed profile for them. Some of the criteria in this table—such as age, gender, occupation, etc.—are self-explanatory, but others may need clarification. Below, I further define these less obvious criteria and explain why they matter.

Political and religious affiliations, and identity

Scientists are trained to view information objectively. Skepticism and independent inquiry are important parts of scientific investigations. Scientists do not accept conclusions unless they see supporting data. They ask for evidence before forming an opinion. That is how most scientists behave anyway. It is important to realize that a nonexpert audience has not been trained to look at the world as a scientist would. As human beings, we do not always make decisions rationally; rather, we engage in motivated skepticism toward new information. Our decisions, opinions, and actions are shaped by many nonrational factors (2.1) including our past experiences, the context in which we form opinions (2.2), our desire to conform to our communities, and to confirm what we believe to be true (2.3). Close to thirty years of research in psychology, social science, and political science, wonderfully summarized by Jacquet, Dietrich, and Jost (2.3), has taught us that as human beings, we engage in motivated reasoning arising from our desire to affirm and preserve three things:

1. Our ego: what we already believe to be true.
2. Our communities and groups: what we believe collectively as a member of a social, political, or cultural group.
3. Our systems of support: status quo systems that support our livelihood.

Our past experiences and communities to which we belong shape our worldviews and identities. Numerous public opinion polls on attitudes about climate change have shown that a conservative political ideology is consistently correlated with rejection of implementing changes toward mitigating climate change (2.3). This may imply that changing minds of political conservatives about taking steps to mitigate climate change is a lost cause, but experience has shown that this is not the case. Research shows that success in changing behavior largely depends on how climate change solutions are communicated. In other words, if you frame your communications in such a way that does not threaten the identity, ideology, and support systems of political conservatives, they are more open to accept and act on them. We will cover this topic in more detail in chapter 3, but suffice it to say, understating the political and religious affiliations as well as the identities of your target audience is paramount for finding success with your communication efforts.

An additional reason for taking into account identity has to do with how human beings tend to make decisions. According to James March, a political science professor at Stanford, one aspect of our decision-making is related to our identity (2.4). In other words, when making decisions, we unconsciously ask ourselves, "How would someone like me react to a particular situation?" Based on the cultural norm of our identity, we tend to go with the herd.

Attitudes on climate change and barriers to taking action

Thanks to outstanding work from YPCCC (2.5) and others (2.6-2.8), we know that attitudes and opinions about climate change are not uniform among the public. This body of work has shown that the public can be divided into six high-level niche audiences based on their knowledge and attitudes toward climate change:

1. **Dismissive**: Most certain that global warming is not happening, or even if it *is* happening, they do not view human activity as an underlying cause. They are quite engaged in the topic and consider themselves well-informed. They do not view global warming as a threat for current or future generations.

2. **Doubtful**: Similar to the Cautious segment (see below) in terms of their level of knowledge about climate change. However, they tend to believe global warming is not much of a threat to people in general. They are more likely to say that global warming is caused by natural changes in the environment.

3. **Disengaged**: This group has the least knowledge and interest in climate change.

4. **Cautious**: Somewhat knowledgeable about climate change, and view it as a distant threat. A large proportion believes that science around causes of climate change is not settled, or that global warming is due to natural causes.

5. **Concerned**: Knowledgeable and concerned about global warming due to human activity, however they do not view it as something that is happening now or affecting them personally. They are not as active as the Alarmed in climate change mitigation efforts, but show an interest to do more.

6. **Alarmed**: Most knowledgeable and convinced about immediacy and causes behind climate change, and they are most vocal and active about climate change mitigation.

An important step toward defining your audience is deciding to which one of these groups they belong, as this will be an important determinant in how you approach them for a conversation about climate change (more on that in Chapter 3).

Table 2.1: Your target audience avatar

Name	Linda Goldberg
Job Title/Profession	Homemaker/Mother
Age	56
Gender	Female
Family income	150,000
Location	Greater NYC area
Identity	Wife, mother
Attitude towards climate change	Disengaged
Influencers	Meteorologists, charismatic news anchors, friends
Political affiliation	Democrat
Barriers to taking personal action on climate change	Lack of knowledge, low interest level
Religion	Judaism
Online activity	Facebook, email with friends, home pages as news source (e.g. Yahoo, MSN)

Once you create an audience profile, similar to Table 2.1, use the information therein to write a story about your niche audience:

> Mrs. Linda Goldberg is the proud wife of a physician and mother of four. Since running a demanding household, she has become completely disengaged with climate change. She always votes Democrat because she finds that Republicans do not care enough about the welfare of all people, but she never considers politicians' stances on environmental issues. Voting is important to her, and she defines herself as truly American especially when comparing herself to her in-laws who have thick accents and cook strange stews.

She sometimes hears stories on the local news about climate change and has begun to recycle since Superstorm Sandy, and now her kids are older and more environmentally aware. She trusts the opinions of the local news anchors and meteorologists because they are also Jewish and come from similar backgrounds.

When Linda goes to the grocery she sometimes finds the packaging of "green" items aesthetically appealing so she will purchase them on occasion. She doesn't care how economical her husband claims energy-saving bulbs are; they don't turn on fast enough when she does laundry in the dark basement. In addition, she has made the change to biodegradable soap, but only because she finds that it is has stopped the irritation on her sensitive skin. She is optimistic by nature and thinks that if climate change is such a big problem, someone very smart and driven will fix it—like the scientists that appear on the MSN online news banner who save baby polar bears, although she doesn't have time to really delve into the issue. Besides, Dr. Oz and the hosts on The View rarely mention it.

Learning from social data

Chances are that the audience profile you have created is largely dependent on your anecdotal knowledge of who they are and based on what they may have in common with *your* values and identity. What if there were a way to refine the profile you have created for them based on data you acquire? This would largely remove the guesswork out of creating your audience profile, and would help position your communication strategy on a solid footing so that you can maximize the chances of success. Facebook Audience Insights provides that opportunity.

Facebook Audience Insights

As of the first quarter of 2016, Facebook had 1.65 billion monthly active users worldwide (2.9). World population is close to 7.4 billion, based on data collected from the most recent census, November 12, 2015 (2.10). So, approximately one-fourth of the world population is actively using Facebook every month, freely sharing data about their interests, identities, values, activities, and group affiliations. Facebook Audience Insights is a tool integrated into the Facebook Ad platform, which allows advertisers to target a relevant audience based on this data. The data in Facebook Audience Insights come from two sources as shown in the following Table 2.2.

Table 2.2

Facebook Native Data	Third-Party Data Partners
Age and Gender	Lifestyle
Relationship Status	Household Income
Education Level	Home Ownership
Job Role	Household Size
Top Categories of Interests	Home Market Value
Page Likes	Spending Methods
Top Cities	Retail Spending
Top Countries	Online Purchases
Top Languages	Purchase Behavior
Frequency of Activities on Facebook	In Market for a Vehicle
Type of Digital Device	

Facebook Native Data is derived from people's activities on Facebook. Third-party data partners—companies such as Acxiom, Datalogix, and Epsilon—provide the information listed in the right column in Table 2.2. These companies collect data about the public based on their online and offline behavior. Here is the process of data collection as explained on Acxiom's website:

"Marketing data about you is collected two ways: data collected offline and data collected online. Offline marketing data comes from publicly available information such as your name, address, birthdate, and census data; information from surveys and questionnaires or product registrations/ warranties; and information from other data providers. Online data often comes from cookies placed on your Internet browser that return information about your online visit to the websites of companies you are shopping with."

What about privacy? In the interest of protecting privacy, none of the data available on Facebook Audience Insights can be connected to any one individual Facebook user. In addition, Facebook does not provide data for an audience of less than 1,000 monthly active users, and data from third-party providers are encrypted (no personal information).

Getting to Facebook Audience Insights

To view the Facebook Audience Insights (FAI) tool, point your browser to www.facebook.com/ads/audience_insights. The FAI interface is broken down into two columns (Figure 2.1):

1 (left column): Filters that allow you to set criteria for a specific audience.

2 (right column): Visual display of data about the selected audience

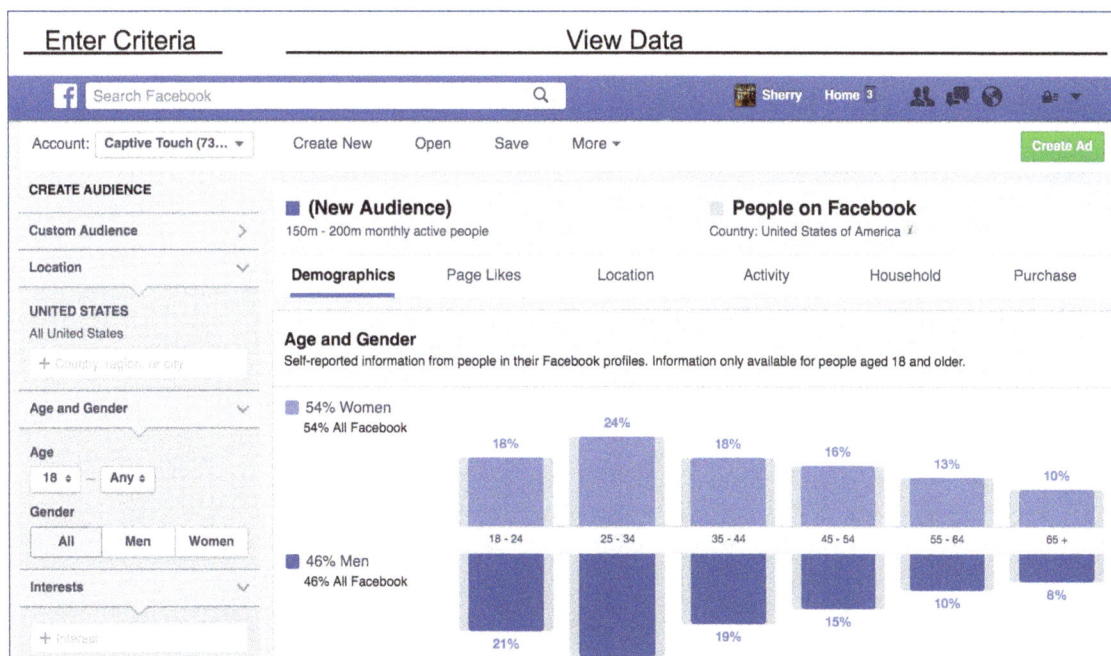

Figure 2.1: Facebook Audience Insights home page

You can set the FAI filter based on a number of categories such as location, age, gender, interests, as well as a number of advanced filters. To illustrate how you can use FAI to learn more about a target audience, we will start with the profile of Linda Goldberg shown in Table 2.1, and refine it based on the information gleaned from Facebook Audience Insights.

Setting up the filter

The first step is to set up the filters (Figure 2.2). Linda falls into the Disengaged group of the public, so climate change is not on her radar, but she has two clear points of passion: her identities as being Jewish and being a mother. For this purpose, the filters chosen in FAI should mirror her identity.

Location: United States

Age: 35–65

Gender: Female

Interests: Passover, Yom Kippur, Rosh Hashanah

Parents: Children ages 4 to 19.

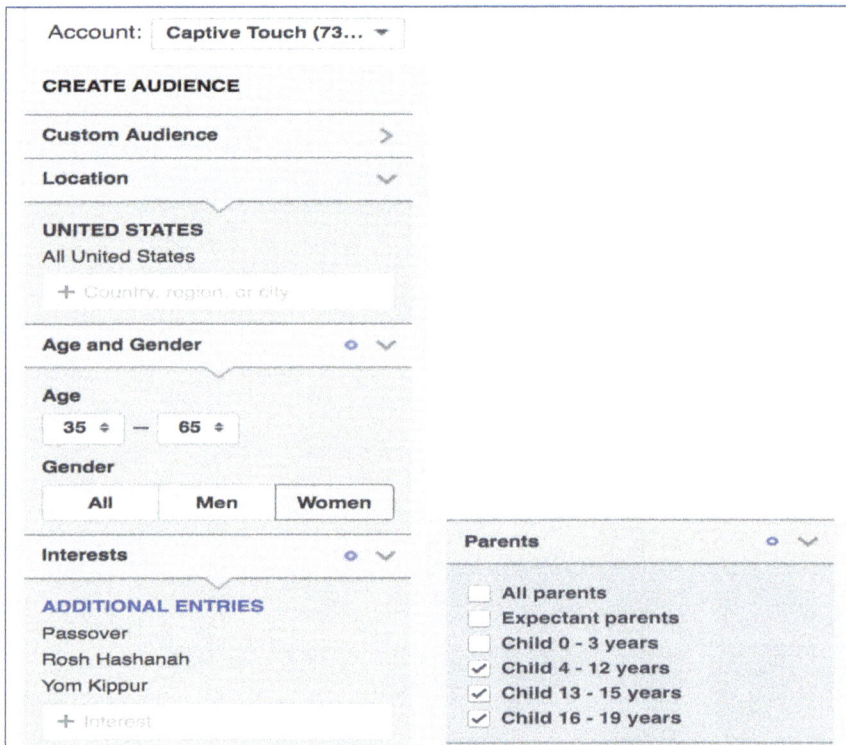

Figure 2.2: Facebook Audience Insights—Setting Linda Goldberg demographics

Obtaining the data

The right column of FAI is a wealth of information (Figure 2.1), which is presented in six tabs:

1. Demographic (Figure 2.3)

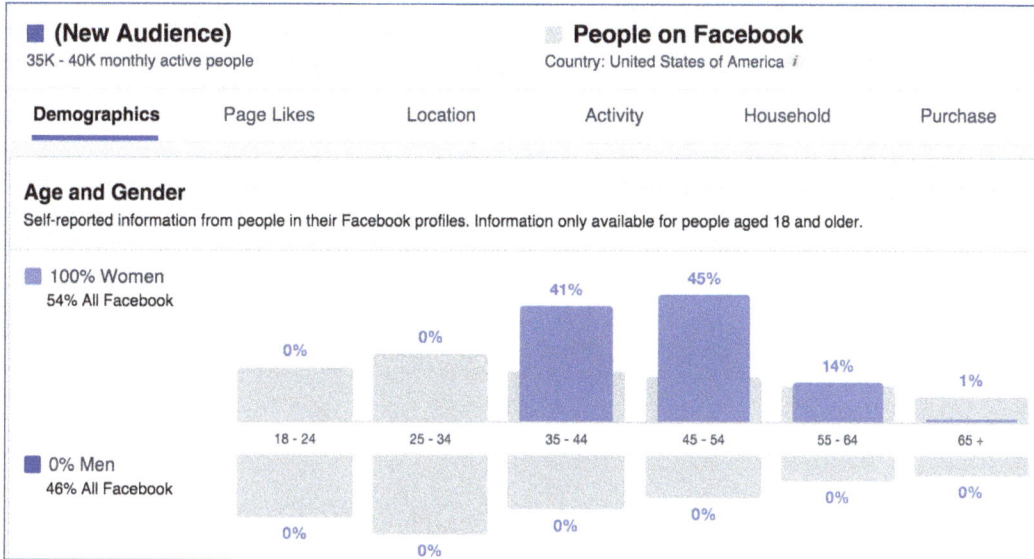

Figure 2.3: FAI—Demographics for Linda Goldberg profile

2. Page likes (Figure 2.4)

Page Likes

Facebook Pages that are likely to be relevant to your audience based on Facebook Page likes.

Page	Relevance	Audience	Facebook	Affinity
Israel Defense Forces	1	8.8K	576K	13.8x
United With Israel	2	11.5K	1.1m	9.8x
GodVine	3	8.9K	4.4m	1.9x
Kohl's	4	9.7K	9.2m	1x
Macy's	5	10.4K	11m	0.9x
Target	6	12.2K	18.2m	0.6x
Amazon.com	7	13.8K	21.3m	0.6x
Walmart	8	11.2K	23.9m	0.4x

Figure 2.4: FAI—Page likes for Linda Goldberg profile

3. Location (Figure 2.5)

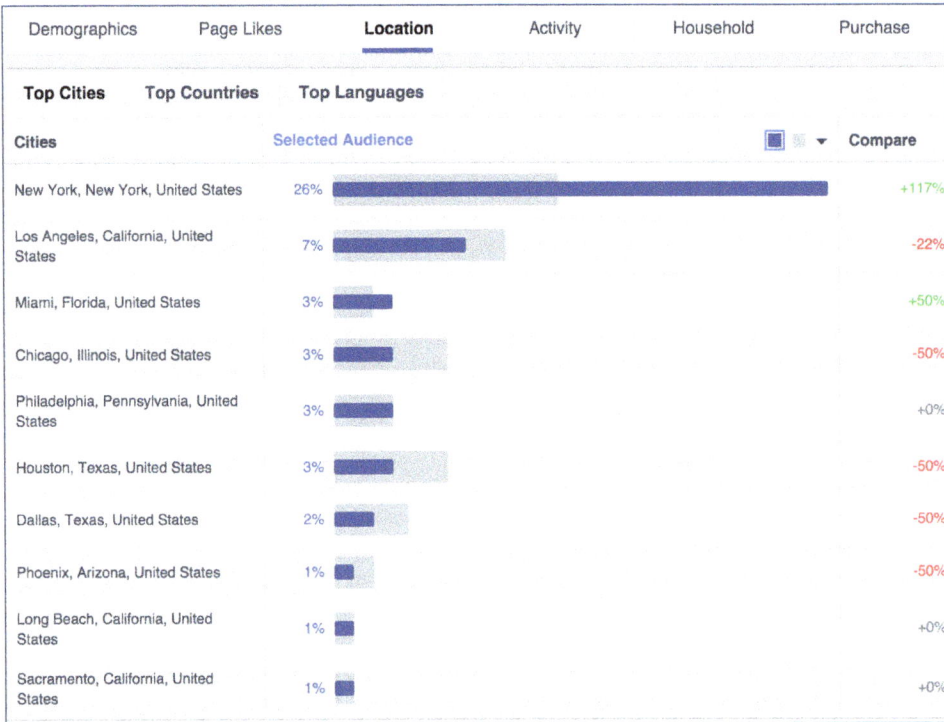

Demographics	Page Likes	**Location**	Activity	Household	Purchase

Top Cities   **Top Countries**   **Top Languages**

Cities	Selected Audience		Compare
New York, New York, United States	26%		+117%
Los Angeles, California, United States	7%		-22%
Miami, Florida, United States	3%		+50%
Chicago, Illinois, United States	3%		-50%
Philadelphia, Pennsylvania, United States	3%		+0%
Houston, Texas, United States	3%		-50%
Dallas, Texas, United States	2%		-50%
Phoenix, Arizona, United States	1%		-50%
Long Beach, California, United States	1%		+0%
Sacramento, California, United States	1%		+0%

Figure 2.5: FAI—Location for Linda Goldberg profile

4. Facebook activity, and devices used to access Facebook (Figure 2.6)

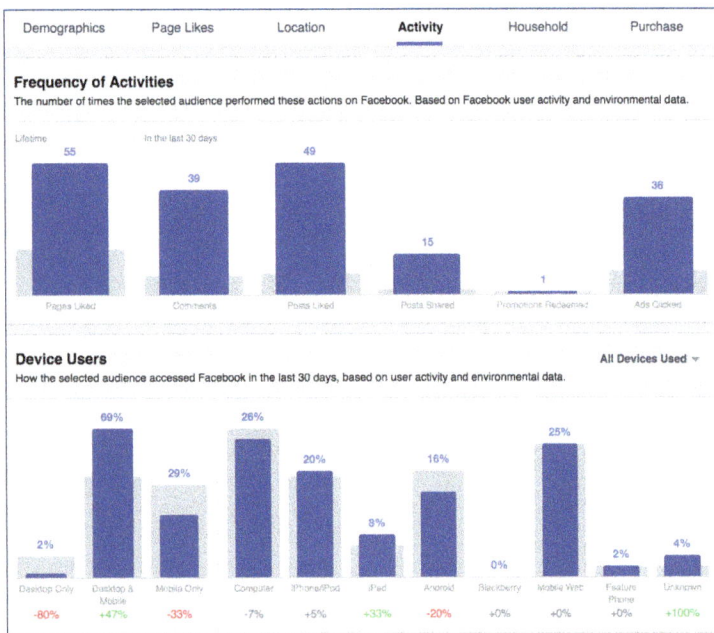

Demographics	Page Likes	Location	**Activity**	Household	Purchase

Frequency of Activities
The number of times the selected audience performed these actions on Facebook. Based on Facebook user activity and environmental data.

Lifetime   In the last 30 days

Pages Liked	Comments	Posts Liked	Posts Shared	Promotions Redeemed	Ads Clicked
55	39	49	15	1	36

Device Users   All Devices Used ▾
How the selected audience accessed Facebook in the last 30 days, based on user activity and environmental data.

Desktop Only	Desktop & Mobile	Mobile Only	Computer	iPhone/iPod	iPad	Android	Blackberry	Mobile Web	Feature Phone	Unknown
2%	69%	29%	26%	20%	8%	16%	0%	25%	2%	4%
-80%	+47%	-33%	-7%	+5%	+33%	-20%	+0%	+0%	+0%	+100%

Figure 2.6: FAI—Facebook activity for Linda Goldberg profile

5. Household income and composition (Figure 2.7)

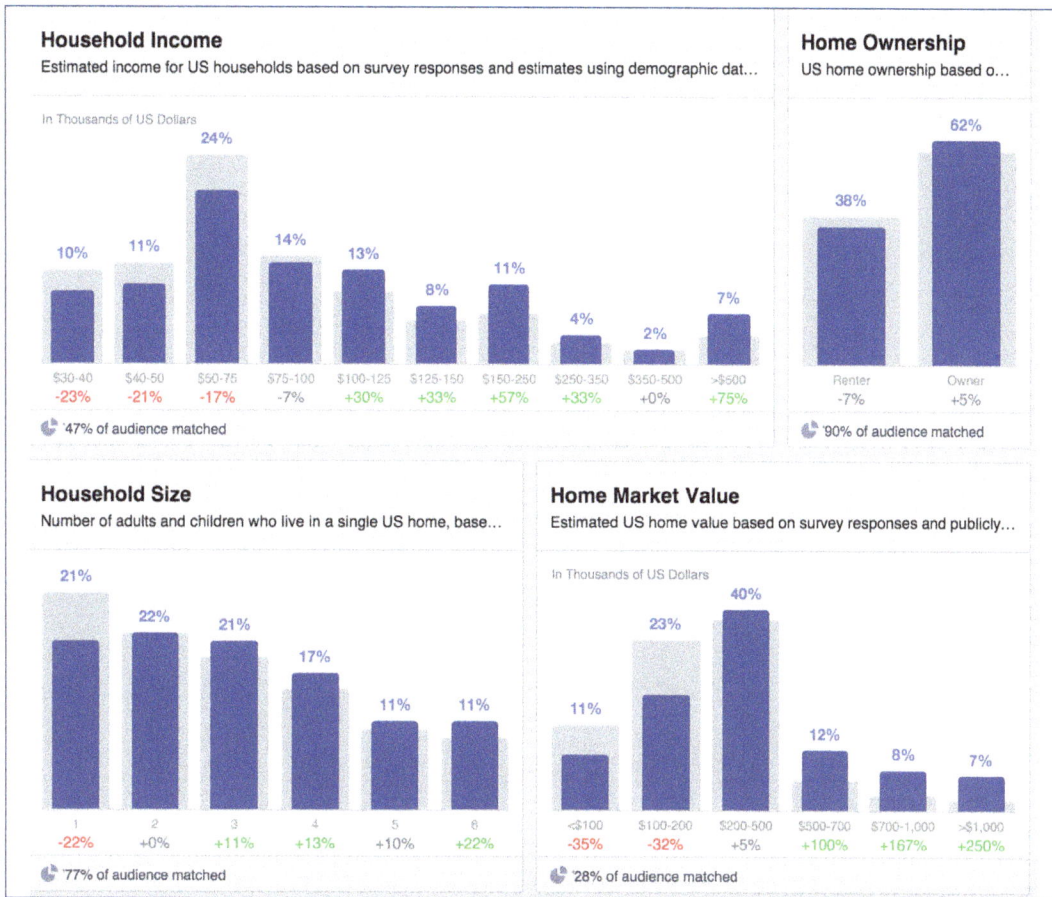

Household Income
Estimated income for US households based on survey responses and estimates using demographic dat...

In Thousands of US Dollars

$30-40	$40-50	$50-75	$75-100	$100-125	$125-150	$150-250	$250-350	$350-500	>$500
10%	11%	24%	14%	13%	8%	11%	4%	2%	7%
-23%	-21%	-17%	-7%	+30%	+33%	+57%	+33%	+0%	+75%

47% of audience matched

Home Ownership
US home ownership based o...

Renter	Owner
38%	62%
-7%	+5%

90% of audience matched

Household Size
Number of adults and children who live in a single US home, base...

1	2	3	4	5	6
21%	22%	21%	17%	11%	11%
-22%	+0%	+11%	+13%	+10%	+22%

77% of audience matched

Home Market Value
Estimated US home value based on survey responses and publicly...

In Thousands of US Dollars

<$100	$100-200	$200-500	$500-700	$700-1,000	>$1,000
11%	23%	40%	12%	8%	7%
-35%	-32%	+5%	+100%	+167%	+250%

28% of audience matched

Figure 2.7: FAI—Household income and composition for Linda Goldberg profile

6. Purchase behavior (Figure 2.8)

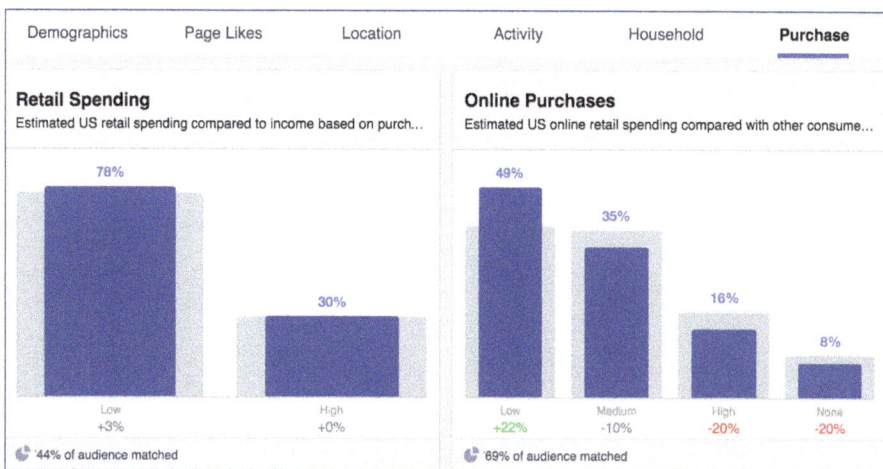

Demographics	Page Likes	Location	Activity	Household	**Purchase**

Retail Spending
Estimated US retail spending compared to income based on purch...

Low	High
78%	30%
+3%	+0%

44% of audience matched

Online Purchases
Estimated US online retail spending compared with other consume...

Low	Medium	High	None
49%	35%	16%	8%
+22%	-10%	-20%	-20%

69% of audience matched

Figure 2.8: FAI—Spending habits for Linda Goldberg profile

You'll notice that the data is displayed in two differently colored charts. The gray-shaded charts denote unfiltered data representing *all* Facebook users, and the blue charts display data filtered for your chosen audience. There is also numerical data presented, which either could be positive (blue) or negative (red). The numerical data shown in percentages is meant to display how closely your chosen audience resembles the overall population on Facebook. This can be useful, for example, if you are trying to decide whether to spend money on serving Facebook ads to your target audience. If your chosen audience is more likely than the general Facebook population to click on ads, then chances are that this investment will be worthwhile, provided that the ad copy, multimedia, and targeting is optimal. FAI data can also be used to fine-tune the detailed profile you have created for your target audience. See how this is done for our example profile for Linda Goldberg in Table 2.3, which is an updated modified version of Table 2.1. (Compare Table 2.1 to Table 2.3.)

Table 2.4: Audience profile matched with communications strategy

Information	Insight from information for outreach
Income: $75K–more than $500K Spending: High end/Upscale	Solar panels, hybrid cars are not out of reach
Location: Greater NYC area> LA> Miami> Chicago> Philadelphia	Offline outreach activities could include locations other than NYC. NYC and Miami residents have personal experience of hurricanes and sea-level rise.
Homeownership: Home owner >$200K	Solar panels should be in reach.
Interests and Facebook pages: Israel Defense Forces, United With Israel, GodVine	Communities for outreach efforts both offline and online.
Education level: College graduate and above	Sets the difficulty level of content.
Mobile device used: More likely to use iPhone and iPad rather than Android. Uses both computer and a mobile device.	Content should be optimized for both mobile and desktop. Outreach via mobile apps should be compatible to the mobile Web and Apple native apps.
Activity on Facebook: Much more likely than other Facebook audiences to like, share, comment, and click on ads. Not likely to redeem promotions.	Outreach via Facebook should include both organic and paid content. Linda is quite active on Facebook! Maybe Facebook should be the primary tool to reach out to her. We shall see.

Table 2.3: An updated version of Table 2.1 based on data obtained from Facebook Audience Insights

Name	Linda Goldberg
Job title/profession	Homemaker/mother
Age	56
Gender	Female
Family income	~~150,000~~ $75K–more than $500K
Location (top 5)	Greater NYC area> LA> Miami> Chicago> Philadelphia
Identity	Wife, mother
Attitude toward climate change	Disengaged
Influencers	Meteorologists, charismatic news anchors, friends
Political affiliation	Democrat
Barriers to taking personal action on climate change	Lack of knowledge, low interest level
Religion	Judaism
Online activity	Facebook, email with friends, home pages as news source (i.e. Yahoo, MSN)
Home ownership	Home owner $200K and above
How does she shop	High-end/upscale, primarily uses credit cards
Interests	Israel Defense Forces, United With Israel, GodVine
Facebook pages they like	United With Israel, Israel Defense Fund, GodVine
Education level	College graduate and above
Mobile device used	More likely to use iPhone and iPad rather than Android. Uses both computer and a mobile device.
Activity on Facebook	Much more likely than other Facebook audiences to like, share, comment, and click on ads. Not likely to redeem promotions.

Now that we have a more detailed picture of Linda's profile, we can be more strategic in terms of approaching her. Table 2.4 is aimed at demonstrating how a detailed demographic for your target audience can help shape your communication strategy.

Mission Statement

By now you have a detailed profile of your target audience with whom you are hoping to communicate about climate change. Now it is time to clearly define why you want to inspire action in your target audience, how you are going to communicate with them, what final outcome you hope to achieve via your communication efforts, and why you think this outcome will make a difference in mitigating climate change. Having clear answers to these questions will help you specifically define a path for your efforts, and you will be armed with clarity of the purpose of your actions. As mentioned in Chapter 1, much of online blogging and social media chatter about climate change has been reduced to rants and uncivil exchange between climate change deniers and activists, or formation of echo chambers by each of these groups. Simply showing up online and sharing information in the hope that *someone* will listen and change their ways is ineffective and can dampen your motivations. Knowing with whom you want to connect, how you are going to communicate with them, and why you believe your efforts will pay off will help your outreach to be built on a solid foundation and based on sound strategy. In addition, clearly defining the outcome expected from your communication efforts will help you measure your success, and knowing the "why" of your efforts will keep you motivated.

To build your climate change communication mission statement, answer the questions below. I have provided sample answers as related to doing outreach with regard to Linda Goldberg.

Writing a mission statement

In order to draft your mission statement, first answer these three questions:

1. **What do you want to do?** Be as detailed as possible about whom you want to reach and what you want them to do.

   Example: *"I want to inspire 'disengaged' Jewish mothers to integrate sustainable living with their interests and daily practices."*

2. **How will you accomplish what you stated above?** Hint: Think about what makes you specifically suited and best positioned to do this. What are some points of common interest that you share with your target audience? If a common interest, how are you connected to your target audience?

   Example: *"I am a young student of Jewish faith, and I am quite active in my community, which includes Jewish mothers—including my own. I know this audience really well—their quirks and sensitivities. I believe I can affect real change in my own community and in the larger online community by being the Jewish voice for climate change. My outreach efforts will include blogging, as well as making connections on social media platforms in which my target audience seems to be active."*

3. **Why do you think the audience you have chosen and your approach will be effective in mitigating climate change?**

   Example: *"My approach will be effective because mothers play a vital part in raising children who will control what happens to our climate in the future. Also, I feel mothers like Linda Goldberg need a channel to become more aware and get involved. They do not put their trust in just anyone talking about climate change. I feel they will be more open to listening if the message is coming from someone of their own faith."*

Take the time to answer these questions as clearly as possible, and don't be surprised to find yourself going back and changing what you have written. It is important that you take the time and do this step correctly, as your mission statement will be the guiding principle for your communication efforts going forward.

Drafting your mission statement

Once you are satisfied with your answers, combine them into two or more sentences. Here is an example:

Example: *"I want to use blogging and the tools of social media to be the Jewish voice for climate change. My goal is to inspire 'disengaged' Jewish mothers to integrate sustainable living with their interests and daily practices because mothers play a vital part in raising children who will impact what happens to our climate in the future. I also feel they need a channel to become more aware and involved in the climate change* *conversation, but they are more likely to listen if a person of their own community is the source of communication."*

Note that you can write this mission statement in many different ways. There is no right way to put your mission into words, but it is crucial to include answers to the *what*, *who*, *why* and *how* of your vision.

Communicating through an influencer

The process for creating a mission statement when working with an influencer remains the same, except that you answer the questions from a different angle. For example, you need to be clear as to why you choose a specific influencer and why you think they will be effective in helping to inform and motivate your target audience toward climate change mitigation.

As you go through the material in this chapter, you may wonder why you spent time gathering data from FAI, when your mission statement could have been well written without them! In the case of Linda Goldberg this may be true, but with others it may not be. In addition, as you will see in the rest of this book, the information in Table 2.4 will help you be more strategic when building content, and searching for online communities of your target audience.

Finally, note that once you select an audience on FAI, you can "save" it for future use in targeting your audience via Facebook ads (note that the how-to of Facebook ads is outside the scope of this book, but I highly encourage you to consider it as an advanced form of outreach). So, investing a good amount of time to play around with FAI and clearly defining your ideal target audience will pay off later for your advertising efforts.

Additional examples

Fans of H. P. Lovecraft science fiction

Opposite page:
Table 2.5: Enhanced audience profile based on FAI.

Name	Paul Draper
Age	30 years old
Gender	Male
Income	$30–$75K
Location (top 5)	Greater NYC area> Los Angeles> Chicago> Houston> Seattle> Philadelphia> San Antonio> Dallas
Identity	Science Fiction/Fantasy/Horror Fan, Writer/Author, Word Nerd, Introvert, Dreamer, Fan of Metal (music)
Attitude toward climate change	Disengaged
Influencers	Lovecraft fan base
Political affiliations	Liberal
Barriers to taking personal action on climate change	Lack of knowledge, low interest level
Online activity	Facebook, online forums, the rest will be determined
Home ownership	Renter
How does he shop	Mostly pays bills with cash, but also uses credit card
Interests	Cthulhu, Stan Lee, H. P. Lovecraft, Dungeon and Dragons memes
Facebook pages they like	H. P. Lovecraft, Cthulhu, Edgar Allan Poe, Dark Horse Comics, Cthulhu Hand Luke
Education level	College graduate
Mobile device used	More likely to use Android rather than iPhone. Uses both computer and a mobile device to access Facebook.
Activity on Facebook	Much more likely than other Facebook audiences to like, share, comment, and click on ads. Not likely to redeem promotions.

Paul is a single thirty-year-old entry-level software developer. He is a die-hard fan of H. P. Lovecraft and the fictional monster Cthulhu, as well as other science fiction works of art. He is an introvert and spends almost all of his free time reading science fiction or chatting with online communities of science fiction fans. Due to being an introvert, he never—or rarely—was invited to parties or student functions in high school or in college. For this reason, he has feelings of powerlessness, disenfranchisement, and loneliness. He is quite removed from the goings-on in the world and prefers to spend his time in the dream world that science fiction creates for him. This is why he does not pay attention to issues related to global warming.

Writing a mission statement

What: Appeal to a preexisting fan base centered around H. P. Lovecraft's *Cthulhu Mythos*, as well as general fans of science fiction/ fantasy/horror. I want to instill in them the motivation to make the fiction they read—such as utopian futuristic civilizations living in balance with nature—come to life. This will be done by providing factual evidence about climate change in a creative way to garner interest and encourage them to learn more.

How: Soft-sell facts using vivid imagery and metaphors, compounded with uncertainty and a sense of impending doom, but with a hint of "tongue-in-cheek" humor. Appeal to the audience's imaginative side, and empower them into taking action by focusing on the detrimental impacts of climate change.

Why: Using art to convey science has potential to grab more attention and catch more interest by relating indirectly with the audience through a common interest—like science fiction—while also providing factual information about the planet's climate.

Example: *"Would there be as many people interested in space if not for* Star Wars?*"*

Mission Statement

When an audience member shares a common bond with a purveyor of information, there is more potential for an audience member to listen. This allows for a greater likelihood of an audience member heeding the informer's data. Targeting a science fiction–based audience—more specifically fans of Lovecraft & Cthulhu—allows me to organically theorize a specific way to appeal to them through a particular writing style.

Table 2.6: Enhanced audience profile based on FAI

Name	Mr. Alex Donovan
Age	35-65
Gender	Male
Income	$150,000.00
Location (top 5)	Greater NYC area
Identity	CEO/Business leader
Attitude toward climate change	Concerned
Influencers	High-profile business leaders
Political affiliations	Moderate
Barriers to taking personal action on climate change	Wants to act on climate in a way that does not hurt revenue, but unsure how to go about it.
Online activity	Professional organizations, LinkedIn
Home ownership	Home owner
Shopping behavior	Prefers fuel-inefficient cars, but is 50 percent more likely to be in the market for a hybrid compared to the rest of the Facebook audience.
Interests	Baseball, he is a New York Yankees Fan
Facebook pages they like	Data not available
Education level	MBA
Mobile device used	Mostly iPhone user, access FB via the mobile app
Activity on Facebook	Responsive to ads, most FB activity is via liking and commenting on posts, relatively little sharing of posts.

Executive Business Leader

Alex is a business leader, the CEO of a successful technology company. He is liberal in social issues, but conservative on fiscal issues. Alex has a keen interest in science and making decisions based on solid data. He understands that human-made climate change is real and needs to be dealt with, but is unsure how to reconcile climate change mitigation with economic prosperity. He wants to reduce the carbon footprint of his company and his personal life, but is unsure how to proceed. Alex says he wants to be active in that arena, but life and the responsibilities of his business keep him very busy. He keeps himself updated about climate change by reading news about new technologies that address this issue.

Writing a mission statement

What: I want to help concerned business leaders like Alex find ways to implement sustainable business practices.

How: Using blogging and networking with Alex on his favorite online business forums, such as LinkedIn, I want to teach Alex green solutions for running his business, and show him that being sustainable can not only reduce the operating costs of his business, but also fulfill his desire to take positive steps toward climate change mitigation.

Why: One of the greatest arguments against climate change mitigation is its negative effects on economic prosperity. However, business leaders are slowly coming to terms with the fact that they need to accept climate change as a reality along with the need to deal with it. Getting the business community on board could be game changing for climate change mitigation, as business leaders usually follow one another. If they can be shown that sustainability can go hand in hand with economic prosperity, business leaders can become a powerful voice for reducing global warming.

Mission statement

Example: *"I want to reach out to online communities of concerned business leaders to communicate sustainable business practices that increase the economic prosperity of their companies. Getting the business community on board could be game changing for climate change mitigation, as business leaders traditionally have been some of the loudest voices of opposition to climate change mitigation."*

Practice what you learned

In this chapter you learned how to do the following:

1. Get to know yourself and your values.
2. Create an audience profile.
3. Use social data to refine your audience profile.
4. Write a detailed narrative for your audience profile.
5. Create a mission statement.

You have seen how the above was accomplished for three sample profiles. Now it is time for you to build the foundation of your communication efforts by creating an audience profile and defining your mission statement. I would love to know your experience using this book, so please join our Facebook group (http://facebook.com/groups/social4climate), share your mission statement, and seek answers to your questions.

Chapter 3

Creating a Communication Strategy

The image of a scientist in popular culture is usually a messy-haired, forgetful, socially awkward individual. The extremely popular television series *The Big Bang Theory* is certainly feeding this stereotype. What many viewers don't realize is that most career scientists don't fit this stereotype and are in fact multitalented individuals:

- **Scientists are excellent writers:** A scientist's job requires concise writing skills due to character limitations of scientific abstracts and publications.

- **Scientists are lifetime communicators:** Obtaining grant money requires scientists to clearly communicate their preliminary data and their vision to reviewing committees. Publishing scientific papers requires scientists to communicate the merits of their findings to the review panel and publishers.

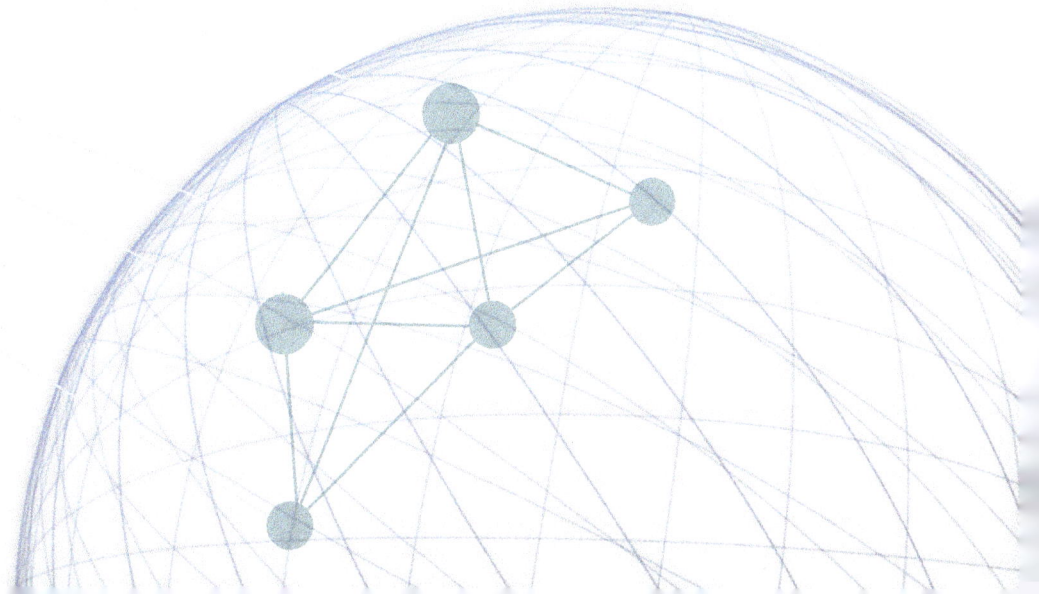

○ **Scientists are lifetime speakers:** In the course of their scientific training and career, scientists have to explain their research to graduate committee members, to principal investigators for their post-doctoral work, to selection committees during their job search, and to conference attendees. Often, scientists have to explain someone else's work when they participate in journal clubs.

○ **Scientists are lifetime entrepreneurs in their own right:** They work long hours, present new and revolutionary ideas, show proof of concepts, raise funding, manage a budget, create jobs, and lead a passionate group of people who share their vision.

With a lifetime experience in communication, why do scientists still need to learn how to communicate? The answer lies in the five factors described below.

Challenges with science communication

1. **Knowledge gaps:** To a climate scientist, data supporting human-caused climate change and the need to act is clear. They've been publishing their findings in scientific journals and discussing them in the context of scientific conferences or blogs. The problem is, a nonexpert audience does not consume information in the contexts just mentioned. There remains a huge knowledge gap that needs to be filled. Thanks to public opinion polls by YPCCC, as far as climate change is concerned, we have some ideas about where these gaps tend to be (2.5). If the point of climate communication is to encourage action on climate change, scientists need to be aware of—and fill—these gaps.

2. **The deficit model:** If we fill these knowledge gaps so that the public understands the science of climate change, they will automatically make the connections. This assumption, referred to as the "deficit model" for science communication (3.1), has traditionally failed. The answer to why simply teaching the facts fails lie in what social science, social psychology, and behavioral economics have taught us about human behavior. A nonexpert audience does not tend to make sense of information and formulate decisions purely based on logical reasoning and data (2.1–2.3). Research has shown that different people process the same piece of information differently, and this difference is largely based on how they have been conditioned, and their tendency to practice unconscious motivated reasoning (2.3). When reaching out to a nonexpert audience, climate

change scientists and communicators must be sensitive to the lens through which a nonexpert audience sees the world, and the motivated reasoning that fuels their decisions.

3. **A language barrier:** In addition to being cognizant of the knowledge gaps and the way human nature shapes comprehension and decision-making, the language that scientists use to communicate climate change is also important. Scientists have been trained to speak and communicate with an expert audience based on data, statistics, and logical reasoning. They speak in percentages and statistical probabilities, and degrees of precision—a language that sounds foreign to most nonexpert audiences. An additional twist is that in some instances, the same words mean two entirely different things in the scientific context and in everyday life. Examples of such words are "theory," "uncertainty," and "very likely," to name a few. Communication with a nonexpert audience has to be spoken in a language, the meaning of which corresponds to what the audience understands.

4. **Doubt in science:** Not only does the nonexpert audience speak a different language and processes information differently, systematic misinformation campaigns have made scientific information confusing for them. Naomi Oreskes and Erik M. Conway have documented in their excellent book, *Merchants of Doubt*, details on how the tobacco and the oil industries funded public relations efforts to raise doubt about harmful effects of cigarettes, DDT, and global warming (3.2). Well versed in the intricacies of communication with a nonexpert audience, their hired PR professionals cherry-picked scientific data, and presented the evidence for a link between lung cancer and smoking, the use of pesticides and wildlife extinction, and the burning of fossil fuels and global warming, as uncertainties (3.2). Similar systematic efforts are also currently propagating myths about vaccines, GMOs and evolution (3.3). These orchestrated PR efforts have been effective in swaying the public opinion to question mainstream and well-established scientific findings. According to a recent survey conducted by the Pew Research Center in collaboration with the American Association for Advancement of Science (AAAS)—on certain issues such as GMO safety, vaccination, evolution, and climate change—there are large gaps between how the scientific community and the public view these

issues (3.4). Learning how to debunk the myths that fuel these confusions will be essential to successfully communicating climate change.

5. **Cynicism:** Cynicism has no place in scientific investigation, but unfortunately it is the newest rhetoric in climate change discussions (3.5). Cynicism could be discouraging to a scientist, because when data supports a finding, cynicism has no legs to stand on. However, when it comes to convincing the public, the industry, or government officials, cynicism may be used as an excuse not to take action, even though they might agree that climate change is real and mostly caused by human activity. So, dealing with cynicism is an essential part of the learning curve toward effective climate change communication.

With these developments in mind, scientists and climate change communicators have much work to do with respect to public outreach and learning a whole new set of communication skills. Thankfully, there is no need to reinvent the wheel. Fifty years of trying to communicate climate science with the public and government officials has taught activists much about effective versus ineffective outreach efforts. The goal of this chapter is to review lessons learned from social science and psychology, and years of on-the-ground communication efforts by climate change activists, so that you can be armed with the right tools and techniques in order to create compelling content.

Filling the climate change knowledge gap

As previously mentioned, public opinion polls have revealed that the public can be grouped into six different segments based on their level of knowledge and attitudes toward climate change (2.5–2.8 and 3.1). One strategy toward effective climate change communication is to tailor the message to the place where a knowledge gap exists. Surveys of the American public (2.5) have identified some of these gaps for each of the six segments of the population. Below, I have aimed to match seven communication-starting points with the six American public segments. Each of these starting points is defined below and visually represented in Figure 3.1.

1. **Awareness of Problem:** Communicating that global warming is a real threat.
2. **Awareness of Causes:** Communicating that humans are causing global warming.
3. **Awareness of Urgency:** Communicating that global warming is affecting us right now, today.

4. **Awareness of Good News:** Communicating all the positive outcomes from efforts toward mitigating climate change.

5. **Ways to Get Involved:** Communicating different ways that everyone can get involved in mitigating climate change.

6. **Action:** Communicating the stories of—and acknowledging—those who are taking action.

7. **Advocacy:** Communicating ways in which action-takers can encourage their communities to join the effort.

The important work of public education must be done, but in addition to just sharing information, we should fine-tune our communication style to go beyond the deficit model of science communication, otherwise our messages will be overlooked, misunderstood, or ignored.

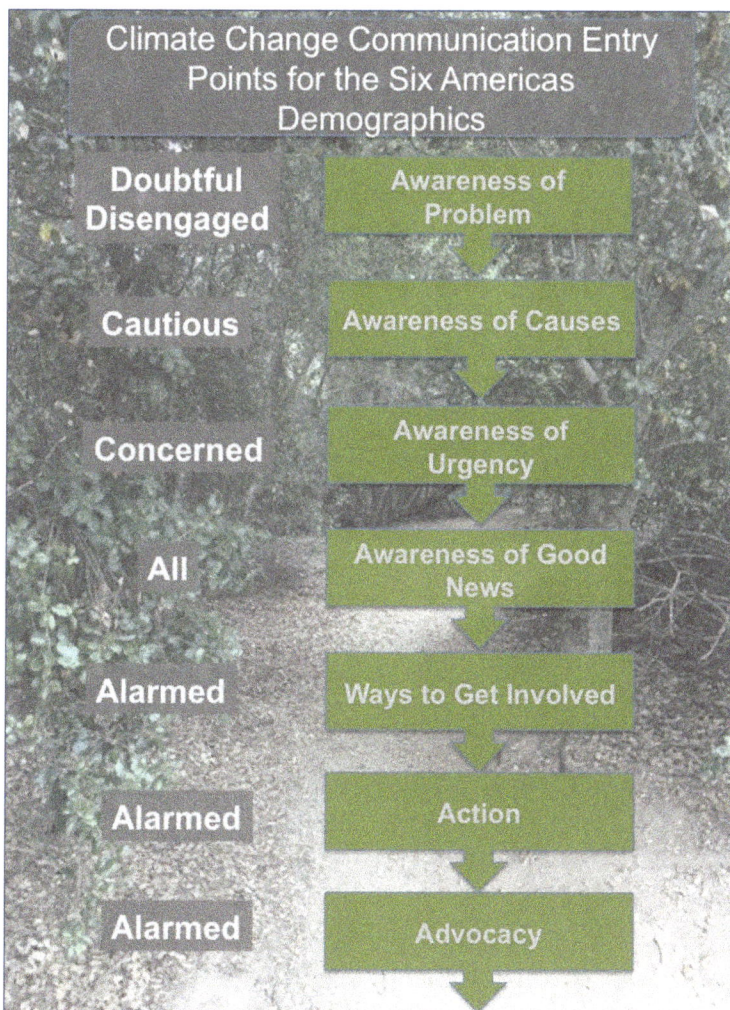

Figure 3.1: Matching the climate change message with knowledge gaps

Going beyond the deficit model

In order to go beyond the failing deficit model of communication, we need to understand the psychology of how human beings make judgments and decisions, and how they react when their reasoning is challenged.

What we know from public opinion polls

Public opinion polls have shown a clear and reproducible positive correlation between a conservative ideology and climate change denial (2.3). What is it about conservatives that make them resist the facts and continue refusing to accept that climate change is happening or that it is caused by burning of fossil fuels? Baffled by this question, climate change communicators have appealed to psychologists and social scientists for answers, some of which was mentioned in Chapter 1. Among the social and psychological findings and theories about human behavior, one stands out in its ability to help understand and explain the roots of denial: The Moral Foundations Theory (3.6).

The Moral Foundations Theory

The Moral Foundations Theory, proposed by Jonathan Haidt, has been formulated based on a large body of research in diverse fields such as evolutionary psychology, anthropology, and social and moral psychology. It is aimed at explaining the basis behind cultural variations in morality: why there is a difference in standard of morality among different populations of human beings. Note that morality in this context does not equate simply with the ability to tell the difference between right and wrong. Rather, it refers to fundamental values that each culture or ideology considers as sacred. Specifically, *"Moral Foundations Theory says that there are (at least) six psychological systems that comprise the universal foundation of the world's many moral matrices"* (3.6). These psychological systems are expressed in a two-word format representing a trigger and the corresponding moral response, as shown in Table 3.1. For example, in the case of the care/harm combination, when harm is inflicted to a member of society, the feeling of compassion is invoked, which leads to a sacred and virtuous response of caring for the harmed.

According to evolutionary psychology, these psychological systems, or cognitive switches, have come about as adaptations to challenges of creating functional human societies. According to Moral Foundations Theory, we all possess these cognitive switches, but genetic differences in combination with our varied life experiences lead us to hold some more sacred than others. Drawing conclusions from his research, Jonathan Haidt suggests that although conservatives find all six of the cognitive switches equally sacred, liberals tend to be mostly driven by their desire to help the weak (care/harm) and to a lesser extent by their desire for freedom of expression (liberty/oppression). This means that although conservatives want to help those in need, they would not compromise loyalty to tradition, nationalism, personal responsibility, and liberty to do so. This difference in moral values, he argues, is the basis of why conservatives and liberals cannot communicate with one another. Each side is bound to the values they hold sacred, which has blinded them to the point of view of the other side (more on that below).

Table 3.1: The six moral foundations and relevant emotions and behaviors—Adapted from *The Righteous Mind: Why Good People Are Divided by Politics and Religion* by Jonathan Haidt. Copyright © 2012 by Jonathan Haidt. Published by Vintage, a division of Penguin Random House LLC."

	Care/Harm	Liberty/ Oppression	Fairness/ Cheating	Loyalty/ Betrayal	Authority/ Subversion	Sanctity/ Degradation
Emotions	Compassion	Oppression	Anger	Rage at traitors	Respect	Disgust
Relevant sacred behavior	Caring	Resistance	Fairness, justice	Patriotism, self-sacrifice	Obedience	Chastity, cleanliness

The interplay between moral foundations and reasoning

Regions of the earth, which have contributed the least to global warming are projected to be the most vulnerable to negative consequences of climate change (3.7). These areas tend to be populated by poor and developing countries. If you are following a liberal ideology—which is safe to say that most scientists do—this very fact, aside from a knowledge of the science of climate change, may be reason enough to convince you to act on mitigating climate change and support regulatory measures to stop carbon pollution, even at the expense of your standard of living. According to Moral Foundations Theory, your Care cognitive switch triggers kindness emotions in your brain, and drives your decision-making. On the other hand, a conservative normally deals with an emotional tug-of-war between all six of their cognitive switches. Experience has shown, in the case of climate change, the Care cognitive switch tends to lose out, leading conservatives to question why we, as developed nations, should change our standard of living and risk our economic competitiveness in order to help the poor somewhere far away on earth. So, even though vulnerability of the developing world to climate change may seem a legitimate reason to regulate carbon pollution to a liberal, it may not convince a conservative to support such regulation.

Thus, according to the Moral Foundations Theory, the primary driver of our decision-making is our emotions, triggered by the cognitive switches in our brains. Once we make that instinctive emotional decision, we then use reason to provide justification for our choices. In other words, according to the Moral Foundations Theory, human beings do not use logical reasoning to find the truth, but they use it as a tool to find ways to explain their emotional—and at times irrational—decisions. This observation is well documented in research done by Jonathan Haidt (3.6). His research has revealed what he calls "moral dumbfounding" where individuals refuse to waiver from their stance on a particular moral judgment, while struggling to provide a logical reason behind that judgment. For example, research subjects would strongly view a particular event repulsive or unacceptable, and stand by this judgment even though they could not provide an underlying logical reason as to why they made this judgment—no matter how hard they tried (3.6).

Other research has revealed a phenomenon, the "worldview backfire effect," which seems to prevent human beings from changing their minds about their firmly believed opinions, even in the face of solid evidence (3.8). During this phenomenon, presenting evidence contrary to a strongly held worldview tends to strengthen rather than weaken that worldview. So, in this case, facts and logical reasoning not only fail to win over cognitive switches; they strengthen them. Examples of these observations include strengthening antivaccine tendencies among parents by showing them evidence that refutes a link between MMR vaccination and autism (3.8). In other words, showing antivaccine parents data that should put to rest their anxieties about vaccine side effects actually reinforces their antivaccine stance. According to these parents' worldview, vaccines are harmful and showing them data that challenges that worldview will not help change their views. Referring back to the Moral Foundations Theory, these parents are so strongly bound to their care/harm cognitive switch that it blinds them to facts that refute their points of view. Research in climate change communication has revealed the same phenomenon. Political conservatives tend to become more skeptical of climate change when they are informed of news about negative health effects of global warming (3.9). In this case, again the care/harm cognitive switch is losing out to other psychological tendencies that conservatives hold so sacred.

Is there hope for reaching conservatives or those on the fence about climate change? How can we win the battle with human cognitive switches that prevent us from taking action on climate change? The answer is not to forge a battle, but to adapt our communication style to the cognitive switches of our audience, which brings me to a powerful communication tool called Framing.

Framing: a tool to go beyond resistance

Framing is a powerful communication tool that has been successfully used by social activists to capture the attention of their target audience, and to get them to understand and *care* about a social issue, including global warming (3.10). Nisbet and Scheufele (3.1) beautifully describe what framing is and how it works:

> *"Frames are interpretive storylines that communicate what is at stake in a societal debate and why the issue matters. . . . Frames help simplify complex issues by lending greater weight to certain considerations and arguments over others, translating why an issue might be a problem, who or what might be responsible, and what should be done."*

The power of framing, as related to accepting a carbon tax, has been clearly illustrated by work of Hardisty and colleagues (3.11). They surveyed a group of self-proclaimed Democrats, Independents, and Republicans for their willingness to purchase more costly products to help reduce carbon emission. A majority of Republicans and Independents did not favor the extra cost when it was presented as a tax, but they did when it was framed as an offset. Such differential framing did not make a difference in the proportion of Democrats choosing the costlier products (3.11); they were generally accepting of the extra cost to reduce carbon emissions regardless of what words were used to express it. Going back to the Moral Foundations Theory, it is likely that the liberty/oppression cognitive switch, due to intense dislike of government intrusion into personal liberties, blinded Republicans and Independents to the benefits of helping the environment. On the other hand, in Democrats, which tend to follow a liberal ideology, the care/harm cognitive switch is the primary driver of their decisions, so framing of the extra cost of products as tax or an offset made no difference in their response.

Table 3.2: Framing climate change messages based on moral foundations and life experiences of sample audience profiles

Persona	Moral Foundations					
	Care/Harm	Liberty/ Oppression	Fairness/ Cheating	Loyalty/ Betrayal	Authority/ Subversion	Sanctity/ Degradation
Linda Goldberg	X			X		X
Paul Draper	X	X				
Alex Donovan		X	X			

Life Experiences	Framing	Analogy/Metaphor
• Had to cancel her son's bar mitzvah due to power outage b/c of Hurricane Sandy. • Her expensive landscaping was destroyed in Miami because of flooding and sea level rise. • Hears her rabbi speak of sustainable living.	• Global warming hurts families. • Threat to American way of life. • Threat to personal health. • Threat to waterfront properties. • Clean energy as alternative. • What will her children's lives be like in 50 years? • Climate Action as a responsibility to protect God's creation: solar energy as a solution. • Climate action=living kosher.	• **Blind spot:** Someone else will deal with climate change. • **Comparison:** Religious ceremonies, or raising children. You play an important role; don't outsource • **Relate back:** Taking care of earth is everyone's responsibility.
• He lives for Comic-Con every year, when he attends wearing a Cthulhu costume. • Was bullied and rejected often at school.	• Global warming threatens the ocean, Cthulhu's home. • Speaking with the voice of Cthulhu, share how global warming makes life in the sea unbearable. • Cthulhu as a witness to the positive effects of clean energy on marine life. • Threat of sea level rise leading to refugee crisis (refugees are often unwanted and rejected).	• **Blind spot:** Disengaged with climate change issues. • **Comparison:** Cthulhu's survival depends on a healthy habitat. • **Relate back:** Ocean acidification is harming sea life, creatures like Cthulhu.
• Endured a power outage due to Hurricane Sandy. • Lost revenue to his company due to outage. • Missed an important meeting due to traffic problems from hurricane. • Is a Yankees fan.	• Global warming as a threat to business operation. • Green energy as a cost-saving measure, increasing revenue. • Sustainable business practices such as video conferencing as opposed to meeting in person, helps the environment, cost-effective for business, no missed meetings.	• **Blind spot:** Value of clean energy for long-term business growth. • **Comparison:** Yankees don't win the World Series in the first game. • **Relate back:** Investment in clean energy is a long-term success plan.

Table 3.2: Framing climate change messages based on moral foundations and life experiences of sample audience profiles

Choosing an appropriate frame

Effective framing makes all the difference, and it starts by knowing your audience, their moral foundations, and life experiences really well. Fortunately, you did most of this work in Chapter 2, so you now can reap the benefit of the hard work you did to clearly define your audience.

Let's go back to the profiles we created in Chapter 2 and list what we think would be *their* moral foundations and life experiences. Then, based on all this information, we will create a list of ways we can frame climate change such that a climate change message resonates with each of them. This exercise has been presented in Table 3.2.

Framing may attract trolls; don't let them discourage you

I mentioned in the beginning of this chapter that I would be reviewing some of the tactics that environmental activists have been using to garner support for climate change. Framing is one of those tactics that ecoAmerica, an NGO, started using back in 2009. An internal memo describing this tactic was accidentally sent to journalists, which led to much negative commentary about ecoAmerica, accusing them of trying sleazy advertising tactics to "sell" global warming to the public (3.12). EcoAmerica's usage of frames was *framed* as a malicious intent to fool the public, which could not be further from the truth. Ironically, the critics found it justified for *them* to use framing, while at the same time criticizing ecoAmerica for doing so. It is important to note that framing is not a tool of deception; rather it is a tool for improving communication. It is used to capture the attention of an audience who is *so* bombarded with information, that one needs the emotional components inherit in framing to capture their attention and help them comprehend the meaning and significance of an issue. So, if you apply framing in your communications about climate change, don't be discouraged by trolls who accuse you of trying to be deceptive.

Frame extension

A note of caution about framing: it can lose its effectiveness with time. Some climate scientists focus their messaging entirely on threats of climate change. Their articles and Facebook posts primarily share bad news, so much so, that I've seen some of their followers criticizing the constant negative tone of their messaging. In a way, it is understandable that a climate scientist, who understands better than anyone the dire consequences of global warming and the urgency to take action, would want to alarm their followers. But this type of one-tone messaging not only will lose its effect with time, it may also lead to feelings of cynicism and powerlessness toward creating change. So it is important for your messaging and framing to be occasionally updated. Social activists understand this phenomenon well; this is why they often employ frame extension (3.10), which means extending the impacts of an issue to other areas of people's lives.

Now, review the Framing column in Table 3.2 and you'll see that for each example audience, I have provided samples of Frame extension. Climate change has been framed as a threat to more than one area of a person's life, and clean energy alternatives have been framed as an optimistic, solutions-oriented view of climate change. When thinking about choosing appropriate framing for your target audience, be sure to implement frame extension so that your message is not monotone and won't lose effectiveness with time.

Communicating with stories and metaphors

A time-tested method of communication in human societies is storytelling. Stories are engaging and they are a familiar way of communication. Scientists may not be characterized as good storytellers, but if you stop and think about it, they have to narrate their research in publications and grant applications in the form of a story. Scientific publications have a beginning that sets the stage and identifies the problem to be solved. The results are the heroes of the story who answer questions about the problem, and the conclusion or discussion is, well, the conclusion of the story. Often, in the conclusion or discussion section of scientific papers, a stage is set for the next story in the sequel of future publications. Scientists are storytellers without realizing it; however, with climate change, the audience is different. Scientists can use that storytelling skill and use it to communicate with a nonexpert audience, except instead of technical language, graphs, and data, they must rely on metaphors and analogies to get their points across.

Why metaphors

Research in the cognitive process involved in learning and memory indicates that we do not simply attain information and store it (3.13). Instead, when we encounter and want to make sense of new information (short-term memory), our brains consolidate it with old information (long-term memory). In other words, we understand new information by drawing help from what we already know. Effective metaphors and analogies speed this process by helping us retrieve a relevant piece of information from long-term memory to make sense of new information (3.13). For example, if you were to explain to Linda Goldberg the harmful long-term effects of CO_2 absorption by the oceans, you could use the metaphor of a mother cleaning up her children's plates by eating the leftovers because she hates wasting food! After awhile, this practice ends up causing the mother her health. By analogy, our oceans may be acting as our atmospheric buffer, absorbing the extra CO_2, but this will make the oceans unhealthy due to acidification. A better solution for preventing food waste is to make less of it! Similarly, a solution for keeping our oceans healthy is to burn less fossil fuels.

How to use metaphors

If you asked me for the best resource to help you learn how to create effective metaphors, I'd refer you to a wonderful book by Anne Miller called *The Tall Lady With the Iceberg* (3.14). In this book, among a host of other valuable techniques, she suggests a four-step process for creating an effective metaphor, which I outline below with some paraphrasing:

1. **Determine your audience's blind spot.** What is their misunderstanding about what you are trying to communicate? What is one mental roadblock that is preventing them from seeing your point?

2. **Understand your audience.** You've already done this in Chapter 2.

3. **Create the comparison.** Based on the knowledge about your audience and their blind spot, find a situation that would be analogous to the message you are trying to convey.

4. **Relate.** Make a connection and show similarities between your message, or the solution you are providing, with the comparison you created in number 3.

You saw an example of this in action above for communicating the value of reducing carbon emission to Linda Goldberg. Here is another example, which is also shown in Table 3.2:

- **Blind spot:** Someone else will deal with climate change.

- **Comparison:** You wouldn't outsource raising your children, or attending important religious ceremonies, because you are an important part of it.

- **Relate back:** Taking care of Earth is as much your responsibility as anyone else who lives on this planet. Don't wait for other people to fix climate change. You play an important role; don't outsource.

Communicating quantitative data and uncertainties

Communicating the meaning and impact of quantitative data may be some of the most challenging things about outreach to a nonexpert audience. How would you help this audience appreciate the significance of the degree of "likelihood" of something or understand the meaning of fractions and percentages? The answer, again, is in framing this information in a way that is familiar to your audience. Proper framing of your quantitative data will help your audience to easily draw comparisons with what they already know. There is a reason that the magnitude of large quantities are often represented in comparison to the size of a "football field" or an "Olympic-sized swimming pool," to name a few. On the other hand, small quantities are often compared to the size of the tip of a needle. The likelihood of rare happenings is often compared to "winning the lottery" or "being hit by lightning." By the time we get to a certain age, based on our life experiences, we already have developed an appreciation of how likely are these events. We can draw upon this past knowledge to make sense of new information about probabilities of other events. Anyone who has had pizza or a pie can appreciate the meaning of fractions when framed in this sense. Framing works, and if you notice, we all use it on a daily basis without realizing it.

Leverage the hard work you have done so far to clearly define your audience, their life experiences and moral foundations, and frame quantitative data in a way that is more familiar to them. Let's see how that would work for our three example demographics:

1. Linda Goldberg: Linda seems to think that others will save the planet from climate change, downplaying her own contributions if she lived a more green life. The impact of the energy-efficient light bulb in the laundry room, about which she often complains, could be communicated to her in terms that are familiar to her. The way a Huffington Post article frames the benefits of efficient light bulbs is a great example:

   "If every household replaced just one light bulb with one certified by the federal Energy Star program, we would save enough energy to light two million homes for a whole year. We would also prevent greenhouse gas emissions equivalent to that of 550,000 vehicles. LED bulbs are a terrific way to fight climate change." (3.15).

2. Paul Draper: Paul is completely disengaged from climate change discussions and lives in the fantasy world of H.P. Lovecraft's work of fiction. He is particularly passionate about fictional creatures like Cthulhu, and attending Comic-Con. One can inform Paul about the impact of living an energy-efficient life in terms of the cost savings to his energy bills. The amount of savings could be expressed in terms of the number of trips he could take to Comic-Con.

3. Alex Donavan: Alex is an executive, so quantitative data is not unfamiliar to him, but he tends to always have a business analytics view to numbers. As a business owner, he is also risk averse, and understands the value of putting in place backup plans and purchasing insurance. Although there are some uncertainties in the exact degree of damage caused to our communities and businesses by climate change, we are confident, as Alex would agree, these damages *will* happen. Alex buys liability insurance for his business, his cars, and his home. Taking energy-efficient measures as a preventative measure can be compared to buying insurance to save one's assets. We purchase insurance to be prepared for damages that may happen, the lawsuit that may be filed, or the fire that may occur. Therefore, the degrees of uncertainties, often expressed in percentages, about climate change could be framed as protecting against future risk. When framed in this way, temporarily forgoing short-term business gains to implement energy-saving measures for a prosperous future may sound more reasonable to Alex.

Spotting and fixing scientific jargon

In 1993 a popular nonfiction book was published that addressed the need for men and women to better communicate with one another. The name of this book is *Men Are from Mars, Women Are from Venus*, by John Gray Ph.D. This metaphor could also be very well used to characterize the wildly different ways in which scientists and the nonexpert audiences communicate—as if they come from different planets! There are at least two things scientists can do to break this communication barrier:

1. **Break free from the curse of knowledge.** Beware of what Chip and Dan Heath call "the curse of knowledge" (3.16) in their best-selling book *Made to Stick*. The Heath brothers remind us that as we move into the upper echelons of higher knowledge, we forget what it is like not to know. This was an eye-opener

for me personally as an instructor of Biology and as someone who regularly enters into heated discussions to debunk scientific myths on Facebook. Until I read *Made to Stick*, I was blind to my own curse of knowledge and deemed anyone who bought into pseudoscience as less than intelligent. Once I became aware of the curse of knowledge, I learned to be more patient with my students, and the nonexpert audience with which I argued, as I realized that they are new to the concepts of which I am fully aware. I have witnessed the curse of knowledge showing its ugly face in heated online discussions between scientists and nonexpert audiences, or in memes and status updates that science-focused Facebook pages share with their followers. Let's make a pledge not to think of a nonexpert audience, who has fallen victim to propagators of pseudoscience, as less intelligent than those of us who have had the training and experience to know better not to be fooled by misinformation. Let's lose the phrase I often see on Facebook: "Do you even *science* bro?" People fall victim to misinformation because of all the psychological factors that we've discussed so far, and due to lack of knowledge. Logical fallacies may seem obvious to a climate scientist, but a nonexpert audience may take them as fact. They do this either because these fallacies validate their moral foundations and life experiences, or they lack the foundations of knowledge required to help them know any better—or both.

2. **Beware of scientific language and jargon.** Have you ever had a nonexpert friend or family member glance over the articles you read and say "this is all Greek to me" or "it goes over my head"? This is not unique to a scientific field; any literature written for or by an audience with an advanced level of knowledge is full of jargon. If we hope to help a nonexpert audience arrive at a level of understanding, the jargon and the big convoluted words used in academic writing needs to be toned down. Audit the words you use and optimize them for understanding.

It is also important to be aware of words that are attributed to completely different things by the public versus the scientific community. Trying to think of all the different words we use in scientific communication and how they may be interpreted differently by a nonexpert audience is an arduous task. Fortunately, some of this work has already been done and documented (3.17) by Andrew David Thaler, a marine scientist (3.18). Inspired by the work of Richard C. J. Somerville and Susan Joy Hassol

(3.19), Andrew has crowd-sourced a list of words used in the scientific literature, their public-perceived meanings, and suggestions for more clear alternative words (3.17).

So far, we've learned the importance of communicating so that people can relate to and understand your message. But for your message to get to that level, it must first be noticed. In this age of information, the public is inundated with messages everywhere they look. What kinds of strategies can you use to get your message noticed and remembered? Keep reading to find out.

Getting your messages to stick

Perhaps the best resource I can recommend for learning how to make your messages heard and remembered is the book *Made to Stick* (3.16). The Heath brothers argue that "sticky" messages have two or more of six crucial elements in common. They summarize these six elements into a six-letter acronym—SUCCES—and here is how they define a sticky message:

1. Simplicity: A message that strips an idea to its core.
2. Unexpected: A message that incorporates an element of surprise.
3. Concreteness: A message that makes an idea tangible.
4. Credibility: A message that is backed up by credible sources. This one scientists should relate to because credibility in science is everything. That's why we have bibliographies, right?
5. Emotional: A message that gets people to care about an idea. This is where everything we learned about social and moral psychology will come in handy. As the Heath brothers so nicely explain:

   "The most basic way to make people care is to form an association between something they don't yet care about and something they do care about" (3.16).

6. Stories: A message that draws in the audience by using stories.

Let's try and apply these strategies to create sticky messages for our example profiles. Remember that a sticky message does not necessarily have to include all of these elements, but I'll try to incorporate as many as I can in my examples.

Then let's try to create a sticky blog article with a catchy title, and imagine that our example audience is scanning his/her Facebook news feed while having coffee in the morning. To help me craft these titles I will refer to Table 3.2 for ideas:

Linda Goldberg: Linda is a dedicated to her Jewish faith and to living kosher. Let's see how we can get her to think about living a truly kosher life. Here is one sample title:

> **"The day I learned I was fooling myself about living a kosher life."** This could be a story from a person of Jewish faith, trusted by Linda, describing how they learned that protecting the environment and the earth is part of living a kosher life. The elements of SUCCES used here would be: Simple, Unexpected, Concreteness, Credibility, Emotional, Stories.

Paul Draper: Paul is obsessed with Cthulhu. The monster talks and Paul listens, and for climate change communication purposes, it helps that Cthulhu lives in the ocean. So let's try to get Paul's attention by showing him that Cthulhu feels uncomfortable:

> **"An open letter from Cthulhu: I am hot and angry, and you don't even care."** This could be a letter from the monster to his fan base describing how the ocean is affecting his well-being and how some people like Paul claim they are passionate about the monster but they don't do a damn thing to save him. The elements of SUCCES used here would be: Simple, Unexpected, Concreteness, Emotional.

Alex Donovan: Alex is really interested in sustainable practices, but he needs to answer to the Board of Directors (BOD) and CFO of his company. Alex's challenge is that the main focus of the BOD and the CFO is profitability. Alex is so busy running his business and keeping it balanced with his personal life that he rarely has time to research sustainable business practices. What if he came across an article that puts the solutions in front of him, titled like this:

> **"The secret to a sustainable business is taking your CFO on a company field trip."** Alex can arrange a tour with his CFO to visit a noncompeting company to learn about their sustainability practices. By using examples of other companies that have done this, this article will demonstrates how understanding the connection between a company's environmental and financial success indicators is the secret to getting the CFO to get on board with sustainability. Elements of SUCCES employed in this message are Simple, Unexpected, Concreteness, Stories. An example of such a blog post is cited in reference 3.20.

I hope you can now clearly see how the SUCCES principle works. I encourage you to read *Made to Stick* for additional examples and scenarios in which you can implement this strategy.

By now, you have learned how to craft your messages in ways that are understood, tangible for your audience, free of jargon, engaging, and memorable. The bad news is that those who financially benefit from climate change denial (and antivaccine/antiscience ideologies) are masters of communication and have been using these very same techniques to spread misinformation and myths about climate change. But the good news is—if done correctly—one can cut through the misinformation, debunk myths, and change minds.

Debunking myths

Rumors, lies, and misconceptions are part of human culture, which is why developing methods to debunking them has been an active area of research. This research has been elegantly summarized by John Cook in Denial101x, a Massively Open Online Course (MOOC) hosted on edX.org (3.21), and in his book *Climate Change Science: A Modern Synthesis* (3.22). Through Denial101x and his book, John Cook teaches us a three-step research-based method for debunking myths, based on what is called the "Inoculation Theory." According to this theory, when trying to debunk a myth, one must avoid putting too much emphasis on the myth itself, as this will have the opposite effect of solidifying it even further. Instead, to effectively address and debunk a myth, one must follow the following steps:

1. State the alternative fact that is being misrepresented by the myth.
2. Introduce the myth along with a warning that it misrepresents the facts.
3. Explain the fallacy in the myth and how it misrepresents the facts.

Let's see some examples:

> *Here is how not to debunk a myth*:
>
> Climate change deniers are at it again claiming that half of global warming can be blamed on the heat coming from buildings and infrastructure in our cities, something called the Urban Heat Effect. It is true that urban areas do emanate heat, but not as much as these authors would like you to believe. In fact, when global temperature is measured in both rural

and industrial regions, scientists find that warming can be detected in both of these regions equally.

Here is how to debunk a myth:

Do you ever wonder how we know that the earth is warming? I mean, how do you take the temperature of a whole planet? When you measure your body temperature you put a thermometer into your mouth. Well, you cannot measure global temperature in just one spot like you do in humans, because the environment on earth is so different in different places. So, what climate scientists do is place weather stations all over the world and measure temperature at hundreds of locations with varying climate and population. The global temperature is the average of all the measurements in these weather stations. These measurements clearly show that global temperature has been rising, and that we continue to experience harsher heat waves due to global warming.

There is a myth going around that says half of global warming is due to heat emanating from buildings in urban areas, implying that the reason the planet is getting warmer is because we have too many buildings. This can't be further from the truth. Why? Because global temperature measurements tell us that both urban and rural areas are showing similar levels of rise in temperature over time. Besides, most of the ~~warming trend~~ rise in temperature on earth has been detected in areas where there is *very* low population. If the Urban Heat Effect had so much to contribute to global warming, the rise in global temperature would not concentrate in areas with low population levels and less buildings.

To find more debunking examples, visit John Cook's website SkepticalScience.com, or take the free Denial101x online course (3.21), as you will be richly rewarded with valuable knowledge.

Cynicism

One of the most challenging roadblocks to creating behavior change toward climate change mitigation is cynicism and apathy. In a way, it is not difficult to see why this happens:

1. Refusal of many lawmakers to either accept climate change or support climate action measures.
2. The scale of change needed to stop global warming.
3. The constant doomsday messages shared by some climate activists.
4. Incidents like the natural gas leakage in Southern California, which makes climate change initiatives by the state seem like a waste (3.23).
5. The unprecedented Supreme Court halting of President Barack Obama's Clean Energy Initiative (3.24).

Luckily, for every discouraging news article about climate change, there are many positive ones, which can be a source of motivation. In fact, public surveys (3.25) and research on the effect of messaging has shown that solution-oriented and optimistic language increases tendency toward taking climate action, even among conservatives (3.25, 3.26). Recognizing the value of good news and focusing on progress rather than a doom-and-gloom climate change message, Citizens for Global Solutions, a grassroots membership organization, has launched a positive messaging initiative, which they call #itshappening. The goal of the project is—in their words—

> "To create 'a call to celebrate how far we've come and, to step up our ambition for the future.' By sharing examples of climate change progress from around the world using pictures and social media, #itshappening tries to '[fight] fatalism with hope, and [counter] apathy with agency.'" (3.27)

So, if your friends, family, and communities are tuning you out because it snowed this winter or they just don't want to hear more bad news, then change to positive messaging. No one can argue with progress and good news—not even deniers.

Practice what you learned

In this chapter you learned:

1. Factors that come between communication from scientists and understanding by a nonexpert audience.
2. Moral Foundations Theory, and how it affects the way people make judgments.
3. How to use metaphors, framing, and stories in order to help facilitate understanding.

4. How to communicate quantitative data.

5. How to create sticky memorable messages.

6. How to debunk myths and guard against cynicism by focusing on good news.

Perfect practice makes perfect. Now, go back to the detailed profile that you created for your target audience, and expand it by defining the moral foundations of your audience. Then, practice writing headlines and articles for them to debunk a scientific myth, communicate a climate change topic, or deliver good news in a way that is sticky, free of scientific jargon, and appeals to your audience's moral foundations.

Be sure to join our Facebook group (http://facebook.com/groups/social4climate), share your progress, and seek answers to your questions.

Now that you have learned effective communication strategies, it is time to use blogging and social media to spread your message.

Section 2

Technology,
Tactics,
and Your Action Plan

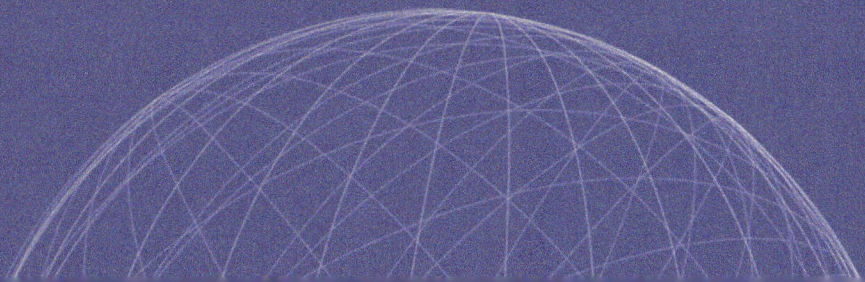

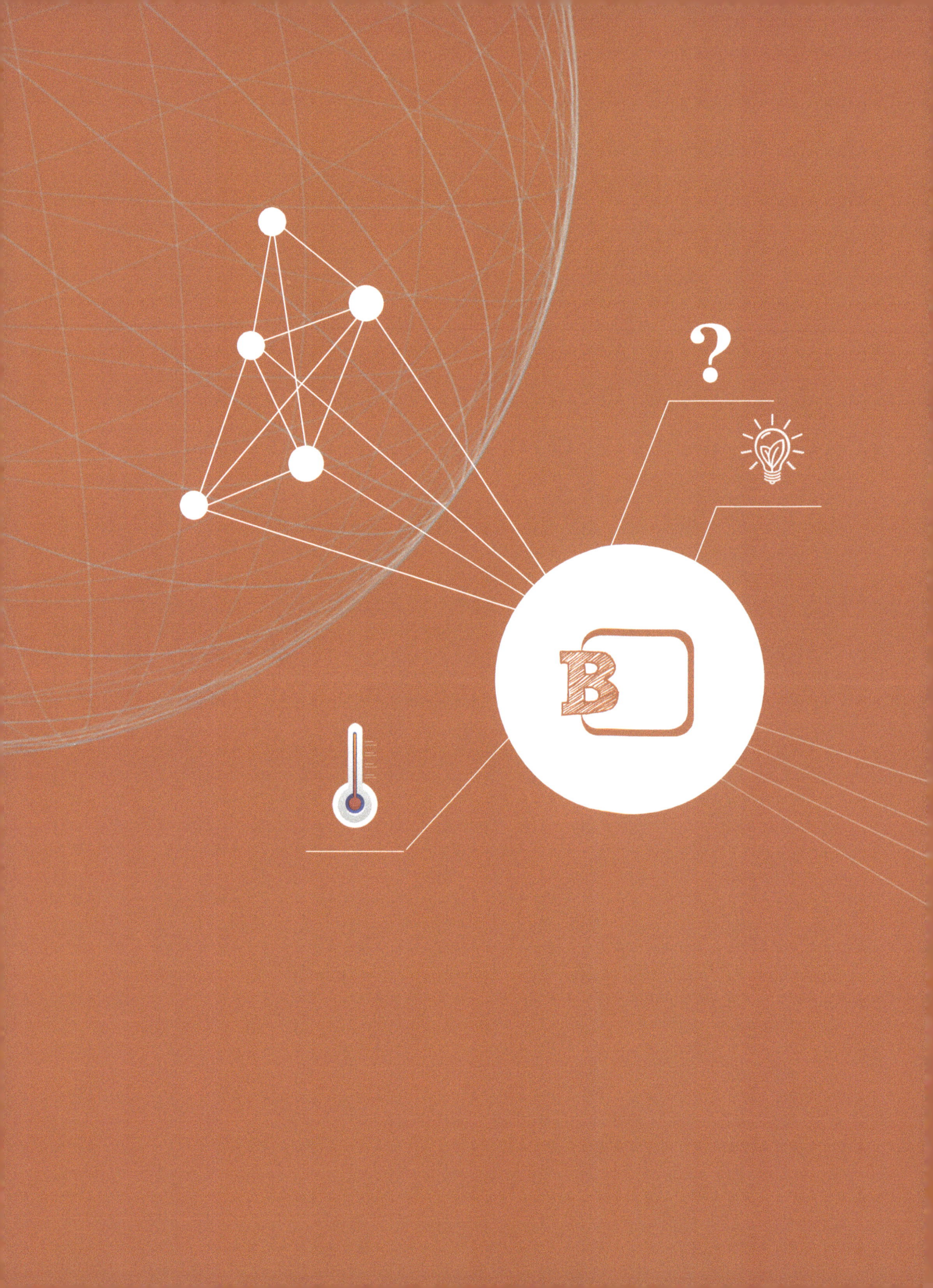

Chapter 4

Blogging

So far, we have discussed setting up the foundation for your climate change communication efforts; now it is time to learn how to take action. We start with blogging, as your blog will be the fountain from where your communication messages emanate.

There are many platforms available for starting a blog, but the simplest, least-time-consuming platform is the free publishing platform offered by Google: Blogger. If you have the budget and access to a Web developer, or if you have the time and desire to venture on a steep learning curve, you are in a position to start blogging from WordPress, which is a far superior platform in terms of options and features. My purpose with this book is to get as many motivated climate activists as possible to start blogging, so I chose the path of least resistance.

In this chapter you will learn how to:

- Set up and customize a blog on Blogger.
- Track the success of your blog.
- Write compelling blog articles for the Web reader.
- Optimize your blog articles for search engines.

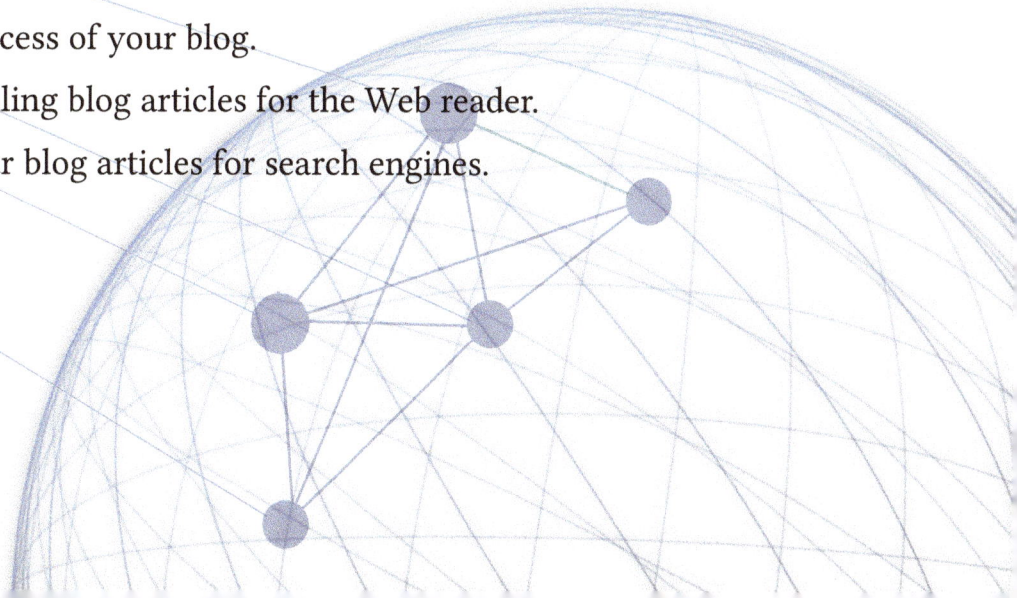

Content can take many forms; we will also review different types of content including images, videos, and infographs, as well as different types of tools and resources to help you create them.

Setting up a blog on Blogger

To set up your own blog, follow these steps:

1. Head over to www.blogger.com/home.
2. Sign in with your Google credentials (Figure 4.1).

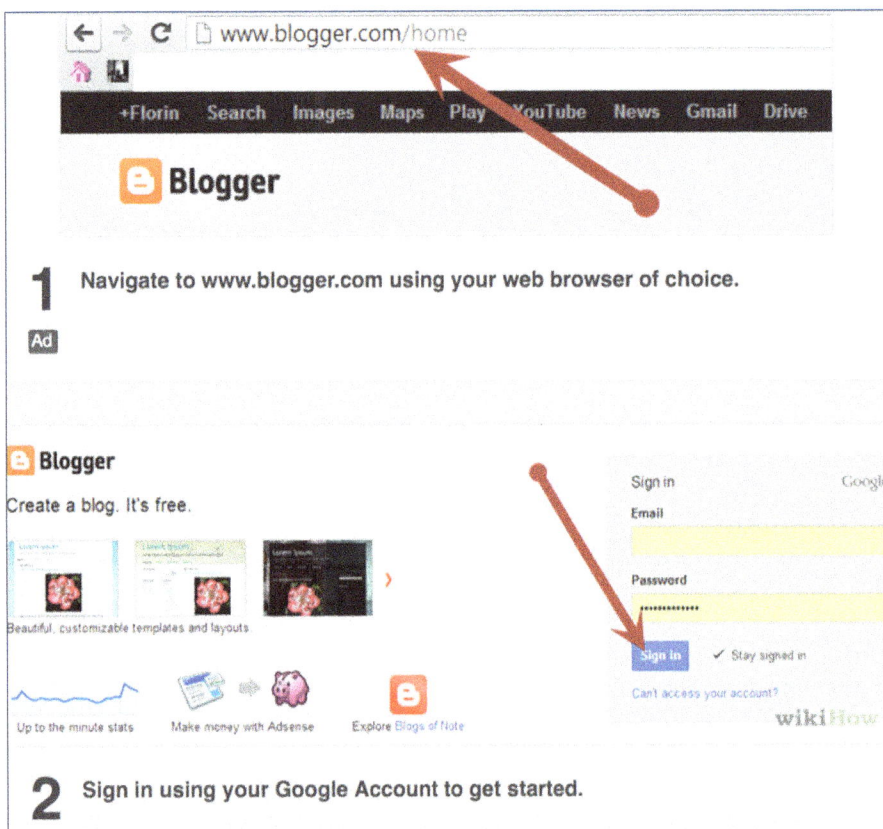

Figure 4.1: Setting up your blog—Starting steps

3. Click on "Create Your Own Blog Now."
4. Select a "Blog Title." This step may take awhile as the title you choose may be already in use. Take care to choose a simple, concrete, and memorable title. The "SUCCES" rules of sticky communication applies to your blog title as well (Figure 4.2).

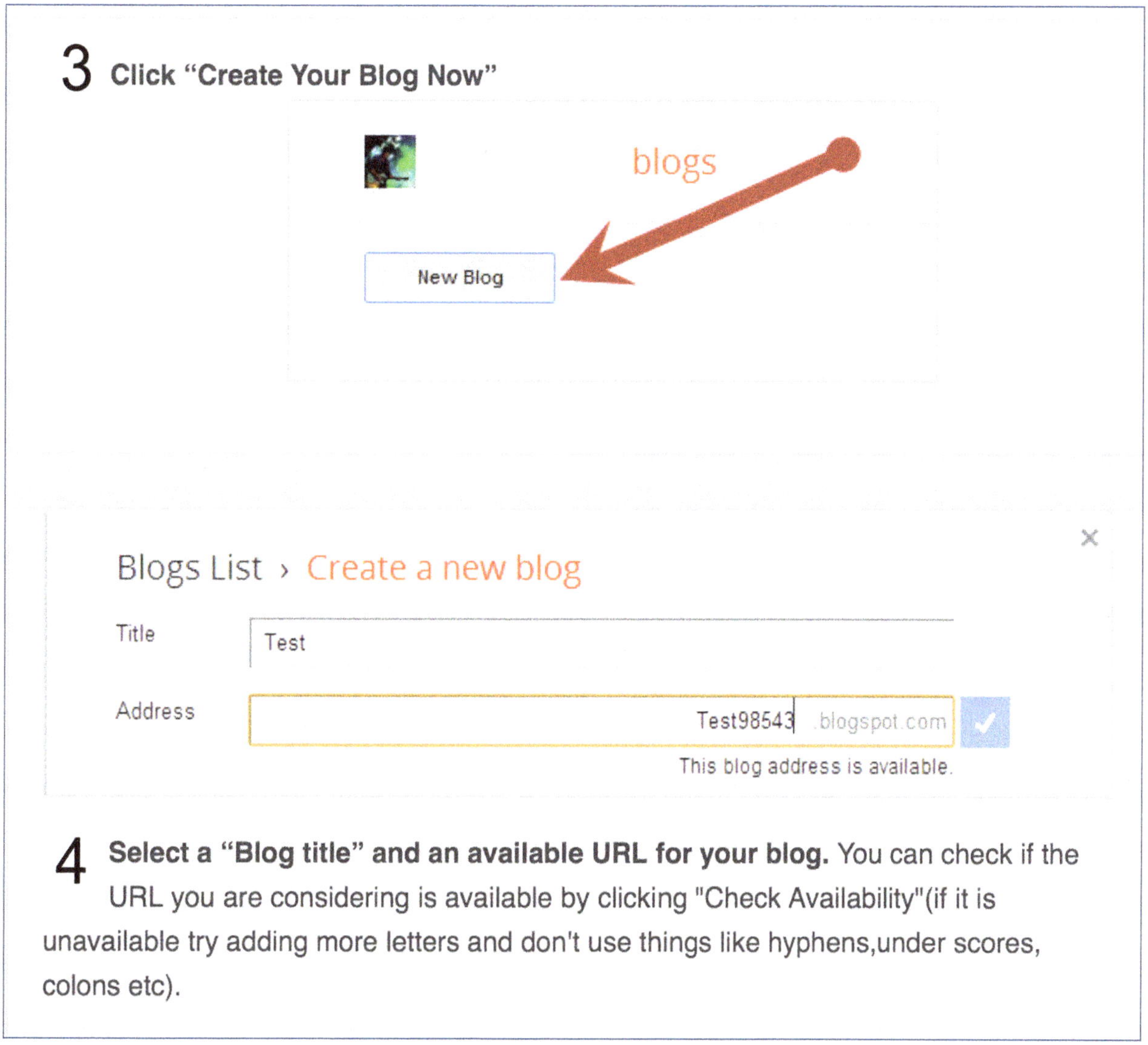

3 **Click "Create Your Blog Now"**

Figure 4.2: Setting up your blog—Choosing a title

5. You may want to choose a domain name and associate it with your blog in the future, but you can skip it for now.

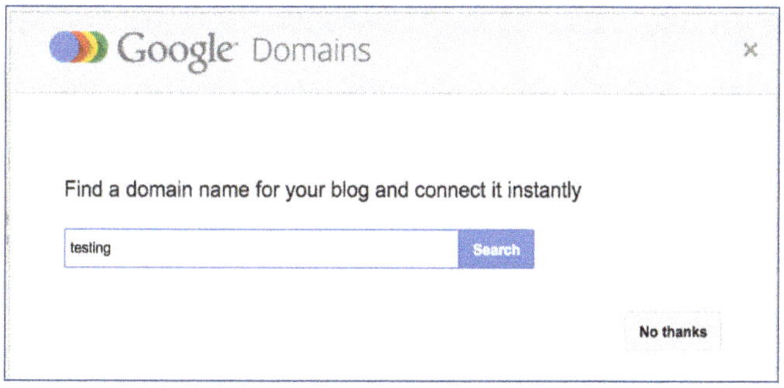

Figure 4.3: Setting up your blog—Choosing a domain name option

6. Choose a template and click "Create Blog" (Figure 4.4).

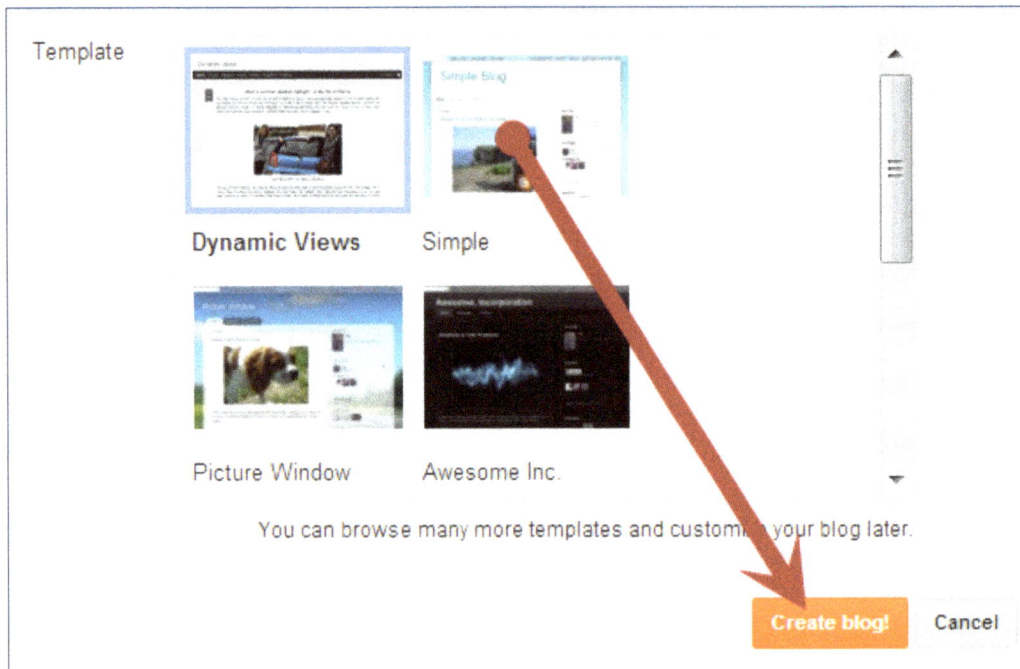

Figure 4.4: Setting up your blog—Choosing a template

You are now the proud owner of your own blog. You can choose from a number of different available templates, but I highly recommend starting with a simple design. You can customize the look of a simple template to match your personal style, color preference, and personal brand, but the most important part of your blog will be the content and how you communicate. Which brings me to one simple change to make your blog article easier to read: Increase the size of the font for your posts. There is nothing more discouraging to a reader than an overly complicated design and an article with fonts so small that it is difficult to read. Follow these steps to change the font size and other customizations.

7. Visit the home page for your blog and click on "Template" in the menu on the left. (Figure 4.5). Once in the Template page, click on "Customize".

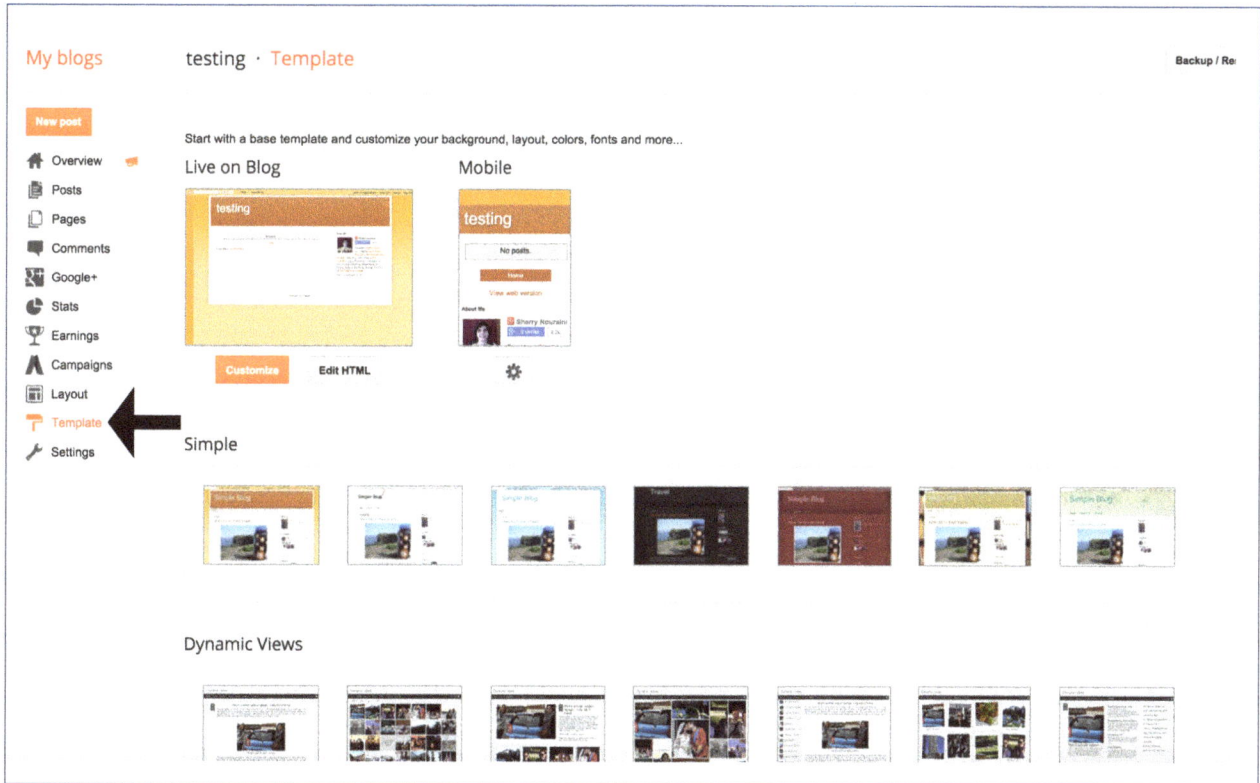

Figure 4.5: Blogger home page—Templates

8. Find the menu on the top left and click on "Advanced" (Figure 4.6).

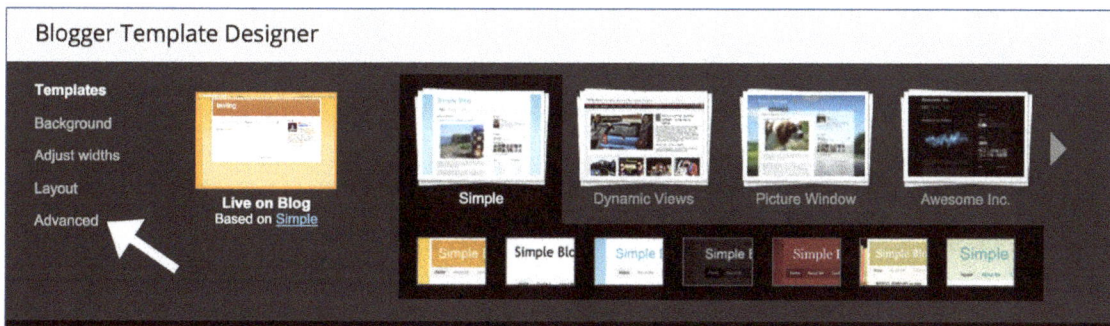

Figure 4.6: Advanced editing of a Blogger template

9. Choose "Page" and change font size to 16. If you don't make this change, your articles will be very small and difficult to read. It is also worth mentioning that many other elements of your blog can be changed here in terms of font type, color, and size, as well as the background color for your page. You can come back to this menu and change things at any time. Remember to click the orange "Apply to Blog" button on the top right so that your changes take effect (Figure 4.7).

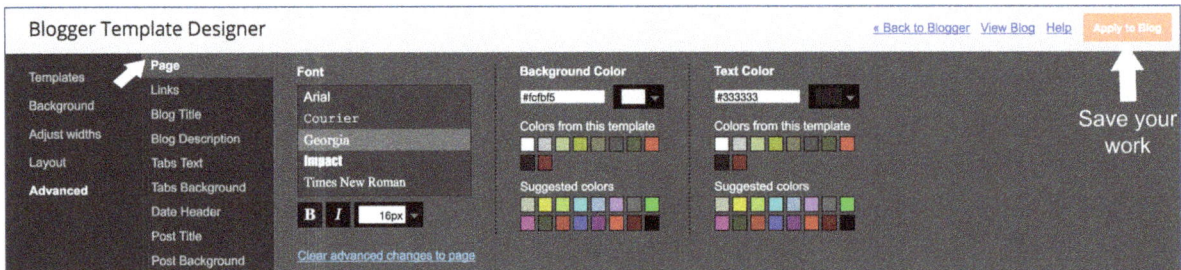

Figure 4.7: Increasing the font size for your blog

10. Writing a blog article. Now that you have set up your blog (wasn't that easy?), it is time to start writing. Visit the home page for your blog and click on the orange button "New post" on top of the left menu. This will take you to a new page with two tabs: Compose and HTML. If you are not familiar with HTML code, you can ignore the second tab. Compose is a "what you see is what you get" text editor, which you can use to write your article (Figure 4.8).

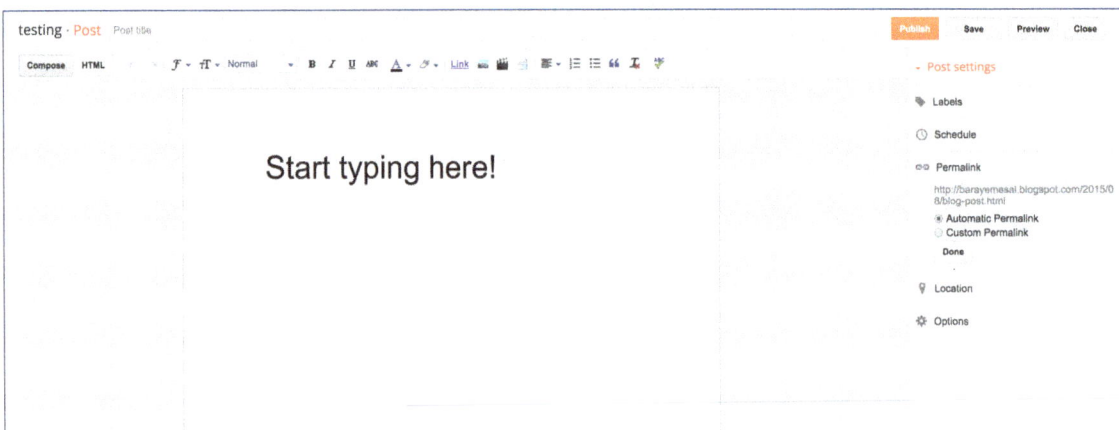

Figure 4.8: Blogger—Writing a blog post

The rest of the menu items on the top are different options for ways that you can enhance or modify your post. Also, on top of this row of editing icons is a box where you type the title of your post, as well as menu items on the right that will allow you to Publish, Save, or Preview your blog post. You can also simply close the page. Always preview your post before publishing to make sure it will look the way you intended.

The right column on this page features these menu items:

- **Labels:** Each of your articles should have a label (or tag) that helps categorize your articles into subject areas.

- **Schedule:** This tool allows you to write blog posts and schedule them for publishing at a later time. This is a great time-saving tool, as you can write a number of articles at the same time and set them up for publishing in sequence.

- **Permalink:** Each of your blog articles has its own URL, which is basically an address and is called a permalink. A permalink is important for two reasons. First, it allows you to direct people to your exact article instead of the home page for your blog, helping your audience find the specific article you want them to see. Blogs are set up so that they always show your latest article as the most recent entry, which means your older posts will become buried and difficult to find unless you can direct your audience to the exact page for a specific post using a permalink. An added benefit of a permalink is that it communicates with search engines what your blog post is about. This is important as you want to increase the chance of your article to be found when your audience uses Google or another search engine to search for information. You can choose either to have Blogger create the permalink or have one custom-made for each article, choosing your own words.

- **Location:** This option is used when you want to highlight a location in your post. If location tracking is on when people search for information using Google, location is one of the factors the search engine uses to match a search result with the person who is doing the search. So, if you want your post to have a higher chance of being discovered by an audience in a particular location, you can turn this feature on.

- **Options:** This section of your blog will give options for three things:

    i. Whether or not you want comments on your blog. I would encourage you to set this to allow, unless you get too much spam or too many trolls.

ii. There will be times when you will post HTML code into your blog post (we will see examples later in this book). Here you can choose either to view the HTML code as is, or if you want the text editor to interpret the code.

iii. How to choose line breaks. You can use either the
 HTML tag in your posts or just use the return button on your keyboard to indicate line breaks. This choice becomes important if you want to look in the HTML tab (or HTML mode) of the text editor to figure out where the line breaks are. I would choose the <br> option just to make clear distinctions where line breaks are in the HTML code (Figure 4.9).

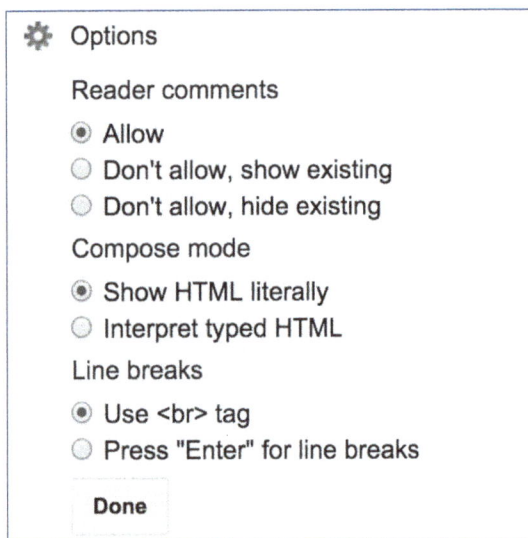

Figure 4.9: Setting up options for your blog

11. Uploading images. You can incorporate images (or videos) into your blog posts by clicking on the relevant icons in the text editor (Figure 4.10).

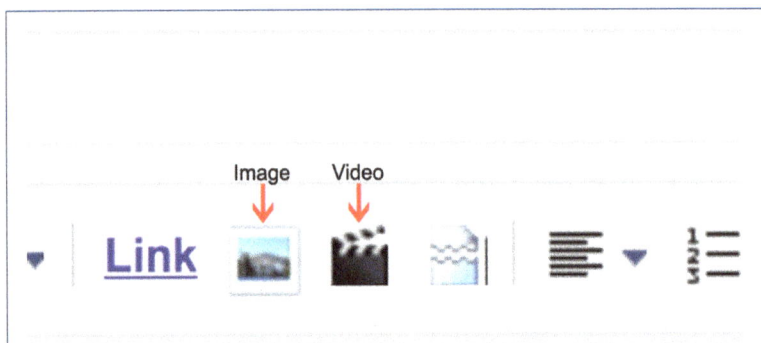

Figure 4.10: Uploading images to your blog post

When an image is uploaded to your post, Blogger will allow you to modify the scale of the image, set the image position in the post, add a caption, and also add text to describe the image properties. I encourage you to take advantage of these features to optimize the proportions and position of the image in the post, and leverage text descriptions to make your images searchable on Google (Figure 4.11).

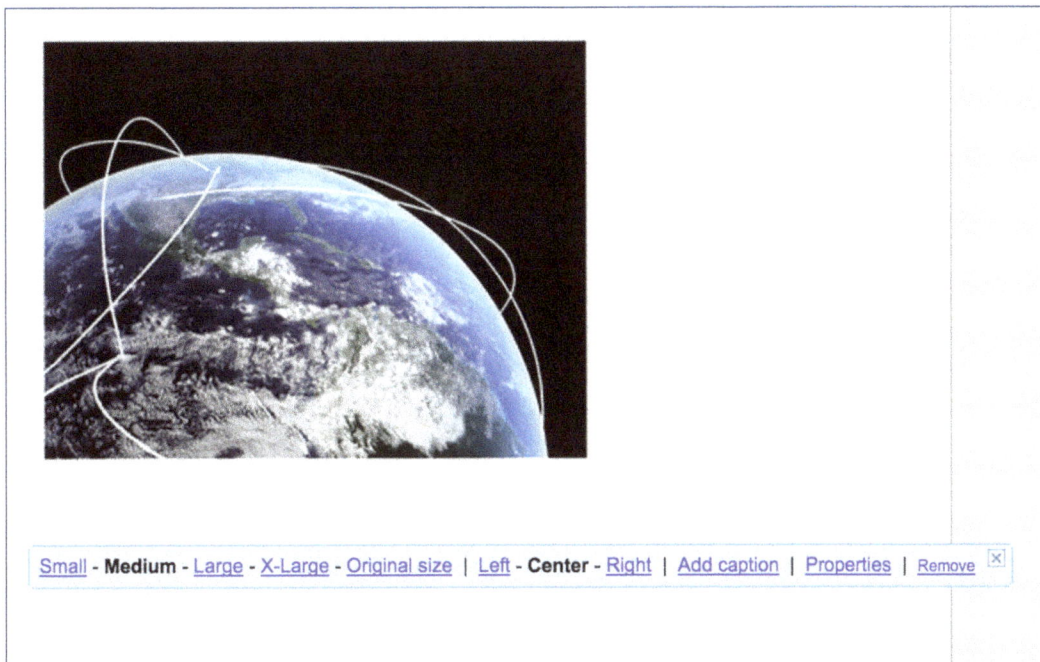

Small - **Medium** - Large - X-Large - Original size | Left - **Center** - Right | Add caption | Properties | Remove ⊠

Figure 4.11: Editing image properties

When you click to open "Properties" you will be given the chance to enter a Title for your image and an Alt Text (Figure 4.12).

> **Title:** This allows you to describe your image and will be shown to your blog visitor when they hover their mouse over the image.

> **Alt Text:** As the name implies, this is an alternative text that provides value in two ways: if the image is not displayed for some reason, alt text will at least provide a description as to what the image contains. It also helps with search engine optimization.

Both of these text descriptions allow search engines to understand what the image is about. Later on, we will talk about using appropriate keywords in your blog so that it can be found when people look for information on search engines. Your image title and alt text should also include these keywords (more on that later).

12. Modifying the layout of your blog. When you preview or publish your blog article, you will notice that an "About Me" box appears on the right column. (Figure 4.13).

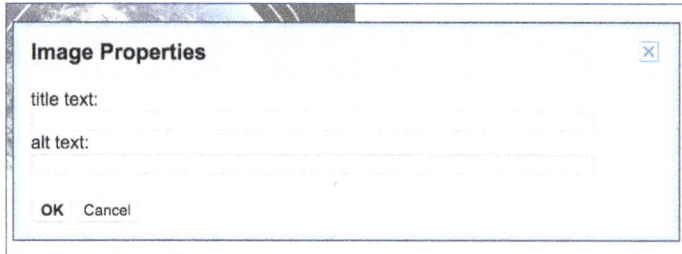

Figure 4.12: Setting Title and Alt Text for your images

The "About Me" widget is set up to show by default in your blog. You can change this setting if you like—or the entire general layout of your blog—by visiting your Blogger home page and finding "Layout" in the left menu. Once you are in the "Layout" page, find the "About Me" box on the right-hand side (Figure 4.14). Edit it to be able to change the title of this box (or Gadget as it is called in Blogger), or uncheck the "Show About Me" box to hide it. However, I would encourage you to keep it there as a personal touch to create trust with your readers.

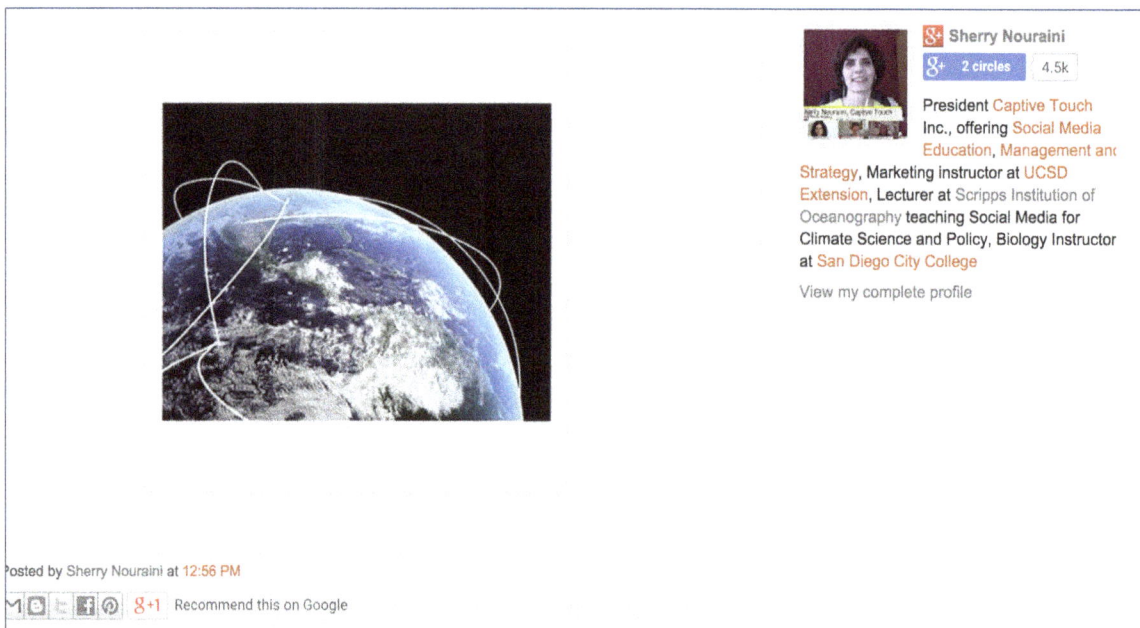

Figure 4.13: Your blog's "About Me" widget

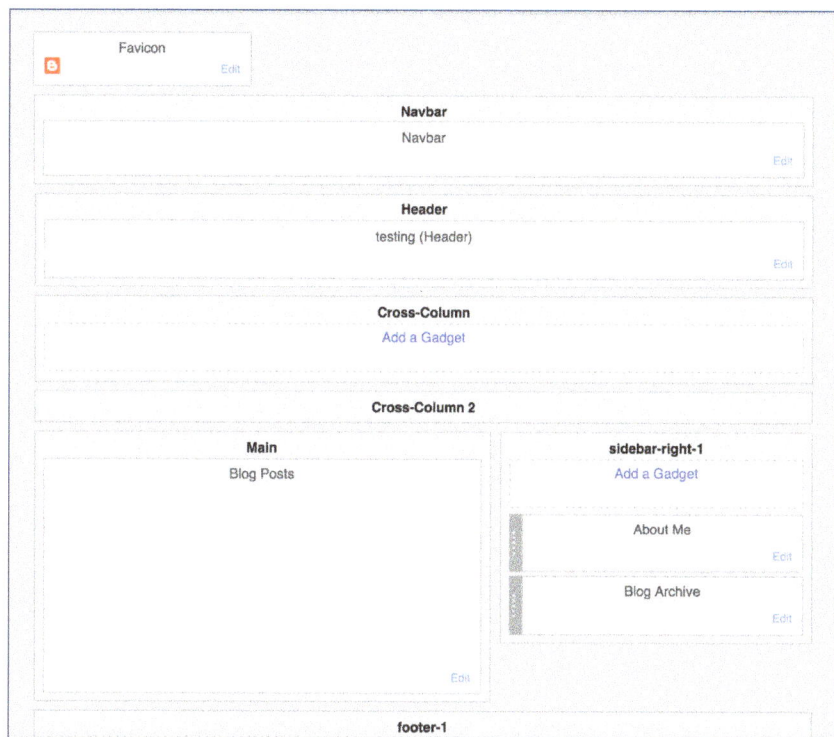

Figure 4.14: Editing the layout of your blog

13. Tracking the success of your blog. Blogger offers a basic analytics dashboard that gives you a general idea as to how your blog is performing in terms of attracting visitors (Figure 4.15). These stats are okay for a beginner, but once you become a more regular blogger, I suggest that you set up and learn how to use Google Analytics (4.1), which is far superior to the Blogger stats.

Here are the metrics for which Blogger provides data in the "Stats" section of the home page:

- **Page views:** This metric displays how many times a particular page has been loaded onto a browser. Be aware that your blog will be viewed by humans and by search engines' robots and spammers. What you care about is the number of humans visiting your blog, which might be as low as 40 percent of the page views that Blogger stats show you. Another source of page views is you, visiting your own blog. You can correct for this by choosing the "Don't track your own page views" (blue text on the right-hand side of the screen) option in the "Overview" section of Blogger stats (Figure 4.15).

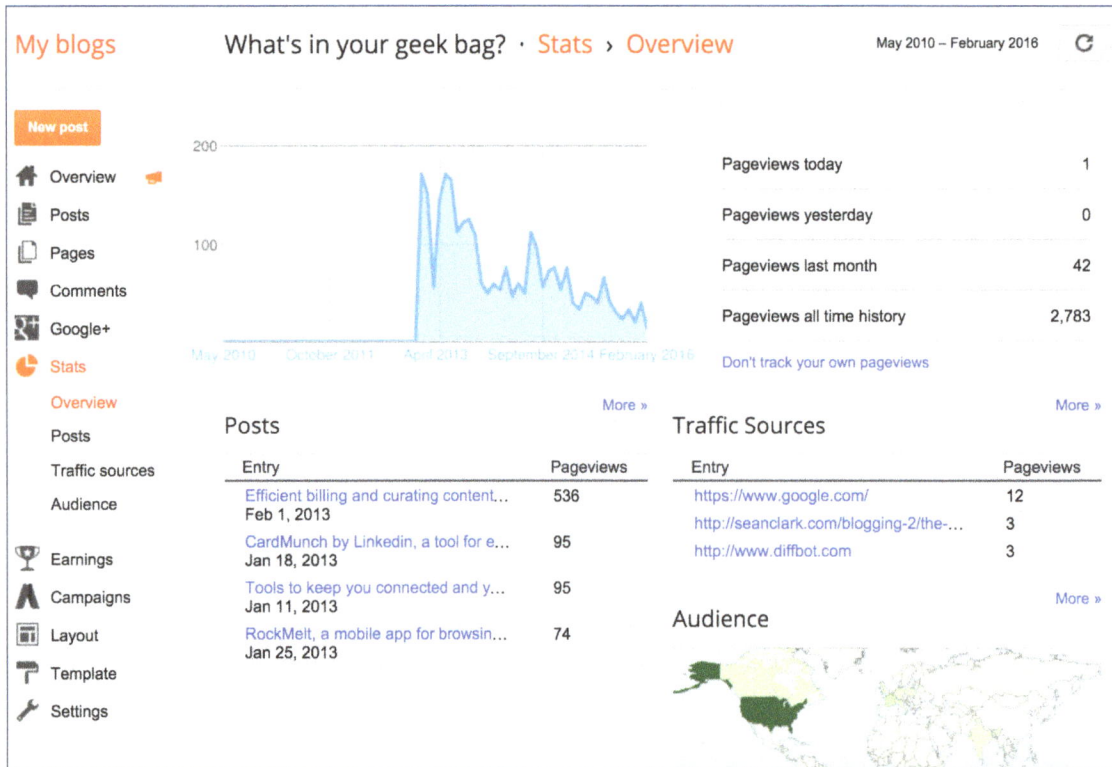

Figure 4.15: Blogger analytics—Overview

○ **Traffic sources:** This section of Blogger stats gives you information about where your blog traffic is coming from and what keywords helped draw Web visitors to your blog (Figure 4.16). Note that Blogger Stats only shows you the top ten performing entities, which is acceptable for a beginning blog. For a complete data set, you will eventually have to set up and work with Google Analytics:

▸ **Referring URL:** These are URLs for Web pages that include a link to your blog. This means someone found the link to your blog on that page and clicked through it to your site.

▸ **Referring sites:** These are websites that contained the URLs mentioned above. For example, if someone (or you) shared a link to your blog on Facebook, the referring site would be Facebook, and the referring URL would be the specific link that was shared on Facebook through which people clicked to get to your site. **A word of caution about spam referrers:** When going through your stats, you might find referring URLs and sites that have nothing to do with the topic of your blog. These are spam referrers who use this strategy to get blog owners to click to their websites. Google makes an effort to block these spammers but it is not always successful. The best way to stop them is not to click through the links. Once your blog becomes more popular among real readers, these spam referrers will be drowned out. The best way to accelerate the rate of gaining real readership is by building your network on social media sites or your niche forums, which is what this book aims to teach. You may also want to use a URL shortener, which I will cover later in this chapter.

- **Search keywords:** Keywords are terms people use to search for information on search engines (Figure 4.17). This is important information, as in addition to gaining readership though your social networks, you also want your content to be discovered by your target audience as they search for information online. Viewing the search keyword stats will help you determine which keywords worked best to help attract readers to your blog. We are going to cover later in this chapter a process for finding effective keywords to include in your blog articles.

- **Audience:** The stats in this section show data about the top ten locations, along with the browsers and operating systems used by the visitors of your blog (not shown).

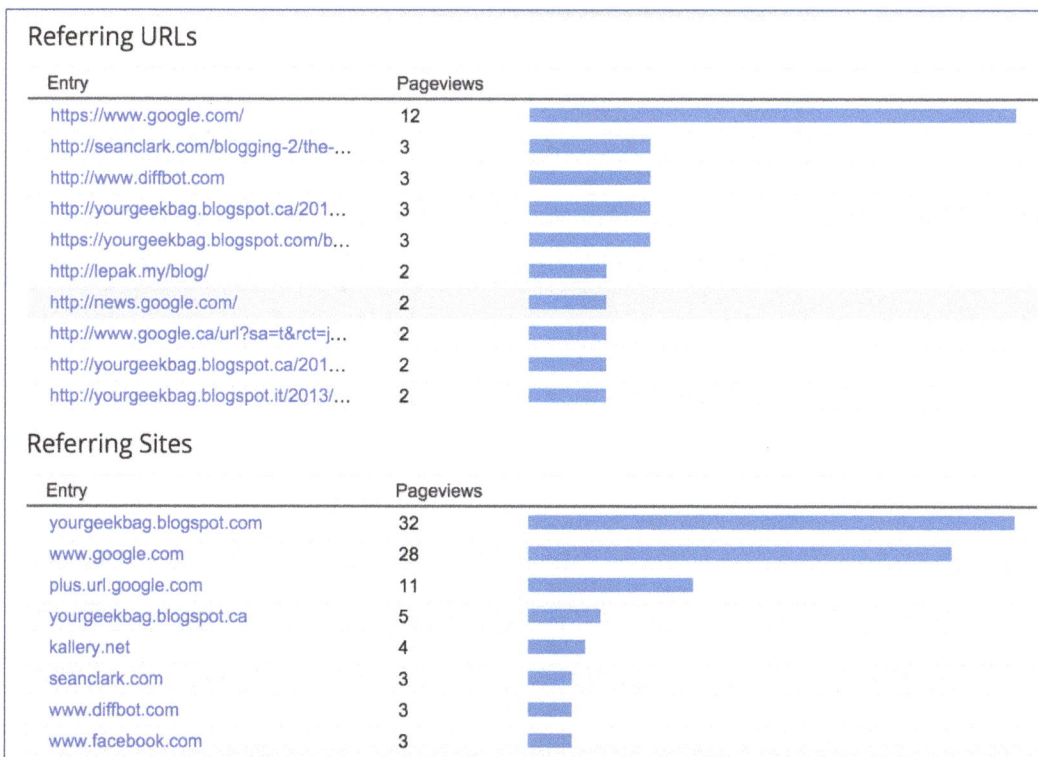

Referring URLs

Entry	Pageviews	
https://www.google.com/	12	
http://seanclark.com/blogging-2/the-...	3	
http://www.diffbot.com	3	
http://yourgeekbag.blogspot.ca/201...	3	
https://yourgeekbag.blogspot.com/b...	3	
http://lepak.my/blog/	2	
http://news.google.com/	2	
http://www.google.ca/url?sa=t&rct=j...	2	
http://yourgeekbag.blogspot.ca/201...	2	
http://yourgeekbag.blogspot.it/2013/...	2	

Referring Sites

Entry	Pageviews	
yourgeekbag.blogspot.com	32	
www.google.com	28	
plus.url.google.com	11	
yourgeekbag.blogspot.ca	5	
kallery.net	4	
seanclark.com	3	
www.diffbot.com	3	
www.facebook.com	3	

Figure 4.16: Blogger analytics—Traffic sources

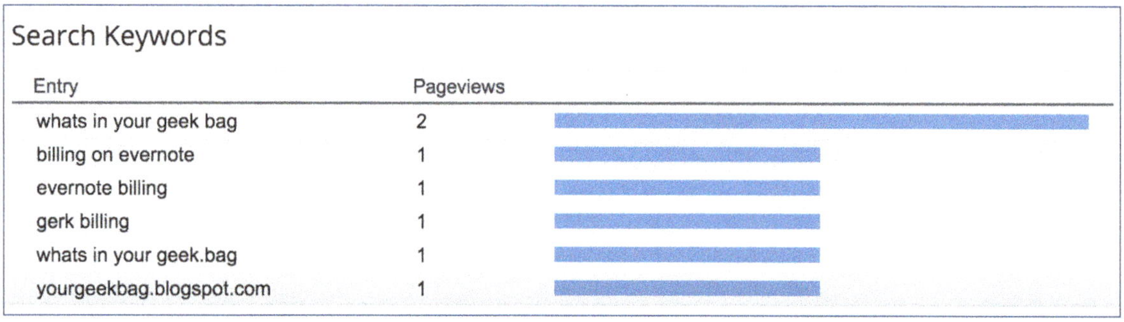

Search Keywords

Entry	Pageviews	
whats in your geek bag	2	
billing on evernote	1	
evernote billing	1	
gerk billing	1	
whats in your geek.bag	1	
yourgeekbag.blogspot.com	1	

Figure 4.17: Blogger analytics—Search Keywords

14. Set your blog to be friendly with search engines. In order to help search engines to make sense of your blog and index it, you'll have to tell them what your blog is about. To do this, visit the "Settings" section on the home page of your Blogger account and do the following (Figure 4.18):

> ▸ Create a blog title.

> ▸ Create a description of your blog.

> ▸ Edit "Privacy" to allow a listing of your blog and let search engines find your blog.

Basic

Title	Social Media Strategy for Climate Change Communication Edit
Description	This blog is aimed at sharing tips on effectively using the tools of social media for climate change communication. Edit
Privacy	Add your blog to our listings? ? ⦿ Yes ◯ No Let search engines find your blog? ? ⦿ Yes ◯ No **Save changes** **Cancel**

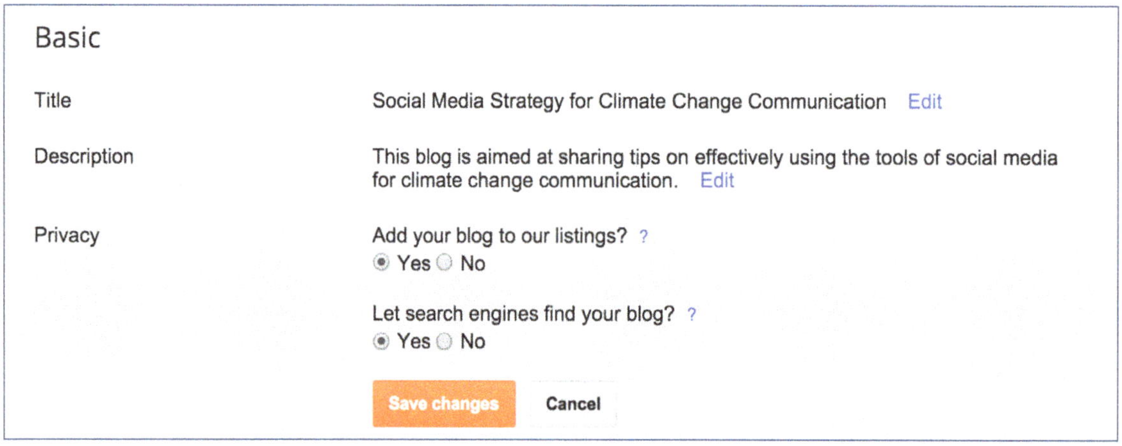

Figure 4.18: Setting up your blog title and description

15. Final thoughts. Increasing your blog readership takes hard work and dedication. It requires creating great valuable content, and proactively promoting your articles in your online communities. Be patient and you will be rewarded.

Deciding what to write about

So, you have set up a blog and are ready to write. Now what? What topics are you going to write about so that your target audience will read? The hard work that you did in Chapter 2 for defining your audience will come in handy again, because by now you have a pretty good idea of what they care about. Now, you need to write articles that connect the dots between what they care about and caring about climate change. Let's go back to the passage I quoted from Heath brothers in Chapter 3: "*The most basic way to make people care is to form an association between something they don't yet care about and something they do care about.*" This is where the framing table you created will come in handy to help you create a list of topics. Once you create a list of topics, think about keywords that your target audience would use on search engines to find information about them. With a list of topics and keywords in hand, write tentative titles or headlines for the articles you would write. I have done this exercise with our three example audiences; let's start with Linda Goldberg and learn why this exercise is important.

Table 4.1: Linda Goldberg—Tentative keywords and blog titles

Topics	Tentative Blog Title	Tentative Keywords
1. Restoring a flooded landscape in a Miami home	Is There Such a Thing as Flood Insurance for Landscaping? Or: Sea Level Rise Makes Flood Insurance Unavoidable.	Flood insurance Flood insurance rates Climate change Rising Insurance rates Sea level rise
2. Living a kosher life	A Checklist for Living Kosher	Living Kosher

Now, your tentative keywords and the corresponding article titles are just that: tentative. You must determine if these keywords are even used in search engines by a human being, and how often they are used.

Keyword research

You can carry out a keyword research investigation using a free tool called Keywords Planner in Google's advertising platform, AdWords. Just like the Facebook Audience Insights, there is no requirement to actually purchase an ad to use the tool, but you do need to have a Google account. The following steps will help you get there:

 I. Visit http://www.google.com/adwords and sign in.

 II. At this point Google will ask you many questions. Skip this step by choosing "Skip the Guided Setup" (Figure 4.19). This will take you to the AdWords dashboard.

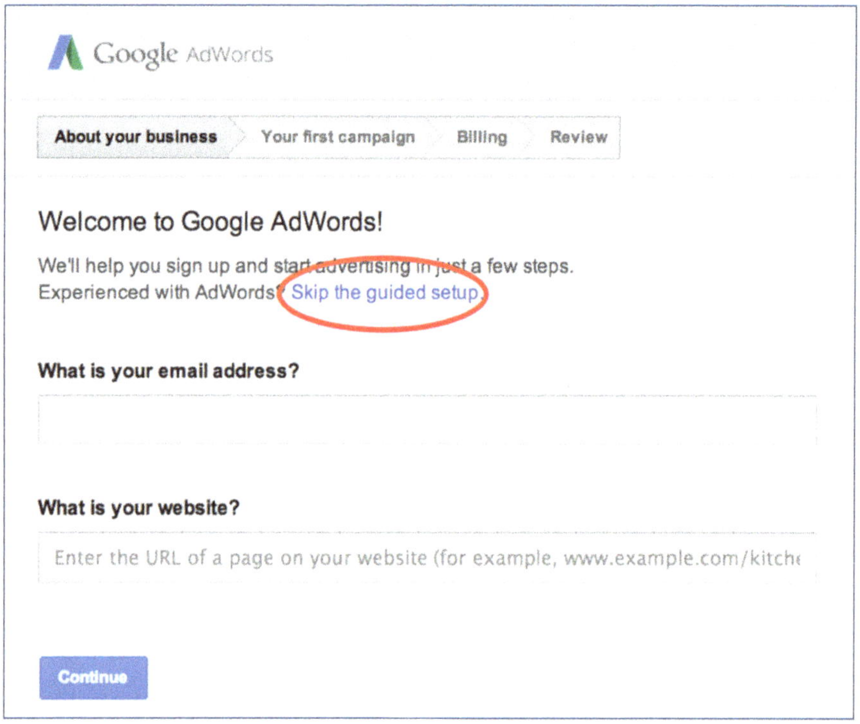

Figure 4.19: Setting up an AdWords account.

 III. Look for "Tools" at the top menu and choose "Keyword Planner" (not shown).

 IV. Under "Find new keywords and get search volume data" choose the second option: "Get search volume data and trends."

 V. Now you can type in your tentative keywords and choose a location, if you like, and click "Get search volume" (Figure 4.20).

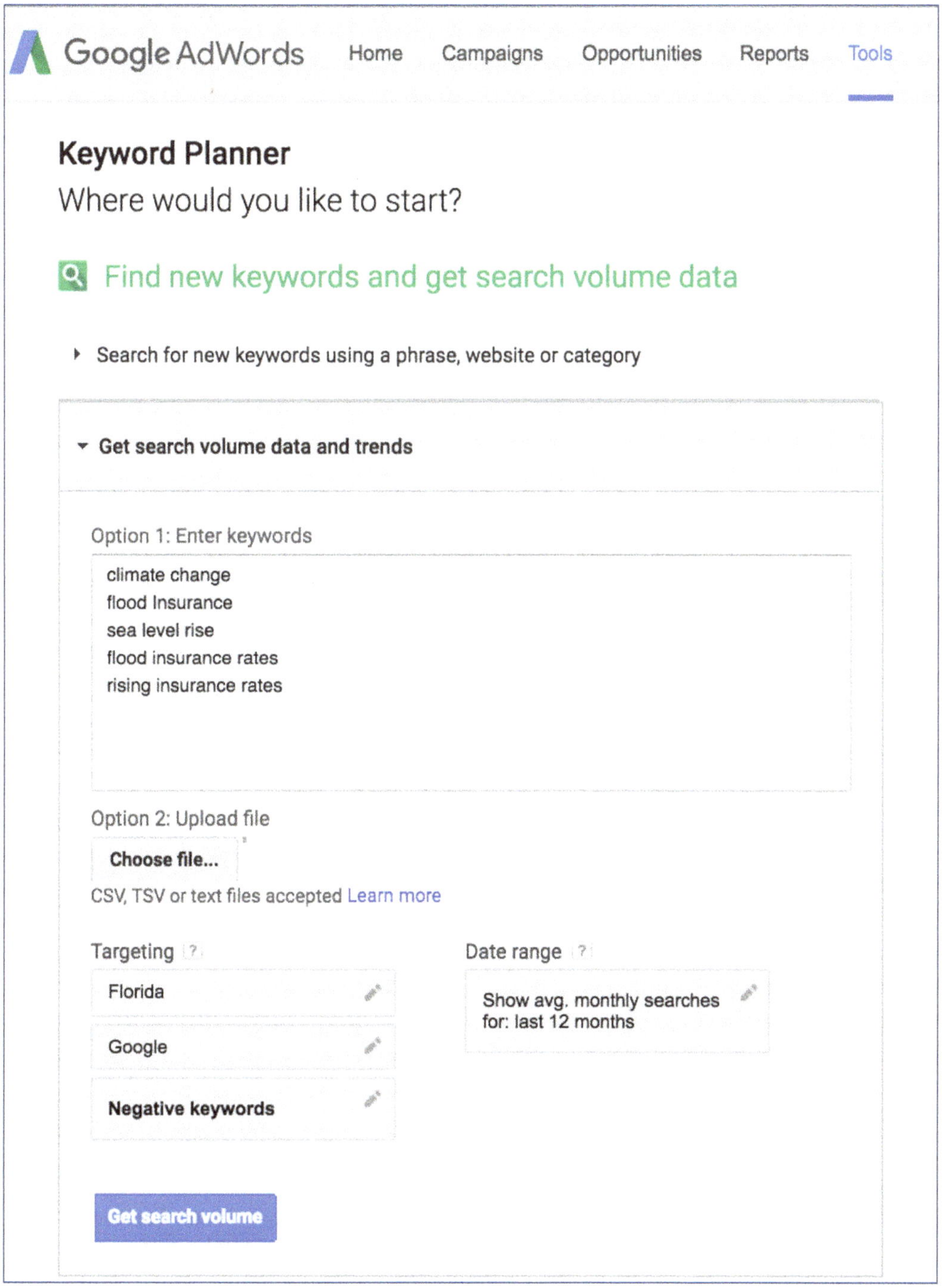

Figure 4.20: Google AdWords—Setting up search

Note that you also had the option of looking for tentative keywords starting with the AdWords tool. You can certainly do this if you cannot think of any keywords based on your audience profile. Once you become familiar with the Keyword Planner tool, I encourage you to experiment with other ways this tool can help you find keywords. You can find links to instructions right on the Keyword Planner dashboard.

Let's see what we can learn from the Keyword Planner tool about the tentative keywords we chose for Linda Goldberg.

- Topic1: Restoring flooded landscaping in Florida. All of our tentative keywords, except "rising insurance rates," have a decent number of monthly searches initiated in Florida (Figure 4.21). We conclude from this investigation that writing about rising insurance rates would probably be unwise because no one is searching for that information, so the keyword is removed from our list.

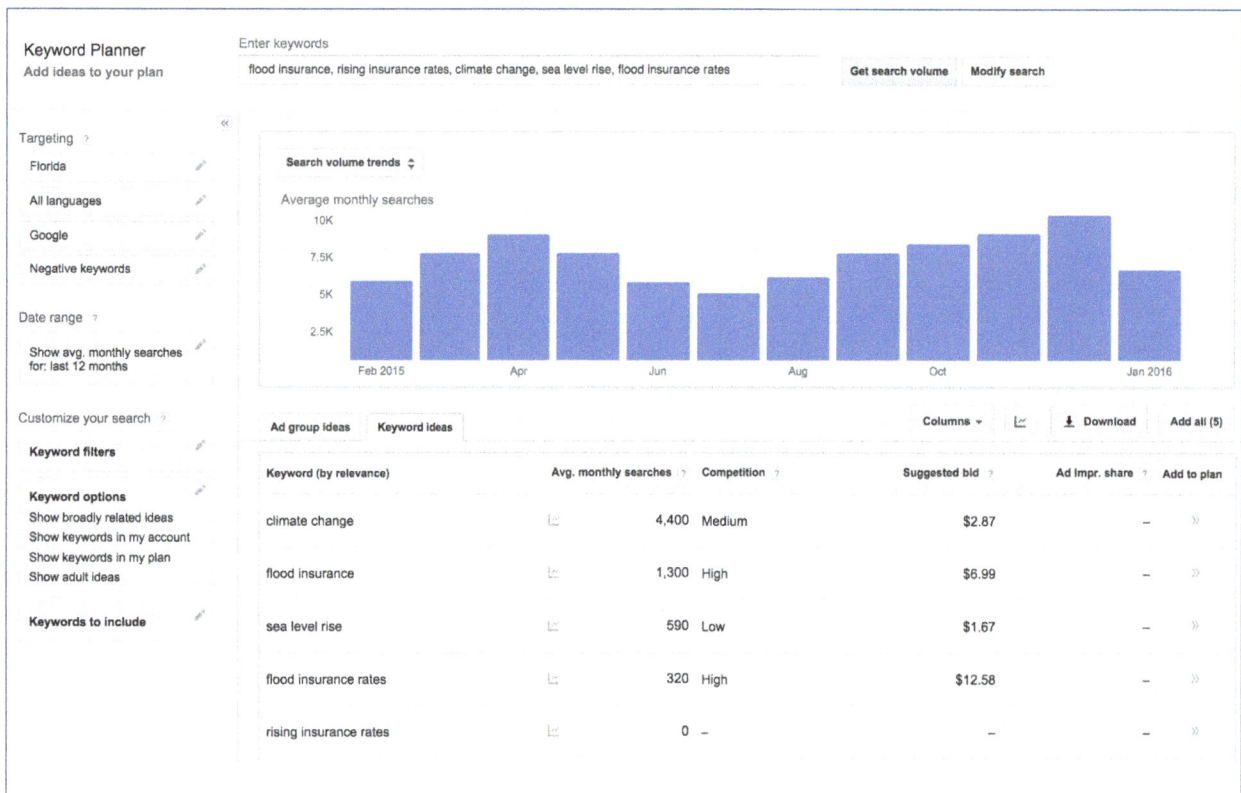

Figure 4.21: Search volume for keywords associated with Topic1

- Topic 2: Living a kosher life. In this case there is only one keyword but there appear to be no searches using this keyword (not shown). One can either try to find alternative keywords using the Keyword Planner, or use an alternative tool to find potential keywords. One such tool that is freely available is Übersuggest (https://ubersuggest.io), which presents a database of all alternative ways that people search on Google for a particular topic. This is a more efficient route as you will see.

Expanding your keywords with Übersuggest

How does Übersuggest work? If you've done searches using Google—which I am guessing you have—you have likely noticed that the search engine offers "suggestions" for other keywords on the bottom of the search results page. These suggestions are not mere guesswork by Google, but they are based on alternative keywords that real people have used when searching for information on a particular topic. What Übersuggest does is scrape all of these suggestions—based on a base keyword—and presents them to the user in a way that is easy to navigate. Let's put the keyword "living kosher" into Übersuggest and see what we can find (Figure 4.22).

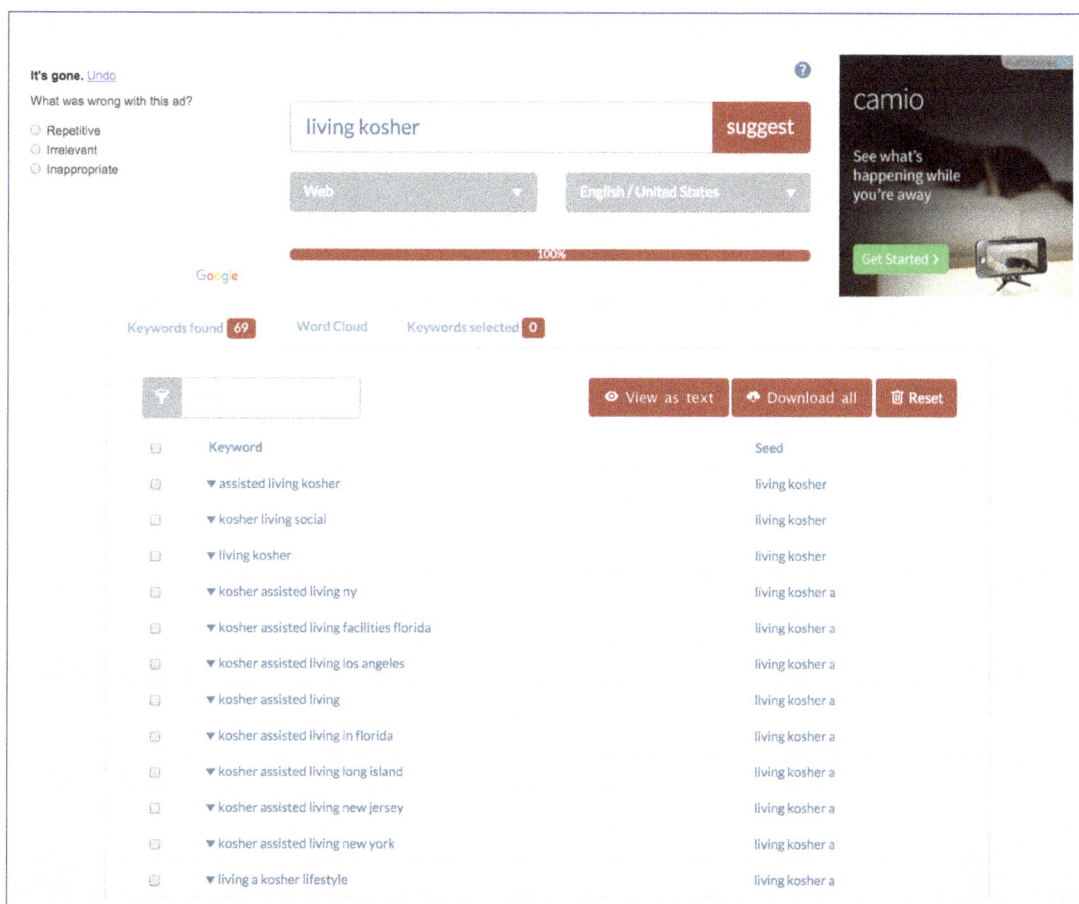

Figure 4.22: Übersuggest: Keywords related to "living kosher"

As you can see, Übersuggest has found sixty-nine alternative ways in which people have searched for living kosher. Going through all these alternatives is quite a monumental task, but the tool gives us an easy way to find possible gems by presenting a word cloud of this data. Just click on the next tab (Figure 4.23).

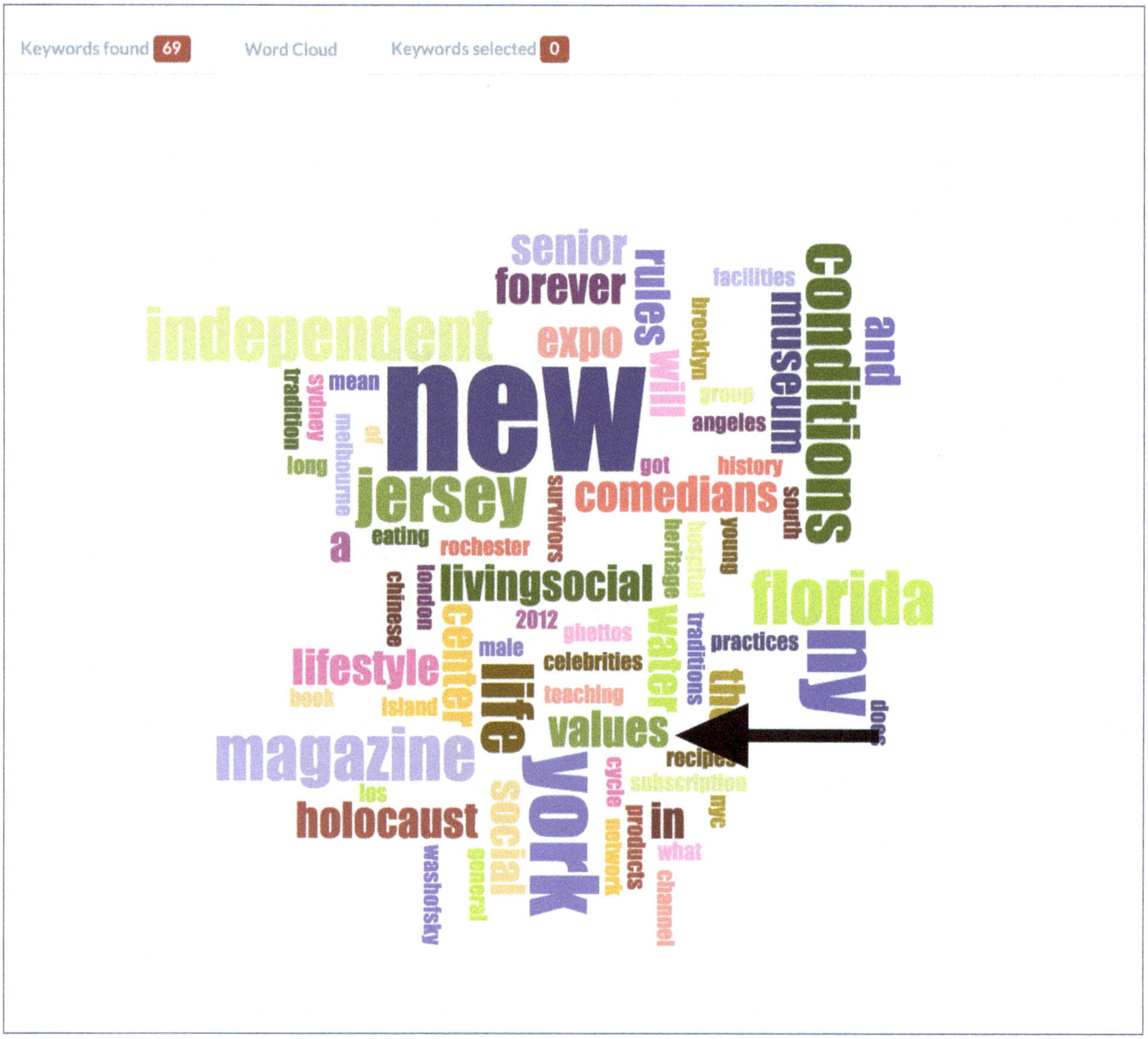

Figure 4.23: Übersuggest: WordCloud for keywords

An investigation of the word cloud reveals all sorts of alternative words that one can use for the "living a kosher life" topic. Let's try the word "values" and filter the keywords found to see if Übersuggest can help us discover a better option. Once we do this, we discover that "living Jewish values" seem to be an alternative for which real people actually search (Figure 4.24).

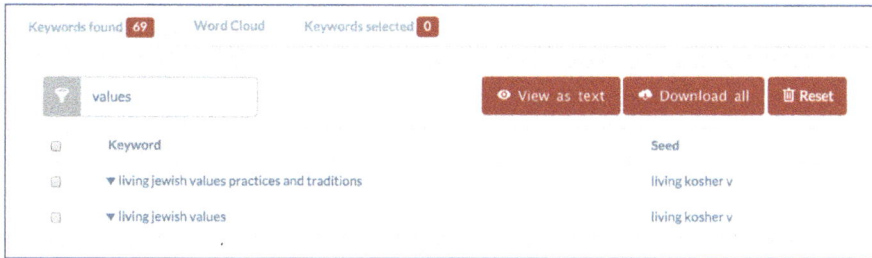

Figure 4.24: Übersuggest—Filtering keyword search results with "values"

Now the question is: What is the volume of search for this new keyword? We can answer this question by running "Jewish values" through the Keyword Planner tool, which reveals that although there are not thousands of searches for this phrase, there is a decent number (Figure 4.25). What other alternative ideas can you come up with using the WordCloud?

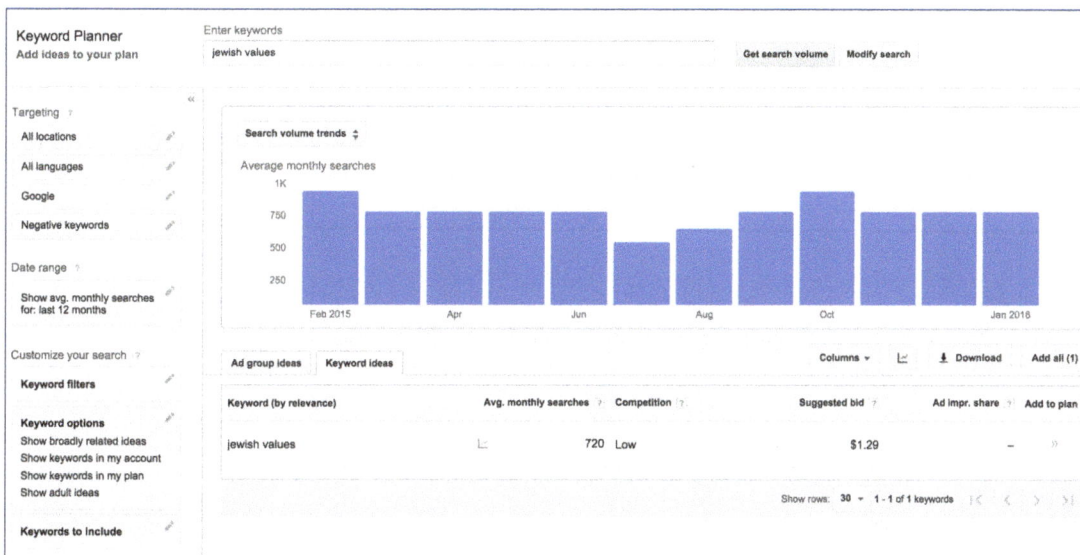

Figure 4.25: AdWords Tool—Search volume for "Jewish values"

After all this investigation, we can edit and create a final plan for writing articles as shown in Table 4.2. Note that using the WordCloud generator I could have come up with other combinations of keywords, but I think with this one example, you get the picture how keyword research works.

Table 4.2: Final keywords and blog titles

Topics	Blog Title	Keywords
Restoring a flooded landscape in a Miami home	1. Is There Such a Thing as Flood Insurance for Landscaping? 2. Sea Level Rise Makes Flood Insurance Unavoidable 3. Flood Insurance Rates to Rise Due to Climate Change	Flood insurance Flood insurance rates Climate change Sea level rise
Living a kosher life	1. A Checklist for Living Jewish Values 2. Choosing Green Energy and Living Your Jewish Values: You Can't Have One without the Other	Jewish values

Now let's see what our keyword research will look like for Paul Draper and Alex Donovan (details not shown, just final products).

Table 4.3: Keywords and blog titles targeted to Paul Draper

Topics	Blog Title	Keywords
Ocean acidification	1. The Call of Cthulhu to Stop Polluting His Home 2. Would Cthulhu Live in the Ocean Had H.P. Lovecraft Known about Climate Change?	The Call of Cthulhu H.P. Lovecraft
High ocean temperature	1. Warming of Oceans Will Be Fall of Cthulhu	Cthulhu Fall of Cthulhu

Table 4.4: Keywords and blog titles targeted to Alex Donovan

Topics	Blog Title	Keywords
Sustainability technology solutions	1. Developing Sustainable Business Practices in Your Office 2. Top Sustainable Technology Companies	Sustainable business Sustainable technology
Incremental steps toward sustainability	1. How to Build a Sustainable Business by Implementing Small Changes	Sustainable business

Alternative sources for writing ideas

What if you draw a blank about what questions your target audience would be asking regarding a particular topic? The solution is searching inquiry-based websites such as Quora.com, Yahoo Answers (http://answers.yahoo.com) or Ask.com for questions that real people are posting online. Here is a list of questions I was able to find on Quora.com using the search term "business and climate change."

- What examples of business-led action on climate change are there?

- What are the major challenges small and medium businesses face when grappling with natural disasters and climate change issues?

- What can an individual, a family, or a business in X (country) do today to prepare and adapt for the coming impacts of climate change?

Searching Yahoo.com with the search term "Judaism and Environment," the following questions were discovered:

- What does Judaism teach about the correct use of the environment?

- How does the environment influence Judaism?

- What does Judaism teach about the environment?

- What does Judaism say about humans' responsibility against the environment?

Note that the value of these findings is not only in the questions you discover, but also the answers that are posted, as they can be a valuable source of information to include in your articles.

Keywords aren't everything

So far, you have learned how to write content that is concise, clear, free of scientific jargon, and memorable. You also learned to create and research for keywords that increase the chances of your content being discovered in search engines. But just because you write well and use the correct keywords doesn't mean your blog will appear on the first page of Google when your target audience searches for information. There are other factors involved, which have to do with the way Google ranks websites for authority and how its smart robot search engines decipher the meaning behind the words in your blog posts. It is impossible to know all the factors involved in Google's search ranking algorithm, but the most common tactics that Search Engine Optimization (SEO) experts offer are the following (4.2, 4.3):

1. **Produce *elite* content.** In other words, just good or great content does not suffice because—in this age of information— your audience is probably inundated with content everywhere they look, so your content has to be *so* good, they can't ignore you.

2 **Get links (citations).** Just like the value of scientific publications is partly determined by how many times a publication has been cited in other high-quality journals, Google gives content a higher value score if it is cited by other high-authority sites (4.3). In SEO language, citation is expressed as links, where within a high authority Web page, there is a link that directs the reader to your content. Authoritative sites include blogs owned by bloggers who influence your target audience's opinions, well-known publications, and major social networking sites. Later in this book we will cover tactics you can use to discover and create relationships with your target audience and their influencers so that your content may be cited.

3. **Have a mobile-friendly website.** More searches are done on a mobile device as opposed to a desktop in ten countries, including the United States and Japan (4.4). If your website is not mobile friendly, it will receive a negative score by Google search robots. Fortunately, Blogger is already mobile friendly, but in case you decide to move to another blogging platform, make sure to use a mobile-friendly template.

4. Have a social-sharing strategy in place. Social networking sites like Twitter, Facebook, LinkedIn, and Pinterest enjoy a very high-ranking authority. In addition, active social sharing of your content—either by you or your followers—can lead to an increased number of citations (4.3). Later we will cover tactics to encourage social sharing.

Text isn't the only form of communication

By now, I have covered creating your own blog, developing a strategy for your content, and SEO tactics to help search engines find them. I've mainly focused on written content, but communication can sometimes be more effective if it is delivered visually via images or video. Let's talk about tools and resources for creating multimedia and best practices for using them.

Infograph as a great data visualization tool

Data can be quite dull and overwhelming. However, one can use the power of visual elements to communicate data or any complex concept. You saw an example of an infograph in Chapter 3 (Figure 3.1) where I summarized data about climate change knowledge gaps among the Six Americas, and offered suggestions for communication starting points for each gap. Let's see another example where information about effects of climate change on oceans is being communicated with Cthulhu enthusiasts like Paul Draper (Figure 4.26). This infograph is Simple, Concrete, and Emotional—three important elements of a piece of engaging content that adheres to the Heath brothers' SUCCES principle described in Chapter 3. It will capture the attention of Paul, who is passionate about Cthulhu (Emotion). Information about sources of Carbon Dioxide (CO_2) and consequences for the ocean are clearly presented (Concrete) and organized separately in two sections of Cthulhu's body parts (Simple).

What if you—like me—lack graphic design experience? How would you create infographs? Infographs and graphic design have become such a popular means of communication that tools have been developed to allow even the design-challenged communicator to put together a decent graphic. Some of these tools are featured in the Appendix.

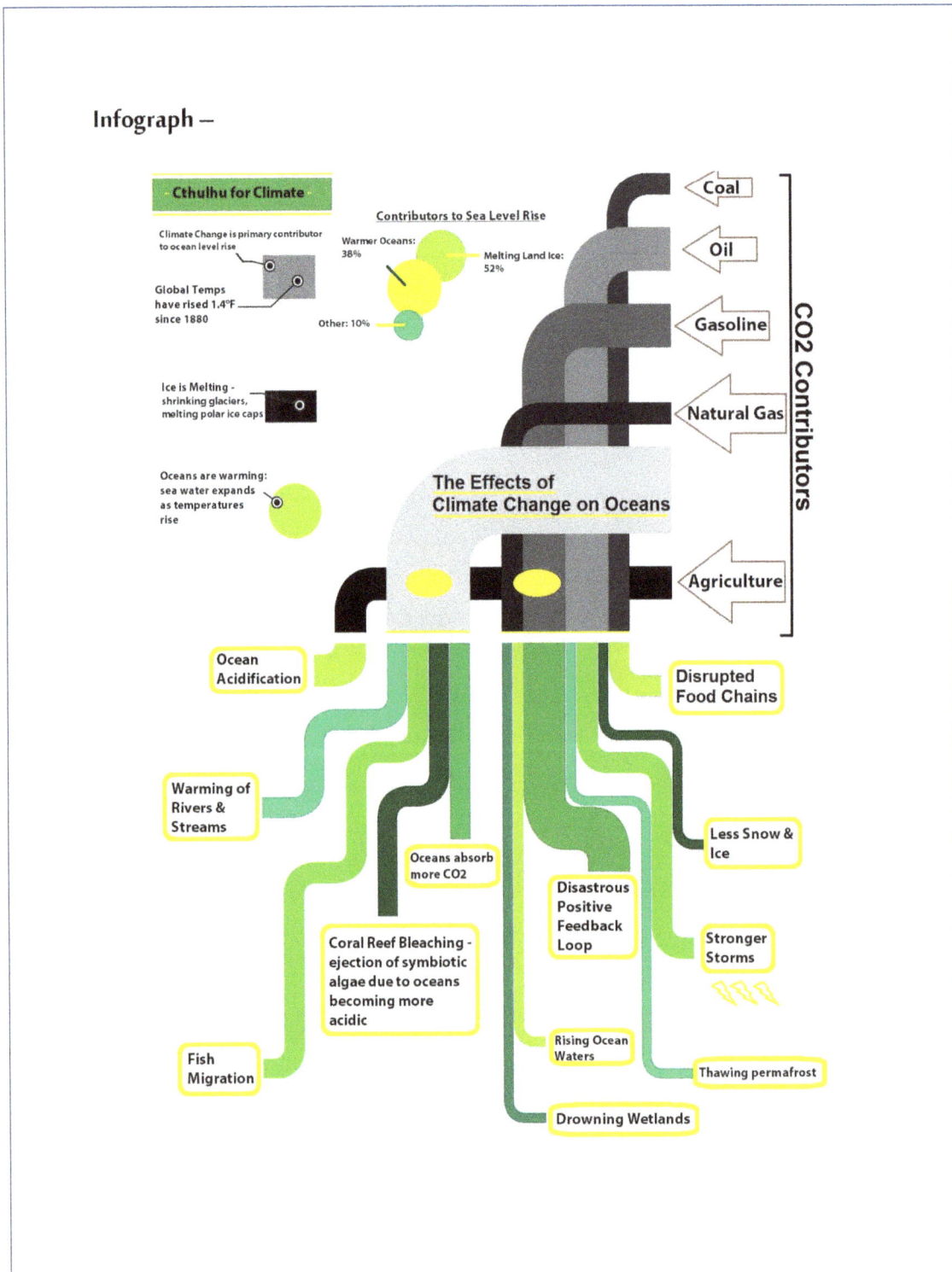

Figure 4.26: Infograph communicating climate change with an audience fitting the profile of Paul Draper. Infograph created by Kris Scarci, printed with permission.

Digital video as a rising means of content consumption

A comprehensive science communication plan must include digital video. There is no escaping it, and here is why: In its 2015 Q3 earnings report, Facebook reported that more than 8 billion videos are viewed on this platform each day (4.5). In addition, in a 2015 case study evaluating digital video viewership, Nielsen reported that while television viewership declined by 10 percent between December 2013-December 2014, time spent on YouTube increased by 44 percent within the same period (4.6). Livestream video production and viewing via mobile livestreaming apps is also a rising trend, although it has not reached the level of adoption as Facebook or YouTube video viewing (4.7). Just this year, Facebook made available the possibility of streaming live vides on its platform, and group video conferencing via its Messenger app. Recording and publishing videos have never been easier, often requiring just a smart phone and an internet connection. Most laptops are also equipped with a camera and video editing software. All you need is a good story, and a will to hit "record"!

Obviously, with a nonexistent barrier to making and publishing videos comes the challenge of competing for your audience's attention. To win this competition, use the strategies you learned in Chapter 3 for creating effective content: what works for written content also applies to video. In addition, communicating via video has its own inherent challenges that are outlined below, along with tactics you can employ to overcome them:

- **Getting them to click "play."** Your first challenge is getting your audience to click the arrow symbol on your video to actually watch your content. Every video has a "thumbnail," a cover image that is rendered when your video is shared online. A well-designed, clear, and concise thumbnail will increase your chances of getting your audience to view your videos.

- **Getting them to stay.** Your next challenge is to keep your viewers watching the video. The first five to ten seconds of your video are key to viewer retention. Make it engaging so that people *want* to keep watching. Video viewers decide early in their watching journey whether or not they want to complete the viewing of your video. One way to get people to stay is to make your original punch-line points in the beginning and provide the supporting material in the remaining part of the video (4.8).

◑ **Getting them to finish it.** In this information age, consumers of Web content have a short attention span. If you can't communicate your point within two to five minutes, they will click away from your video (4.8). You may be wondering how you can possibly communicate a big topic like climate change in less than five minutes! If you've had to write a scientific abstract, then you already have the skills to be concise. In addition, you have the option of breaking down your message into topic points and creating a short video for each point. Note that you don't have to reinvent the wheel every time you create content. Your blog articles, which tend to be more comprehensive than short videos, are a great source for the above mentioned "topic points." Create the content once, and then repurpose it from text into images, a slideshow, an infograph, or a video.

◑ **Keywords are important here too.** If you spend the time and effort to create videos, you might as well set them up for discovery in a search engine. You have already done the work of finding keywords for your blog articles. You can use the same keywords in the title and description of your videos. It is also a great idea to upload a transcript of the video to the platform on which you are uploading the video, and create captions for your videos. Creating captions are especially important for a mobile audience or those who watch videos on Facebook, as captions allow the viewer to consume the content without sound: they could be *anywhere,* but they still can watch the video and absorb the message therein. An added benefit of captions is they make your videos accessible to those with hearing disabilities, and compliant with the Americans with Disabilities Act (ADA). ADA requires accommodation for the disabled by all places of public accommodation (4.9). To ensure compliance with this law, some websites will not allow posting a video without captions. If your videos are not captioned, you risk the chance of reducing the range of audiences you can reach.

◑ **What if you don't like cameras?** You may be hesitant to appear in front of a camera, but you don't have to! Video content can be prepared using a slide show or an animated story. There are a variety of video creation tools (see Appendix) that allow you to tell a story via cartoons and illustrations, and make your point in a more artistic way.

◑ **Example videos.** There is no shortage of video content that attempts to explain climate change, but there are only a few that I have been able to find that come close to meeting the criteria for the potential to capture and retain a nonexpert audience. I list them here:

▸ Steroids, baseball, and climate change (4.10)—a great analogy between baseball batting average and global warming. This video is short and uses a great analogy to explain temperature variations as related to global warming. However, there is quite a bit of scientific jargon in the script, and the video thumbnail along with the first few seconds of the video could be more engaging (4.10).

▸ Animated videos explaining science of climate change (4.11). Videos are short and concise. These videos would make an excellent resource for teaching in a science class, but they would not be suitable for engaging the public (4.11).

▸ The Climate Change Playlist (4.12). The videos in this playlist are hosted on the "It's OK to be Smart" YouTube channel, and they are perhaps the most engaging climate change education videos I have seen. However, they tend to be about seven to eight minutes long, which is a bit longer than optimal (4.8), and better suited for a science class. Nevertheless, this set of videos is a great model to follow for creating engaging videos (4.12).

Practice what you learned

In this chapter you learned:

1. How to set up your own blog.
2. How to monitor the success of your blog.
3. How to develop content ideas.
4. How to create and research suitable keywords for search engine optimization (SEO).
5. Tactics to employ to position your blog for better ranking.
6. Alternative forms of content.
7. Best practices, and resources for creating engaging videos.

Now it is time for you to practice what you learned and start blogging. With your audience profile in hand, do your keyword research, come up with titles for your blog posts, and write an article. Then practice repurposing your article into other forms of content. Be sure to join our Facebook groups (http://facebook.com/groups/social4climate) and share what you have come up with, and ask questions.

Content creation would be a waste of time if no one finds and consumes the content. We've already covered tactics to make your content findable on search engines, but a more targeted and effective means of generating readers is sharing it with an audience who shares your values or trusts you. No other tool is more suitable for this task than social networking sites. In the next few chapters we will cover tactics for finding and building a community of followers and readers using the tools of social media.

Tools of social media

Tools of social media are always in flux, and features change on a daily basis. New social networking sites enter the market frequently, and some old ones die or dwindle in user base. The continuous change in the digital world is the reason why it is difficult to write a "how-to" book about social media tools without it becoming obsolete the day it is printed. However, there are major social media channels that have staying power, and all share elements that will always be relevant. In this section I focus on selected mainstream and enduring social media tools, and basic features within that will help you find and form a community with your target audience. These basic features are fivefold:

- **Profile:** Your identity. Most social media profiles have a number of elements in common: your biography (bio), a profile photo, a link, a location, and a banner image. Your profile basically tells the world who you are. When it comes to creating your bio, think about including keywords that define you and your interests, as well as those for which your target audience may be searching. The keyword research that you did in Chapter 4 can be a starting point.

- **Organization:** Organizing your interactions. When it comes to efficient social interactions, managing your social connections is everything. Most social networking sites offer tools for contact management, and for those who don't, there are tools to help you organize.

- **Search:** Filtering social data to find what you want. Millions of people use social media every day, and by doing so, enrich these networks with tons of data. This richness of data is a valuable feature of social networking sites that most people tend to ignore. The search engines built into these social networking sites will enable you to find your target audience based on the words they use in their profiles and status updates.

- **Status updates and terminology:** Your message and language. Each social network offers a variety of different options for the types of content you want to share. Some options are more restrictive than others, but they all allow you to post text, images, and videos. The language of social media networks is universal, with few variations.

- **Community:** Where users gather based on shared interests. We are social creatures and tend to hang out with our own communities. Each social network offers features that allow their users to find and connect with their tribe.

- **Analytics:** Measuring outcome. Engaging in social networking sites without measuring your effectiveness is futile. Fortunately, all offer tools to help you measure your success; however, it is easy to get overwhelmed by the number of things you can measure. What you need to do first is to decide what metrics to track, which depends on your objectives and your tactics. I will provide examples for measuring what matters, and I will also provide an overview of the metrics that you can measure for each social network.

Your overall social media strategy

There are two ways to go about building and communicating with your intended audience, and the choice depends on how closely you are connected based on shared values. People tend to be open to— and consider—ideas when delivered by their trusted opinion leaders who share their values (4.13). If there are few or no common values between you and your target audience (for example a humanist versus an Evangelical Christian), consider delivering your message through an influencer: someone who is on the same page with you on climate change and shares values with—and is trusted by—your target audience. A great example of this approach is how the message of climate action was communicated with the Catholic community through Pope Francis. Data has shown that reaching out to the Pope made a measurable change in the attitude of the Catholic community toward climate change mitigation (1.15). The flowchart in Figure 4.27 summarizes the steps toward connecting and communicating with your target audience.

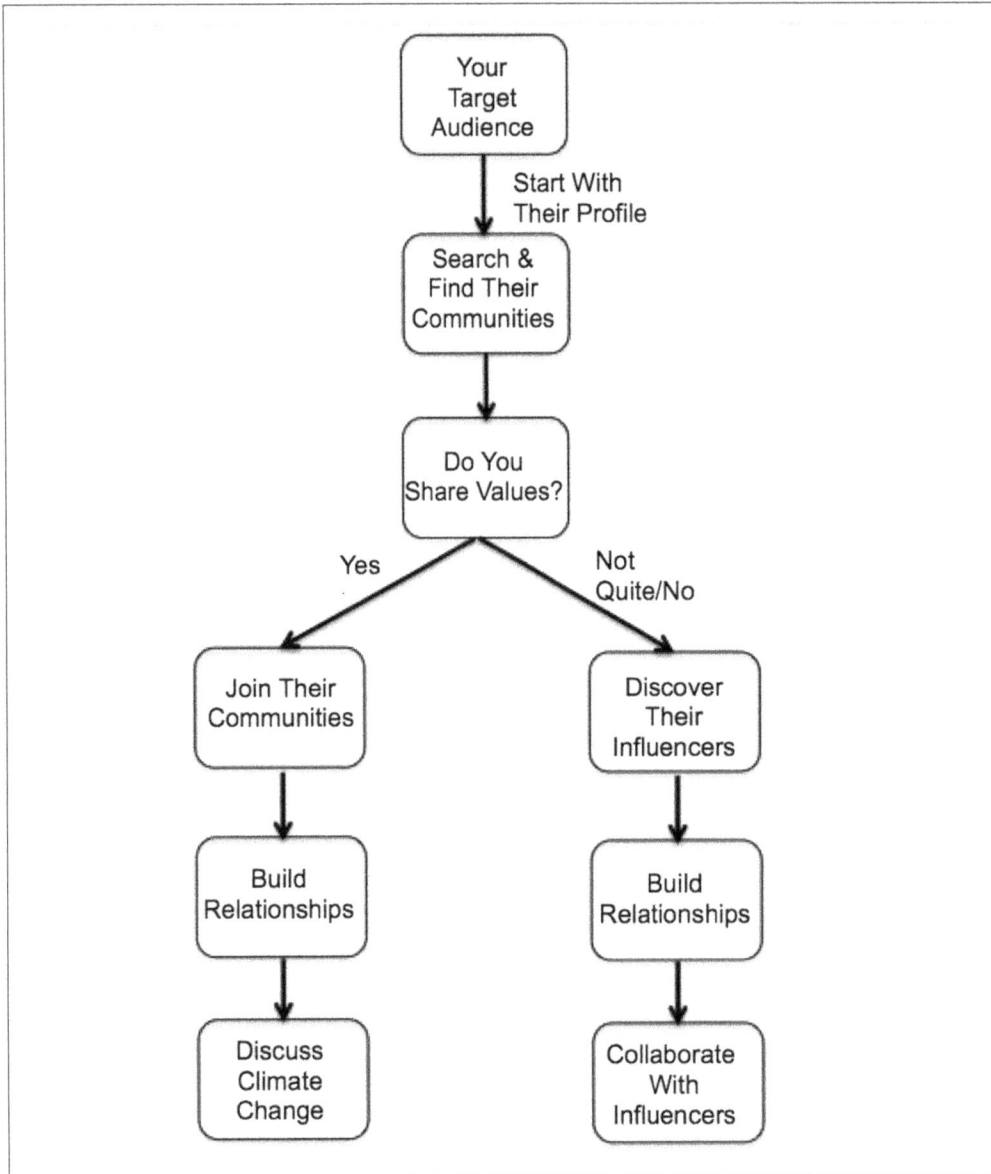

Figure 4.27: Flowchart for strategic communication with a target audience

Regardless of your chosen communication approach, your social media strategy will rely on comprehensive research into where your target audience tends to congregate online, and identification of their online opinion leaders. In the upcoming chapters, we will cover how to use the search functionality of social networking sites for this research. But first let's focus on building your identity.

tweet!

t

Twitter

Twitter is a micro blogging platform, allowing only 140 characters per status update, and serving 310 million monthly active users. Twitter offers the fastest means of finding and connecting with a target audience, provided that you use this platform strategically with the aim of making personal connections. In this chapter you will learn how to use different features offered by Twitter and other tools to make strategic connections—a task that hinges upon presenting a good first impression via your profile.

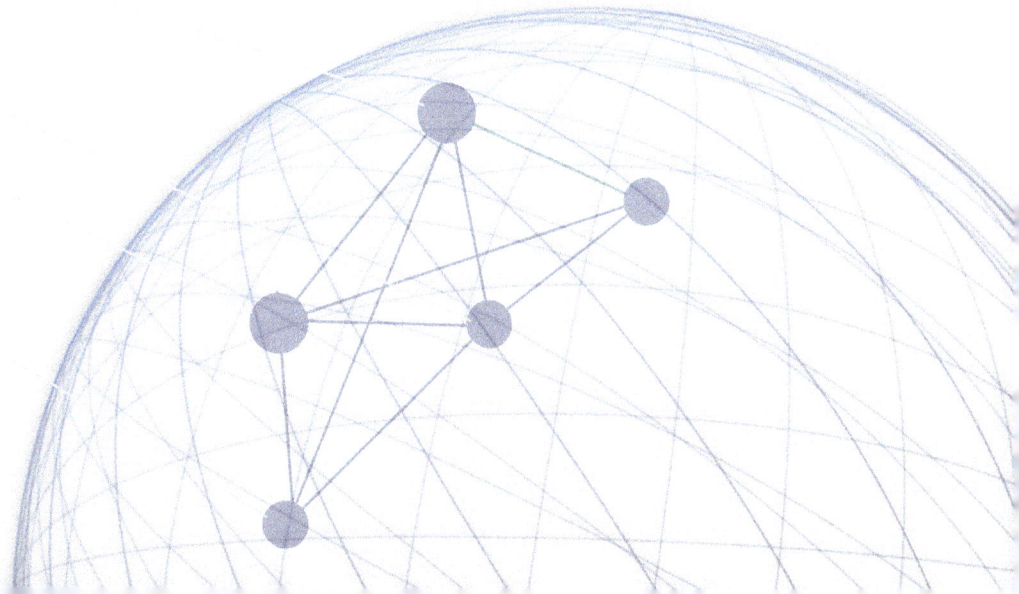

Profile

To set up a Twitter profile, visit www.twitter.com/signup. Type in your real name or the name of the organization or entity you represent. Engagement always is more effective when it is done through a personal profile, but you can also find success if you approach people through an organization or group persona. If you are setting up a Twitter account for an organization, put verbiage on your bio pointing to the Twitter profile of the person who is in charge of managing the account (for example "Tweets by @snouraini"). This way, you can make a human connection with your profile visitors in an attempt to build trust. You also have to choose a username, which will represent your identity. For example my username is "snouraini" and my identity on Twitter is "@snouraini." Every time someone wants to communication with me or get my attention or cite my name, they would include "@snouraini" in their message. Your username also forms the basis of a URL for your Twitter profile; for example, mine is www.twitter.com/snouraini. There is a fifteen-character limit for the username you choose. Be sure to choose a username that is easy for you and your connections to remember.

During the sign-up process, Twitter will offer suggestions as to the accounts you may want to follow. Skip these steps and visit your profile page. Add a profile image, a banner image, and then complete your biographic information (Figure 5.1). Include a link to your blog, and share your location if you are comfortable with this being known to the public. You have a maximum of 160 characters for your bio, so be strategic about how you write it. Resist the temptation to create a cute or fancy biography; it is best to be concise and clear while incorporating keywords. When people visit your profile they should be able to tell who you are, what you care about, and with whom you are affiliated; don't confuse them by providing vague descriptions.

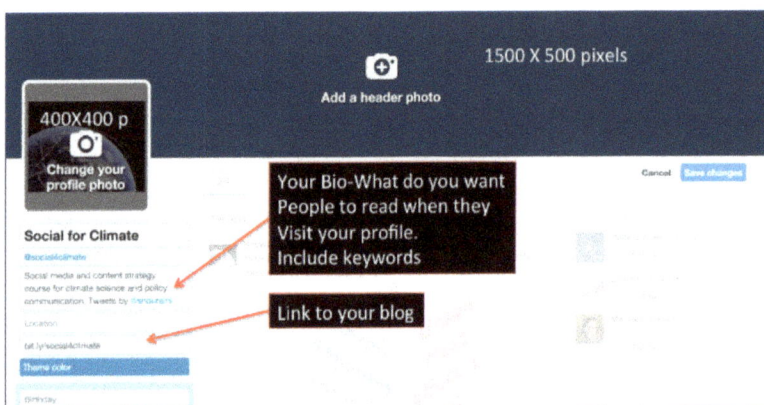

Figure 5.1: Completing your Twitter profile

Organization

Before starting your Twitter journey, get organized. You are going to start following a few accounts, but as time goes by, the number of accounts you follow on Twitter will grow to thousands. When you follow so many, it will be difficult to manage and keep up-to-date with everyone in your community. Twitter shows you updates from the accounts you follow on a chronological basis, so whether or not you can see an update about a particular topic depends on when you choose to visit your stream. To address this issue Twitter offers organizational features such as lists and hashtags.

Twitter lists allow you to assign the accounts you follow into categories so you can be selective about the groups of people with whom you want to interact at a given time. Twitter hashtags allow users to assign a category to their status updates. For example, if a status update relates to the topic of climate change, hashtags #climatechange #globalwarming #gw or #climate are incorporated into the message. In this way, Twitter users tag their status updates, which makes it easier for those who are interested in the topic to find them.

Creating lists

Begin with your audience profile and the keywords you have already created. Make a list of topics, stakeholders, strategic partners, organizations, and opinion leaders with whom you may want to connect. Twitter allows you to create up to 1,000 lists—each of which can be assigned up to 5,000 accounts. To create lists, view the top right hand corner of your Twitter page, and click on your photo. Click on "Lists" on the drop-down menu to visit the relevant page and start making lists (Figure 5.2).

Let's see how lists can help you remain focused in your outreach efforts. Say that your target audience is Linda Goldberg. There are a number of different ways you can begin raising awareness about climate change with this demographic. You can

Figure 5.2: Creating lists

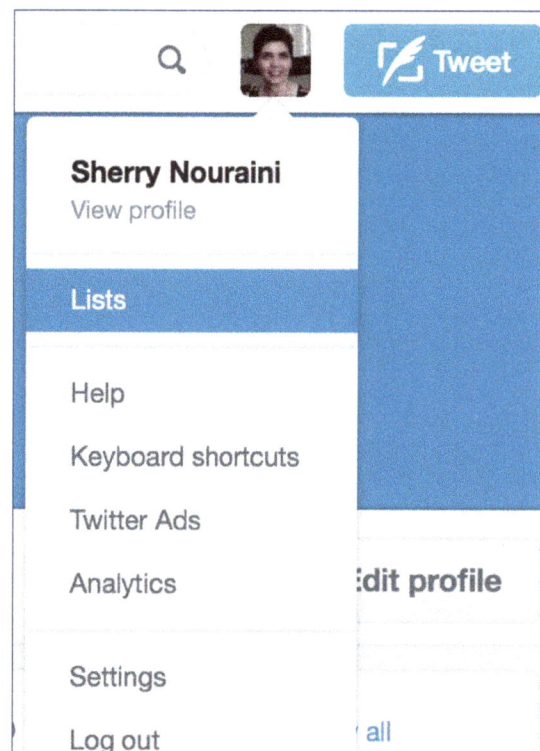

search for and build relationships with Jewish mothers on Twitter. Alternatively, you can approach Jewish religious thought-leaders who influence the community of Jewish moms. I think you'd agree that the nature of conversations with these two groups could be quite different, but you'd want to make sure to build relationships with both. You can leverage the list feature of Twitter, and create a separate list for each of these groups to focus your outreach efforts. Joining Twitter and trying to build relationships with different types of users at the same time is like going to a party and trying to have conversations with two different crowds of guests about two different topics at the same time: quite overwhelming and fruitless, wouldn't you say?

Hashtags

By using hashtags, your target audience is essentially erecting a sign to help you find them. But how do you know which hashtags they are using? Here is where you have to do some hashtag research, which begins again with your keywords. Your hashtag research will include both searching on Twitter, and using a great hashtag discovery and analytics tool called RiteTag (ritetag.com). RiteTag also suggests hashtags related to your keywords, some of which you may not have considered.

Let's use one example from the previous section of the book to see how you would go about building lists, discovering hashtags, and doing searches on Twitter:

Linda Goldberg: Let's recall our mission statement for reaching Linda Goldberg:

> *"I want to use blogging and the tools of social media to be the Jewish voice for climate change. My goal is to inspire 'disengaged' Jewish mothers to integrate sustainable living with their interests and daily practices because mothers play a vital part in raising children that will control what happens to our climate in the future. I also feel they need a channel to become more aware and involved in the climate change conversation, but they are more likely to listen if a person of their own community is the source of communication."*

A communicator guided by this mission statement would benefit from focusing on the online Jewish community. Based on this premise, let's create a list of categories and hashtags.

Categories: Jewishmoms, Jewishorganizations, Jewishschools, Jewishspiritualleaders

Now create a Twitter list for each of these categories. Going forward, once you find a Twitter user who falls into any of the above categories, assign them to the relevant list. This way, you are organized from the outset. Now, let's see how you go about discovering relevant Twitter accounts. As mentioned above, your journey starts by doing some hashtag research.

Hashtags: #jewish #jewishvalues

Let's focus on the word "Jewish." Searching RiteTag will not only reveal useful analytics about this specific hashtag, it will also uncover other hashtags that contain this word (Figure 5.3, not all discovered hashtags are shown in the image). I have labeled the icons in this image so you would know what each of them mean. Each hashtag entry is shown with performance metrics (level of engagement), contents of the tweet containing the hashtag, and link to further statistics related to each hashtag. The button named "Connections" leads to an array of hashtags (Figure 5.4) that normally appear on the same tweeter update with the seed hashtag (in this case #jewish).

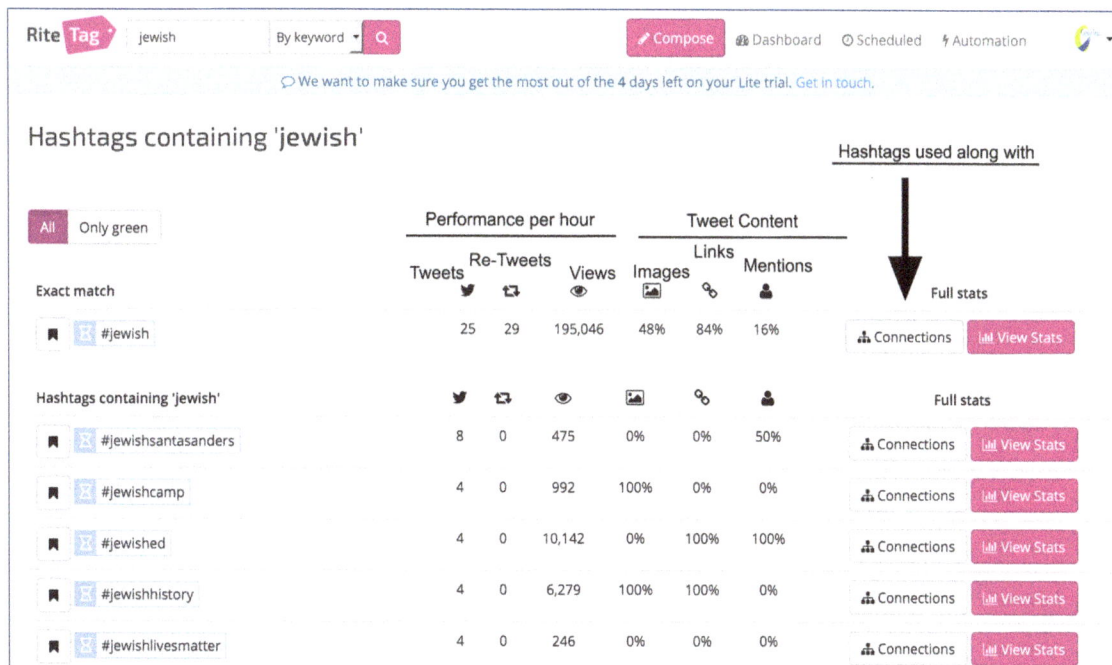

Figure 5.3: Using RiteTag for hashtag research

RiteTag has a color code for the hashtags:

Green: This hashtag is popular in real time.
Red: This hashtag is too generic.
Blue: This hashtag is used frequently over time.

No color: This hashtag is not used or not followed very frequently.

RiteTag discourages using "no-color" hashtags, but I think some of them may be worth exploring because they may point you to exactly the niche audience you are trying to find. Reality test RiteTag's assumptions, and always judge your discoveries based on quality rather than quantity. For example, even if a seemingly unpopular hashtag helps you find just one Jewish mom who lives in New York, and who is active in her community, she may be all you need to reach a community of Linda Goldbergs.

Reviewing the search results on RiteTag, I put together a list of hashtags that I felt would be relevant. I also dug deeper into the analytics of these hashtags by following the "View Stats" menu (see Figure 5.3). This analysis allowed me to find a list of other hashtags that are related to my original list (Figure 5.4). RiteTag also reveals Twitter users who tend to use these hashtags in their status updates (not shown). Then I checked the profiles of these people on Twitter and chose the ones who seemed relevant to our target audience and our categories of Twitter lists (see "Relevance" column in Table 5.1).

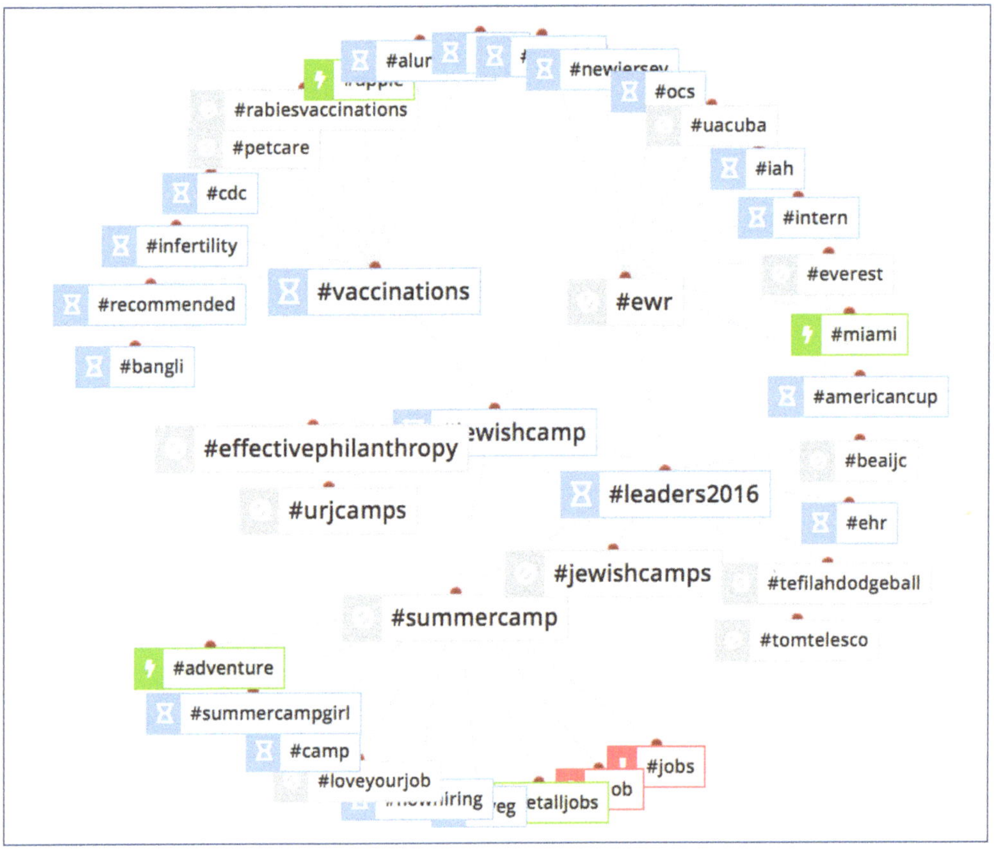

Figure 5.4: Hashtags related to #jewishcamp

Table 5.1: Relevant hashtags and profiles for outreach with Linda Goldberg sample profile
* People's usernames have been modified to respect the privacy of these users

Hashtag	Connections	People	Relevance
Being Jewish	None	@kweiner98 *	Sociologist mom, CT
Jewishcamp	#leaders2016 #summercamp #camp #vaccinations #newark #newjersey	@JewishCamp @Tawonga @BowTieDad * @CampZeke @Deanberg98 * @sbbEZas123	Camp for kids, NY Camp for kids, CA Camp supporter NY-based camp Exec Director of NJ-based community Jewish mom and teacher, ATL
Jewishwisdom	#wiedu #edchat #cpchat #txeduchat #wischat	@Jill_plat *	Mom, Teacher

Thanks to RiteTag, we have a set of hashtags that people actually use and a number of accounts that can help us connect with the Jewish mom community. But this is just the tip of the iceberg! Using the hashtags that we've discovered, we can use Twitter's search function to find additional Twitter accounts, which can be a source of a wealth of additional information. Let's see how that works.

Search

Twitter offers a powerful search engine that provides search results in a variety of formats. The search function also allows you to narrow down your searches to a particular location, uncover Twitter posts associated with a particular sentiment and language, and set a time range for the Tweeter status updates you want to find.

To get the most bang out of your search efforts, head directly over to advanced search using this URL https://twitter.com/search-advanced. See Figures 5.5 and 5.6 for a screen shot of this example search: #jewishcamp, Tweets in English, near New York.

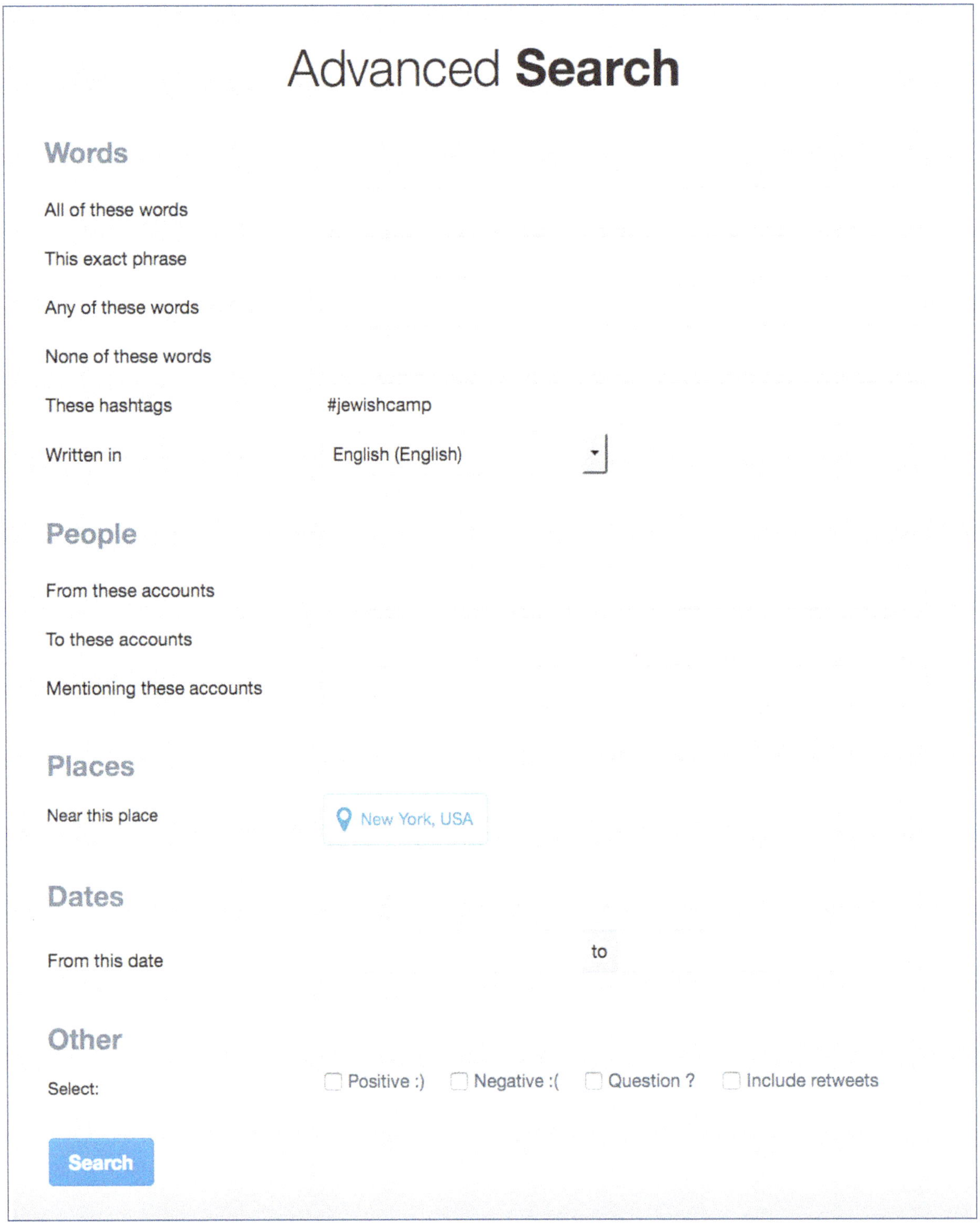

Advanced **Search**

Words

All of these words

This exact phrase

Any of these words

None of these words

These hashtags   #jewishcamp

Written in   English (English)   ▾

People

From these accounts

To these accounts

Mentioning these accounts

Places

Near this place   📍 New York, USA

Dates

From this date   to

Other

Select: ☐ Positive :)   ☐ Negative :(☐ Question ?   ☐ Include retweets

[**Search**]

Figure 5.5: Twitter Advanced Search

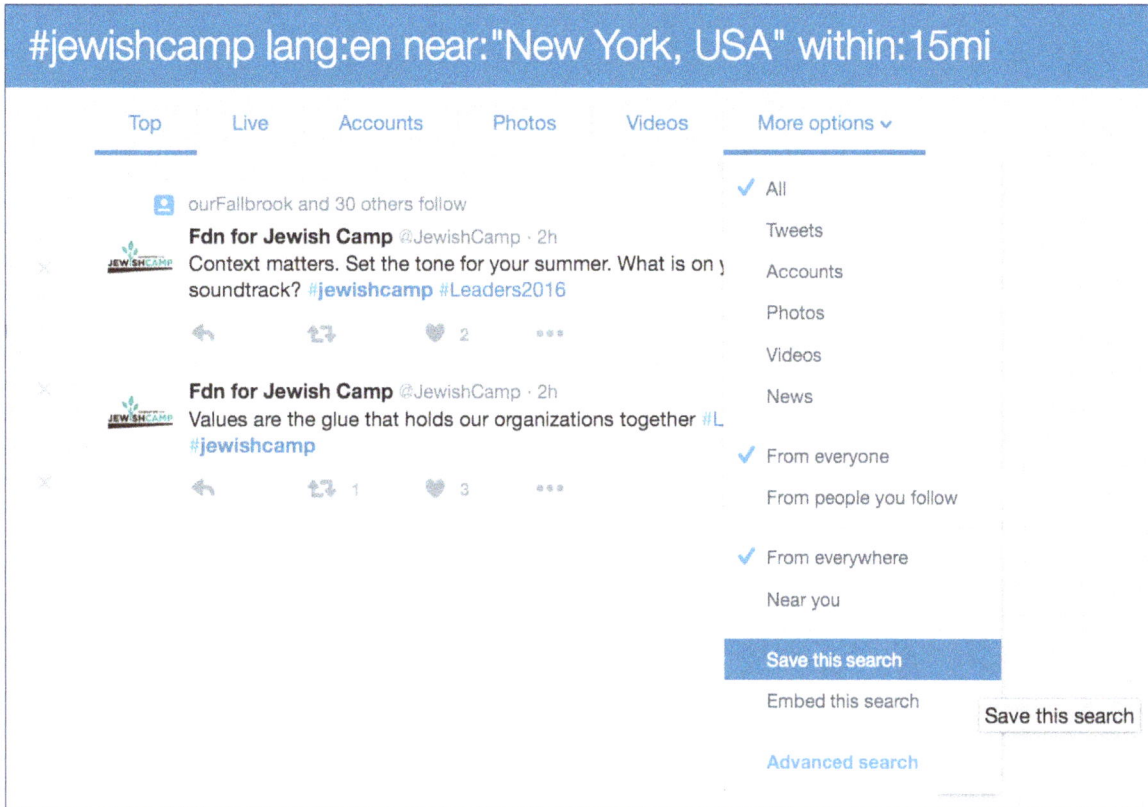

Figure 5.6: Twitter Search Results page

Note that you can save this search on Twitter and come back to it for fresh results. The "Search Results" page is a gold mine of information as you can discover Tweets, accounts, photos, and news about a particular topic. For example, if you view the "Accounts" tab you will see that this search has uncovered people who were not revealed in your RiteTag search (Figure 5.7).

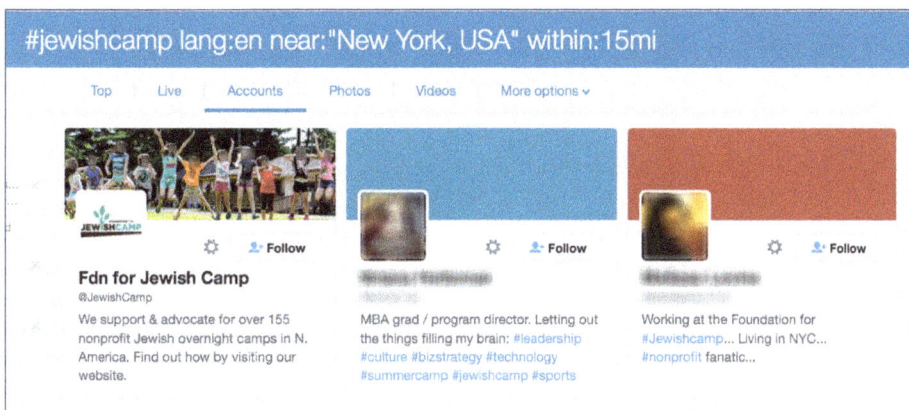

Figure 5.7: Accounts discovered through Twitter search

A word of caution

Doing hashtag research is a great way to reality test your assumptions, as I often have to do. If your intention is to search for and find communities of Jewish moms on Twitter, you may be inclined to choose the #jewishmom hashtag based on the assumption that this demographic would be using it to create an identity around being a Jewish mom. Interestingly, searching Twitter with the hashtag #jewishmom reveals that this phrase is primarily used by the Twitter community for satirical updates about the stereotypical overprotectiveness of a "Jewish Mom" rather than as a signal for identity. This may be one of the most valuable lessons you can learn about social media: the meaning and the context of the language used is not determined by you, but by the members of a social network. You are not in control of the message; they are. Either respect it, or fail at communicating with them.

A secret weapon: other people's lists

Earlier I introduced you to lists as an organizational tool. However, they are also wonderful discovery tools! Serious Twitter users are avid list creators: they spend the time to create lists of relevant accounts, which means they have already done some discovery work for you. So, when you find an account that is highly relevant to your target audience, check their lists to see if you can find additional relevant accounts in *their* lists. For example, our latest discovered account, Foundations for Jewish Camp, has six lists, all of which can connect us with additional relevant stakeholders in the Jewish community, one way or another (see Figure 5.8).

To connect with these additional accounts, you can either visit each list to find new accounts to follow, or just subscribe to each list. Note that subscribing is not the same as following; it will allow you to only see Twitter updates from these accounts without the account holders being aware that you are doing so. If you want a "Hey, I am following you" tap on their shoulders, you need to follow each account individually. One particular list that may be valuable in our example is the "jewishmedia" list, in case you wanted to reach out to journalists, with readership in the Jewish community (see Figure 5.9).

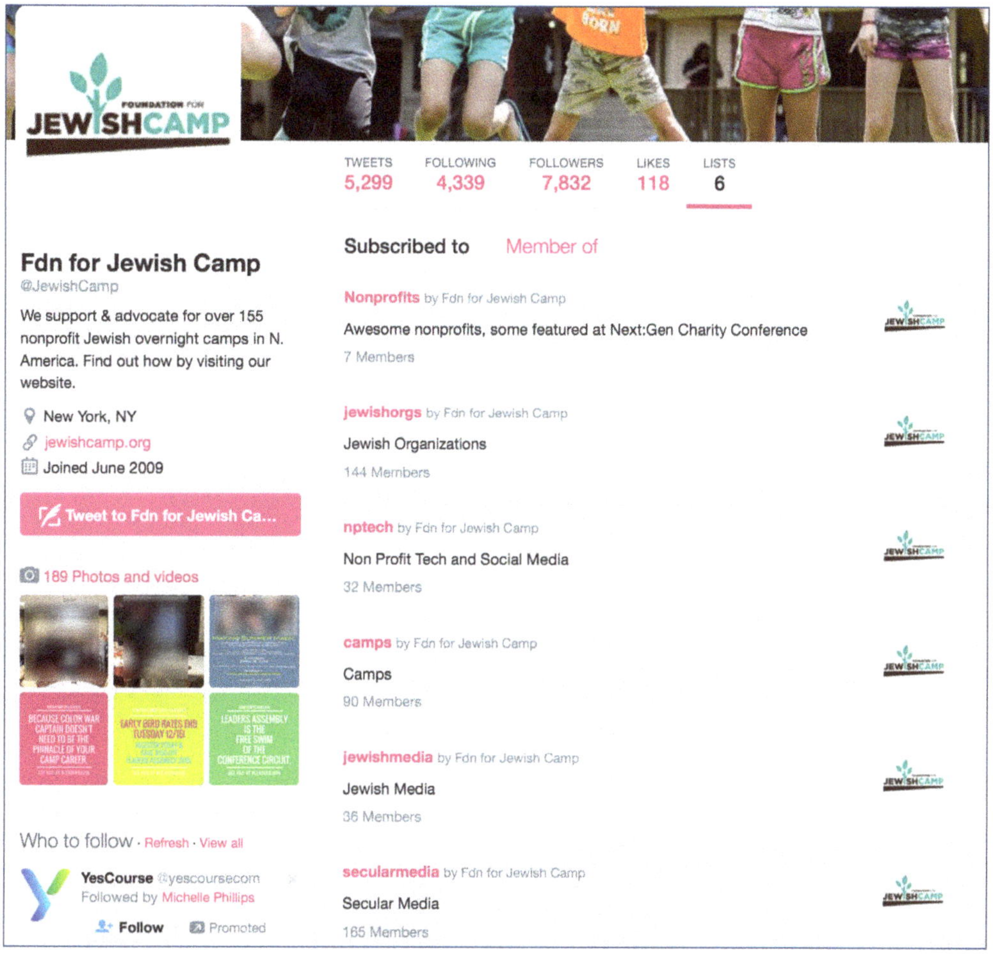

Figure 5.8: Exploring other people's lists

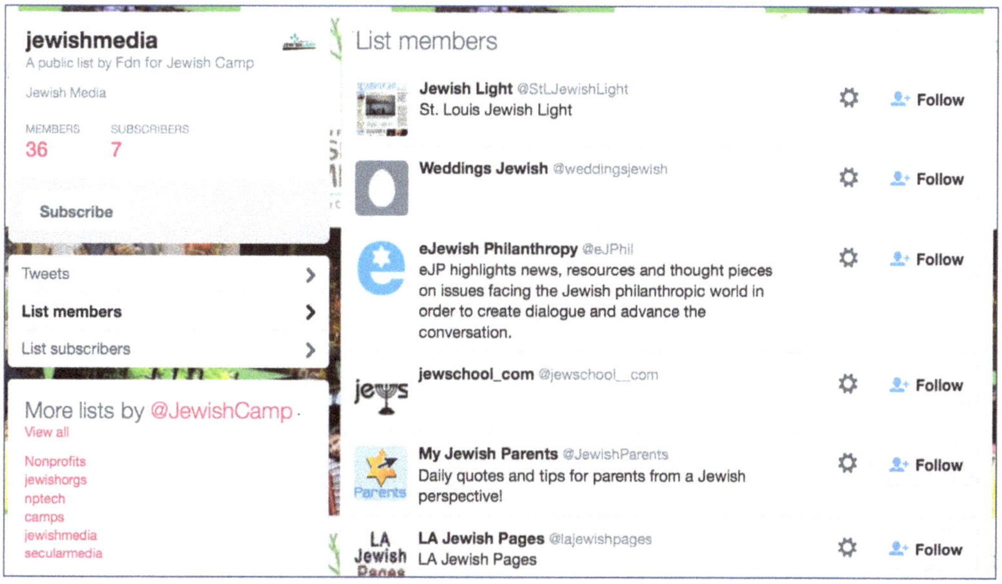

Figure 5.9: Jewishmedia list created by @jewishcamp

Status updates and terminology

The status update box is where you type your message, but it also provides a range of useful features (see Figure 5.10).

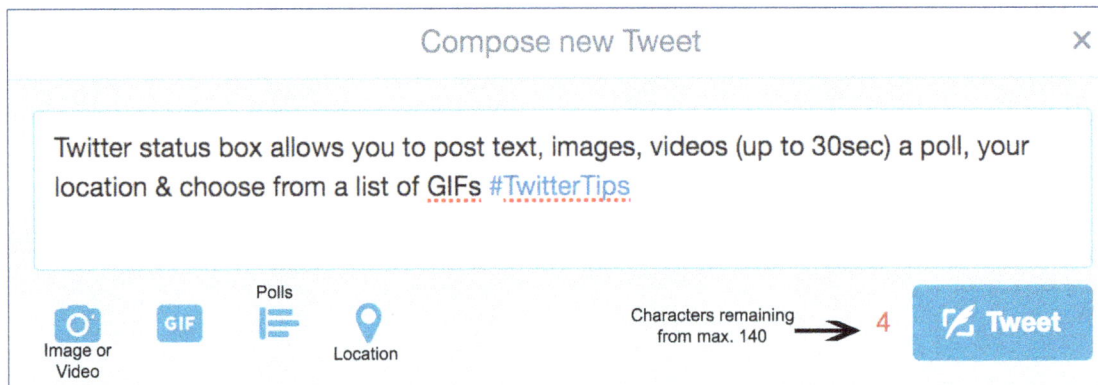

Figure 5.10: Twitter status box, before posting

In Figure 5.10, I have labeled the various features of a status box, most of which are self-explanatory. The GIF (Graphic Interchange Format) option provides you with a library of animated GIF media to choose from depending on your topic. An animated GIF is an animated clip created from multiple images, which are played back in sequence.

Once you post a status update, you will have a host of other features that takes a Twitter update to the next level (Figure 5.11).

As of the writing of this book, Twitter status updates have a limit of 140 characters, but the company has announced that this limit will soon apply only to the text portion of a status update (5.1). More specifically, media attachments like photos, GIFs, videos, and polls will no longer "count" toward the 140 character limit. In addition, when replying to a Twitter user by including their Twitter handle (for example @snouraini), @names will no longer count toward the 140 character limit (5.1). It may be that by the time you read this book, these changes are in effect.

Here is a list of various Twitter features and the functions they provide:

1. **Reply.** Like a face-to-face conversation, the "Reply" feature allows your followers to start a conversation with you based on the message you have posted.

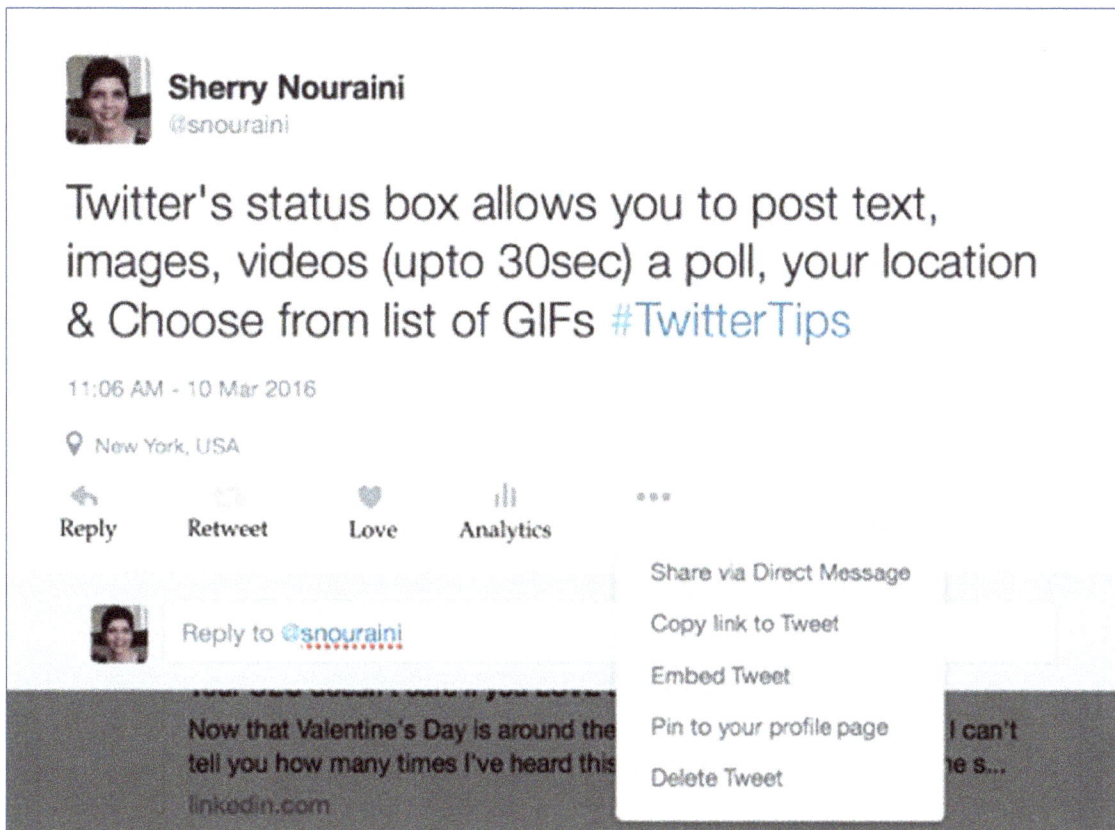

Figure 5.11: Twitter update, after posting

2. **Retweet.** This is one of the most valuable and powerful features of Twitter. If your status updates resonate with someone who is following you, they can share your update with their followers. This is the power of word of mouth and has the potential of amplifying the reach of your message. If someone Retweets your status updates, be sure to thank them and use the occasion to start a conversation. Retweeting and replying is also a powerful means of building relationships with stakeholders that we discovered in the previous section, so don't be shy to use these features.

3. **Love.** This is a feature similar to what you may be used to seeing on Facebook as the "Like" button. In addition to showing "Love" for status updates posted by those you follow, this feature is used by the Twitter community to bookmark particular Tweets for future reference.

4. **Analytics.** Clicking on this link will provide data on engagement with your status updates. We will cover analytics in more detail later in this chapter.

5. **Share via direct message.** As opposed to a public Twitter update, direct messages are private conversations between Twitter users. This feature allows your followers to share your status updates privately with their connections.

6. **Copy link to tweet.** It may not seem so, but each and every one of your status updates is a Web page with its own URL. What is even more amazing is that Twitter updates are indexed by Google, which means that they can be discovered when people search for information on this search engine. This is why it is important to make an effort to incorporate the keywords that you discovered in Chapter 4 in your status updates. Your followers can copy the link to your tweet (status update) to reference elsewhere on the Web, or even in a publication.

7. **Embed tweet.** Your status updates can also be transplanted on to a Web page outside of Twitter, for example as part of a blog post. This feature as well as the one above allows your messages to travel outside of Twitter and onto the greater cyberspace.

8. **Pin to your profile page and delete tweet.** These are only available to you, the author of the status update. Pinning a tweet to the top of your profile prevents it from moving further down on your profile page as you generate more status updates. This is useful when you have a status update that you want your profile visitors to see and by which to get to know you better. These may include an introductory video, a link to a slideshow, a poll, or a form for signing up for a petition or a newsletter.

9. **Mute, block, and report (not shown).** If you want to disconnect from status updates of certain users, you either have the choice of muting their updates without unfollowing them (mute), stop them from following you (block), or report them to Twitter if you think their activity breaks Twitter's terms of service.

Community

As of the first quarter of 2016, Twitter has 310 million monthly active users worldwide (5.2). It is impractical and unrealistic to think that one can connect with and develop meaningful relationships with everyone on Twitter. Like any other large community, the Twitter user base connects with one another based on their mutual interests, expressed in the form of hashtags. But building community on Twitter does not stop at including a hashtag in a status update. Here are some of the ways the Twitter user base has been building communities, which offers you a great means of finding your target audience:

1. **Twitter chats.** This is a live conversation, which is moderated, centered around a topic, and held on a set schedule. The moderators of Twitter chats create a unique hashtag, which allows users to search for, find, and participate in conversations. Twitter chats are great for meeting like-minded individuals and having online substantive discussions. If you refer to Table 5.1, you'll notice that our hashtag discovery work on RiteTag revealed a number of Twitter chat hashtags that are connected to #jewishwisdom: #wiedu, #edchat, #cpchat, #txeduchat, #wischat. In this case, I would not be exaggerating if I said that this is a gold mine, as Twitter chats will directly connect us with a community of educators who value the Jewish faith and have mothers among their members (see relevance column in Table 5.1).

   But relevant Twitter chats aren't always so easily discovered, so how do we find them? Fortunately, there are a number of curated databases of Twitter chats that can be used to search for those relevant to our needs:

   - **RiteTag.** Twitter chat hashtags usually include the word "chat," so searching RiteTag with the keyword "chat" will reveal about 200 hashtags relating to Twitter chats. Although time-consuming, you can monitor engagement statistics for chats on RiteTag to see if they are active and have a good following. These statistics will also reveal Twitter users who are connected to these chats.

   - **Twitter chat schedule by TweetReports.com.** This is a useful inventory of Twitter chats, which is sortable by a number of criteria. This inventory also lists the moderator of the chat, which is great for connecting with influencers on a particular topic. Chat moderators are usually well-known and trusted within their community.

   - **Twubs.com Twitter chat schedule.** This is a comprehensive list of live Twitter chats, which you can browse to find relevant communities. Although time-consuming, Twubs will show you Twitter accounts of chat participants, which again, is a gold mine.

2. Conferences and webinars. An additional, powerful means of connecting with your target audience is through participating in online conversations about conferences and webinars. Most of these events have a designated hashtag set by the organizers. Event hashtags serve many functions for event attendees and organizers, some of which include sharing event updates, helping to find information, networking, sharing experiences at the event, and continuing discussion of relevant topics after the event. To leverage the power of conferences or webinars, you should research and create a list of past and future events that your target audience is likely to attend.

You may be wondering how to find such conferences and events. Sometimes a simple search on Google is all you need. For our example audience, Linda Goldberg, I searched for "Jewish parenting conferences" and found a list of events on this topic, most of which happen in New York—one of the locations where people like Linda reside (Figure 5.12).

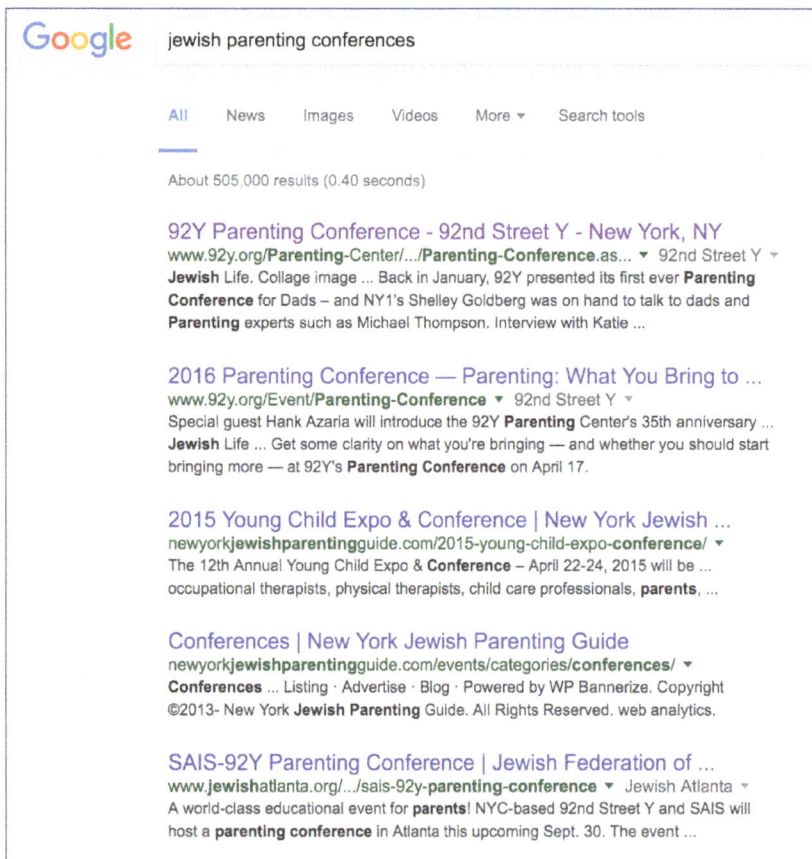

Figure 5.12: Google search results for Jewish parenting conferences

Visiting the website for one of these search results, "92Y Parenting Conference," led me to their Twitter account and discovery of the hashtag for this conference, #92Y. Gold mine! Following the hashtag for this conference can help one find Jewish mothers, learn what topics are on their minds, and start conversations with this community. Also, conference organizers and speakers at this event would be potential influencers for outreach efforts.

In closing, using the tools of social media does not have to limit you to online interactions. It is important to be holistic in your outreach efforts and take every opportunity to connect online and offline.

Analytics

To access the analytics section of Twitter, visit the advertising dashboard: http://ads.twitter.com. There is no requirement to buy ads for accessing Twitter analytics. Click on "Analytics" in the top menu and you will see a number of items, each of which I describe below, with the exception of Twitter Cards (Figure 5.13). I consider Twitter Cards a more advanced feature, which is outside of the scope of this book.

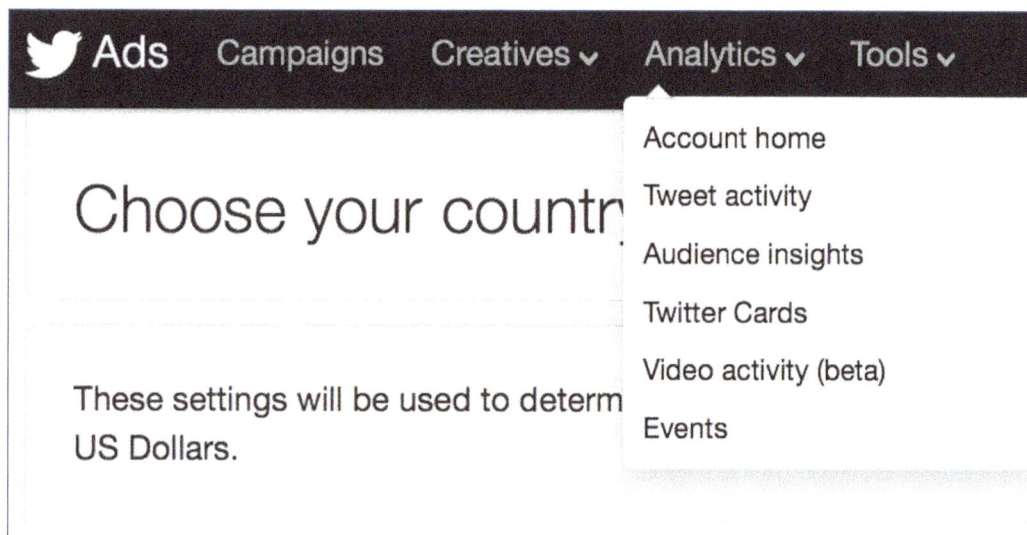

Figure 5.13: Twitter analytics dashboard

Account home

On this page, Twitter analytics provides performance data for your account for the most recent twenty-eight days, and also compares this performance to the previous period, represented by a numerical percentage value (Figure 5.14).

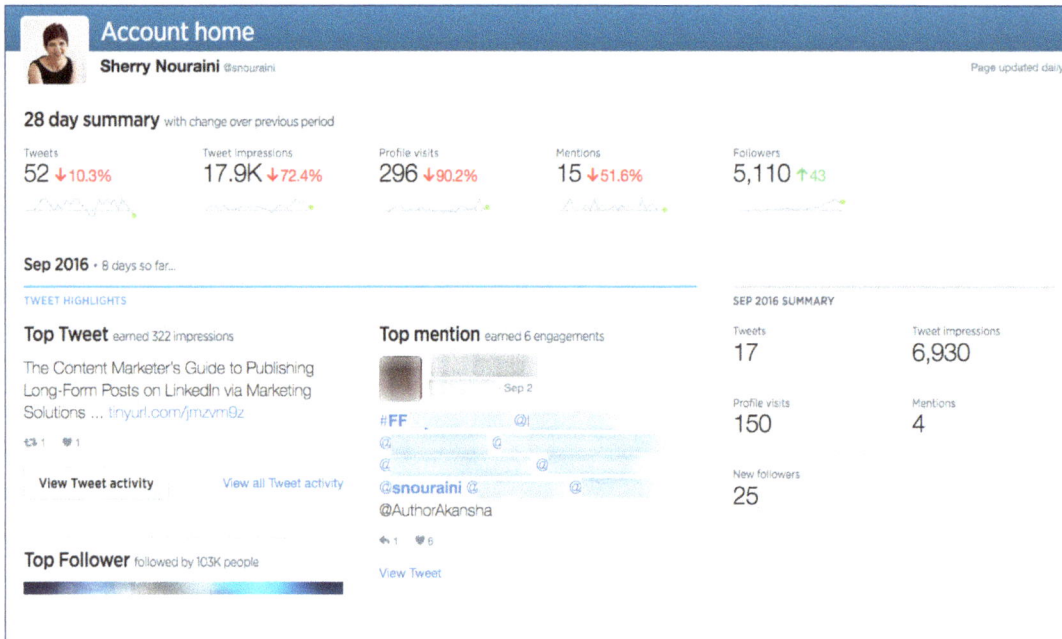

Account home
Sherry Nouraini @snouraini
Page updated daily

28 day summary with change over previous period

Tweets	Tweet impressions	Profile visits	Mentions	Followers
52 ↓10.3%	17.9K ↓72.4%	296 ↓90.2%	15 ↓51.6%	5,110 ↑43

Sep 2016 · 8 days so far...

TWEET HIGHLIGHTS

Top Tweet earned 322 impressions

The Content Marketer's Guide to Publishing Long-Form Posts on LinkedIn via Marketing Solutions ... tinyurl.com/jmzvm9z

View Tweet activity   View all Tweet activity

Top Follower followed by 103K people

Top mention earned 6 engagements

Sep 2

#FF   @
@
@   @
@snouraini @   @
@AuthorAkansha

View Tweet

SEP 2016 SUMMARY

Tweets	Tweet impressions
17	6,930

Profile visits	Mentions
150	4

New followers
25

Figure 5.14: Twitter analytics Account home

You can use this data to gauge your success in attracting attention to your profile. For example, if you are gaining fewer followers than the previous twenty-eight day period, you should ask why and consider what you can change to turn that tide. Conversely, if you are gaining many more followers than last month, you should investigate to find out whether you can attribute this success to something you did so you repeat it.

On the Account home, Twitter analytics also highlights the best-performing activities on your account (Figure 5.14 bottom): Top Tweets, Top media shared, Top followers, etc. Pay attention to your top-performing activities to learn what resonates with your followers. In addition, if you notice "Top followers" (users with a large following), investigate whether or not they could be influencers with whom you can partner. Note that a large number of followers in and of itself is not an indicator of influence, as followers can be purchased. However, not everyone buys followers, so some of your top followers may be of value.

Tweet activity

Twitter provides performance analytics in twenty-eight-day chunks for up to 3,200 Tweets going back as far as October 2013. The "Tweet Activity" page provides an overview of number of impressions and engagement rate (likes, Retweets, replies, link click, etc.) for your tweeter posts for the past twenty-eight days, as well as detailed performance data for individual Twitter status updates (Figure 5.15). "Impression" refers to the number of times a user is served a Tweet in timeline or search results.

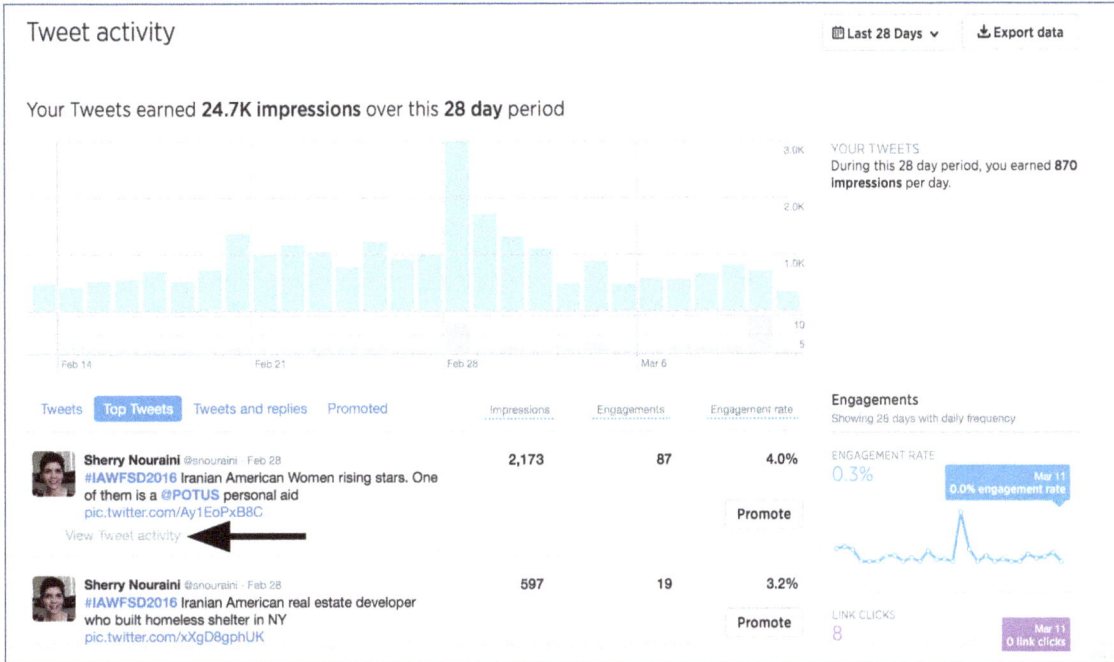

Figure 5.15: Twitter analytics dashboard, Tweet activity

You can also view details about how your followers interacted with your status updates by clicking on "View Tweet Activity" (shown by arrow in Figure 5.15). See Figure 5.16 for a screenshot of the details this action reveals. For a definition of each of these metrics, view reference 5.3.

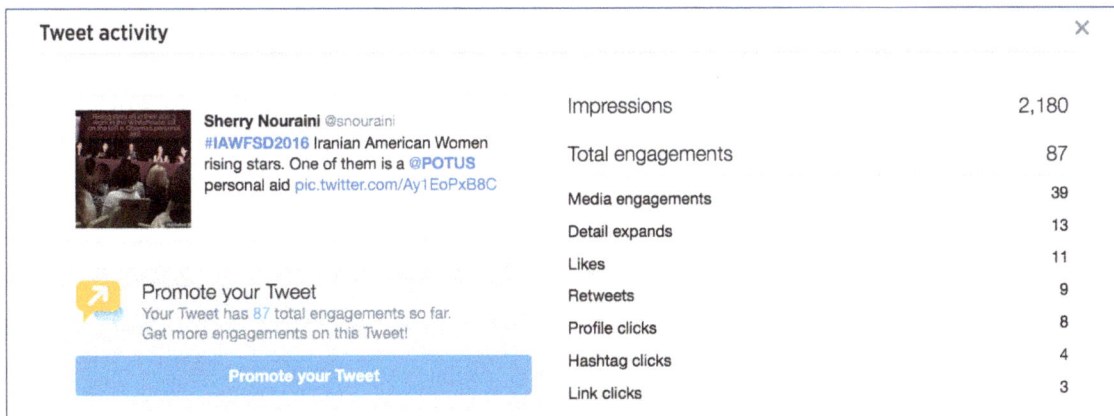

Figure 5.16: Twitter analytics for an individual tweeter update

By reviewing your "Tweet activity" you can discover status updates that most resonated with your followers, judging from whether they viewed your profile, clicked on the hashtags and links you included in your status updates, and Retweeted or liked your status update.

Audience Insights

The "Audience Insights" section of Twitter analytics is similar to the Facebook Audience Insights (FAI) tool that we discussed in Chapter 2. The source of data for Audience Insights is the information that Twitter users share on the platform, as well as data collected by a third-party data partner. In addition to data about the entire Twitter user base, Audience Insights also provides information about your followers, as well as your organic audience. An organic audience is a collection of Twitter users who potentially viewed your Twitter updates (as opposed to those who viewed your updates as a result of an ad you placed on Twitter). Note that organic audiences may or may not include your followers; these are users who potentially saw your updates because they followed you, or because someone they follow shared your status updates.

Similar to FAI, data in Audience Insights can be filtered based on demographics, interests, and a host of other criteria; however, this can only be done for the entire Twitter user base. There is no option to filter data about your followers or organic followers (Figure 5.17).

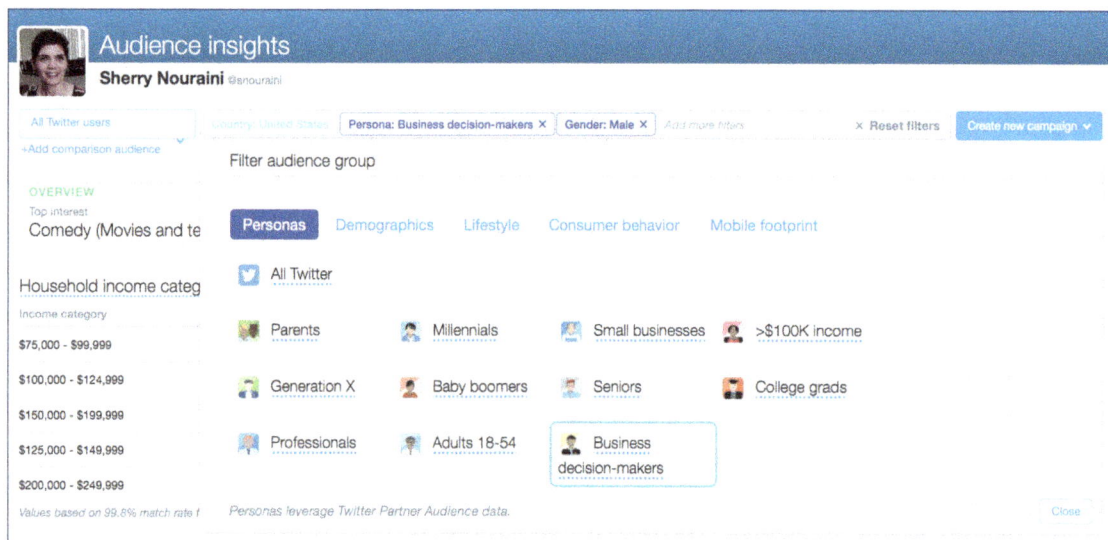

Figure 5.17: Twitter analytics, Audience Insights—All Twitter users

How can this feature be useful? For one thing, you can gain intelligence about a particular group of the Twitter user base. Say for example that we wanted to find out the television genres in which our Alex Donovan profile is interested (to use as a conversation starter). We can choose filters "business decision-makers" among the "Personas" category and "male" within "demographics." Performing this action reveals (among other things) the top television genre in which this demographic is interested: Sports (Figure 5.18).

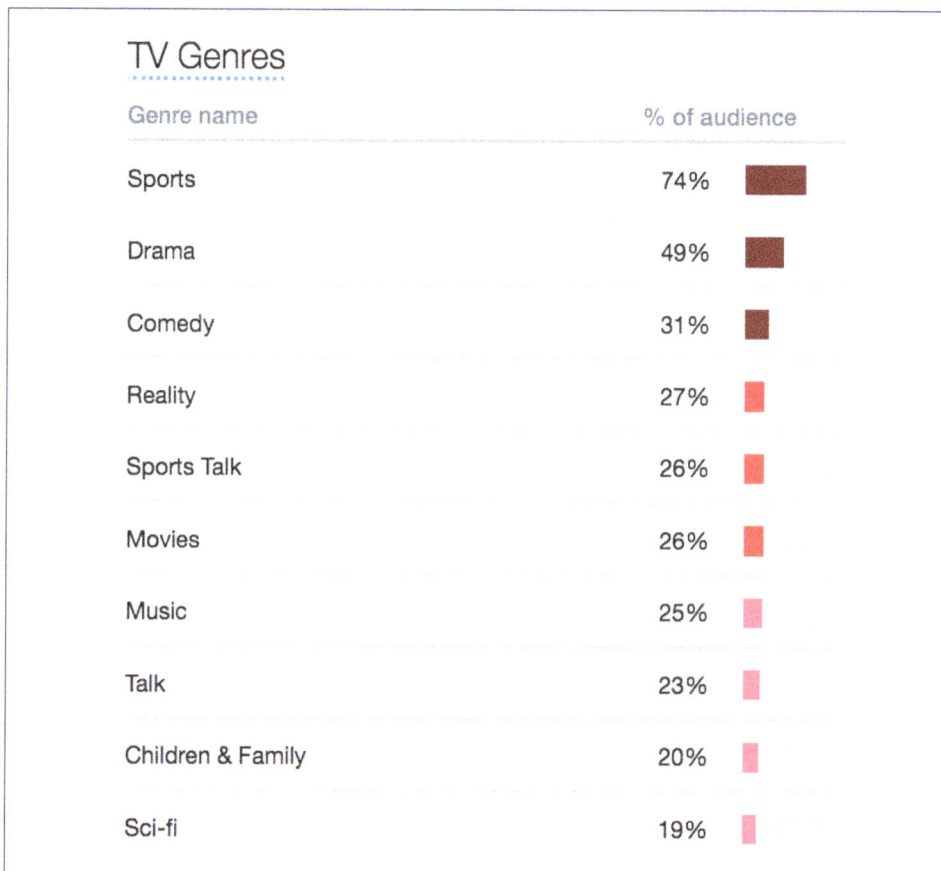

TV Genres

Genre name	% of audience	
Sports	74%	
Drama	49%	
Comedy	31%	
Reality	27%	
Sports Talk	26%	
Movies	26%	
Music	25%	
Talk	23%	
Children & Family	20%	
Sci-fi	19%	

Figure 5.18: Top television genres of male business decision-makers on Twitter

When data in FAI is filtered for a particular demographic, this tool automatically presents data both for the entire Facebook user community as well as the data filtered for that demographic. On Twitter "Audience Insights" you have the option of looking at these two datasets separately or viewing them as a comparison just like FAI (Figure 5.19).

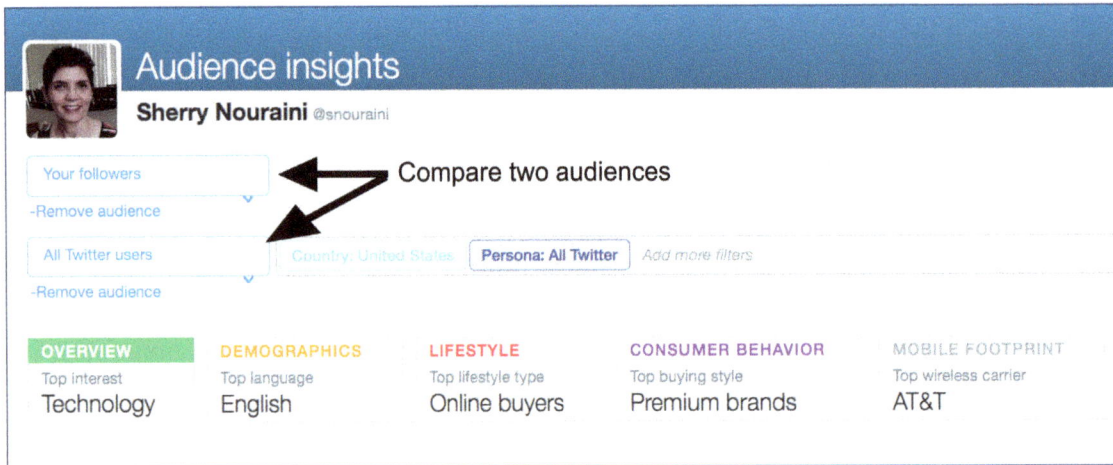

Figure 5.19: Twitter Audience Insights—Comparing two audiences

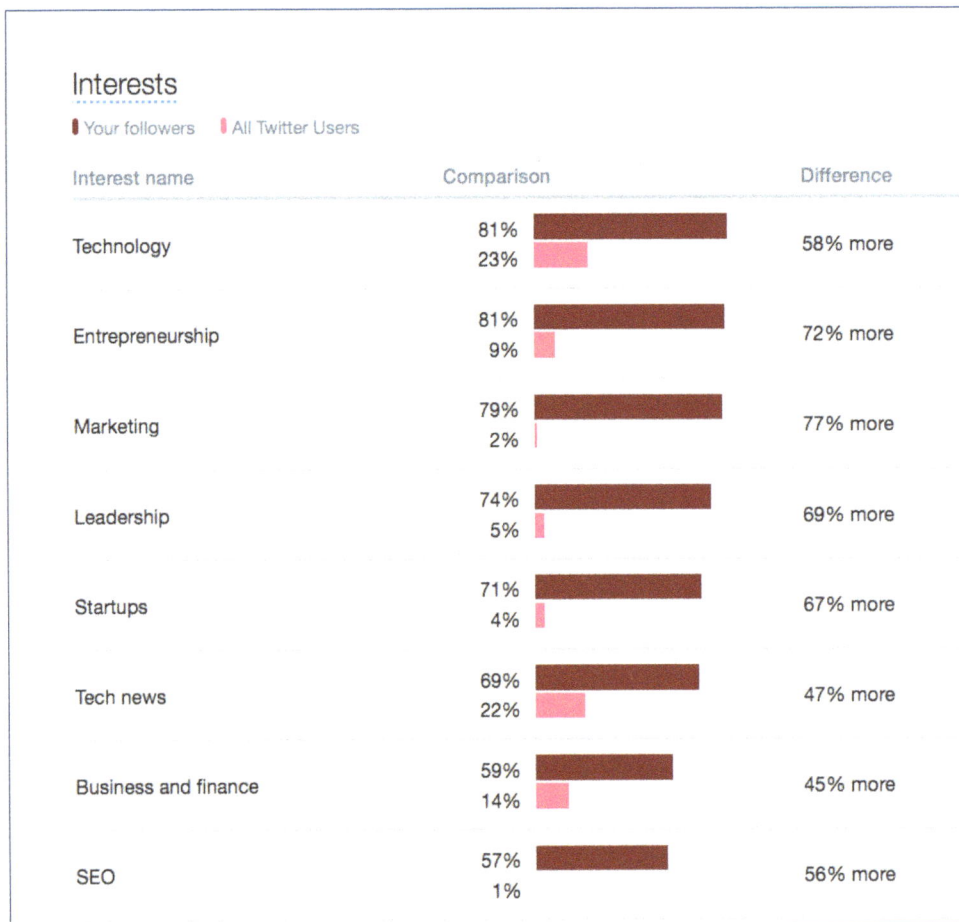

Figure 5.20: A comparison of my followers versus the entire Twitter community

However, I highly recommend that you view data about your followers—in addition to data on the entire Twitter community—for two reasons. First, research on behavioral economics has shown that we better understand data when it is presented to us as a side-by-side comparison, rather than when shown as a stand-alone (2.1). Second, when creating a community on Twitter (or any other social network), it is important to focus on building a niche audience who best matches your target audience profile. The side-by-side comparison of the data for your Twitter followers and the larger Twitter community will act as a check-and-balance system for deciding whether or not you are on track with building a niche audience. For example, for my personal Twitter profile, I focus on building a community of users who are interested in marketing, technology, and entrepreneurship. As shown in Figure 5.20, I am right on track with this goal. Examining the data in Figure 5.20 reveals that even though a large percentage of my followers are interested in marketing, technology, and entrepreneurship, only a small fraction of the entire Twitter user base fit these criteria, indicating that I am focusing on a niche audience.

Note that if you are new to Twitter, there will be insufficient data for you to analyze your followers in the beginning. However, you can still use the Audience Insights tool to analyze a particular demographic of users on Twitter, and gain insights about the larger Twitter user base. When you build enough of a follower community, be sure to check the Audience Insights feature to learn about your community's interests and habits.

Video activity

As mentioned earlier in this chapter, Twitter allows uploading videos of up to 140 seconds long. The "Video Activity" section of Twitter analytics provides data on the number of times that each posted video has been viewed, along with the view completion rate: what percentage of the one hundred forty-second video was watched (not shown). This latter metric is important because it is an indication of whether your videos are capturing the attention of your audience, and if you are keeping them interested.

Events

This is a relatively recent feature in Twitter analytics, and leverages the fact that events—whether offline or online—tend to extend into conversations in Twitter. As discussed in the "Community" section of this chapter, event hashtags play an important role in bringing conversations and networking on events into Twitter.

Twitter monitors conversations about events, and for some events, presents data about the demographics of users who are holding those conversations. It also presents top tweets and samples of live Tweets related to each event (Figure 5.21).

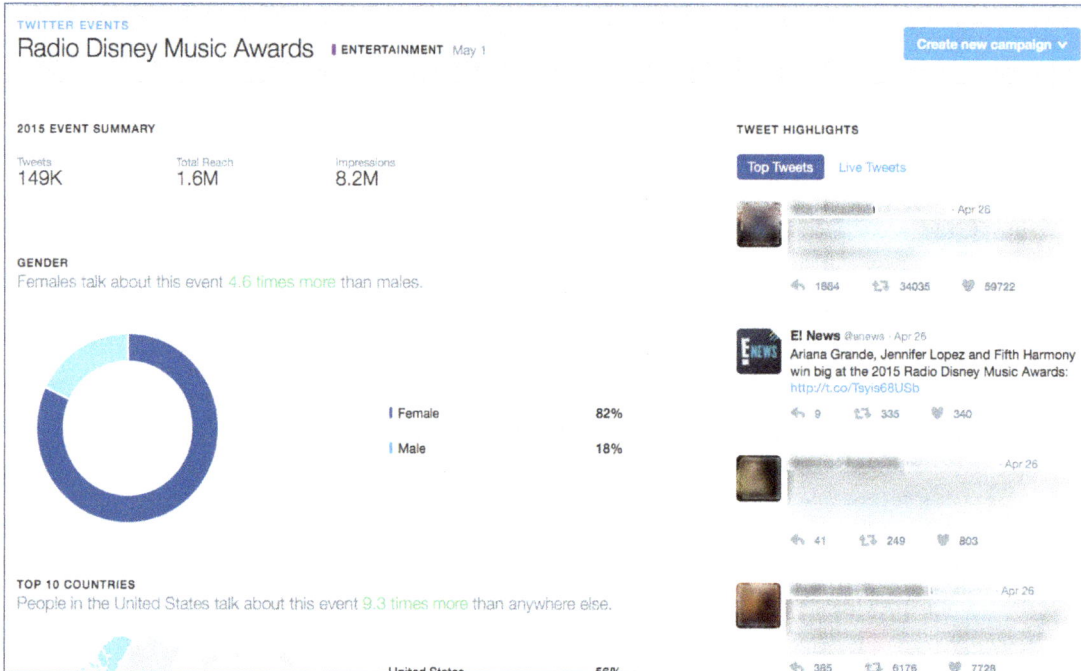

Figure 5.21: Twitter analytics Events—Audience demographics

In addition to a list of all events, Twitter analytics has a separate listing for sports or movies that are discussed on Twitter (Figure 5.22).

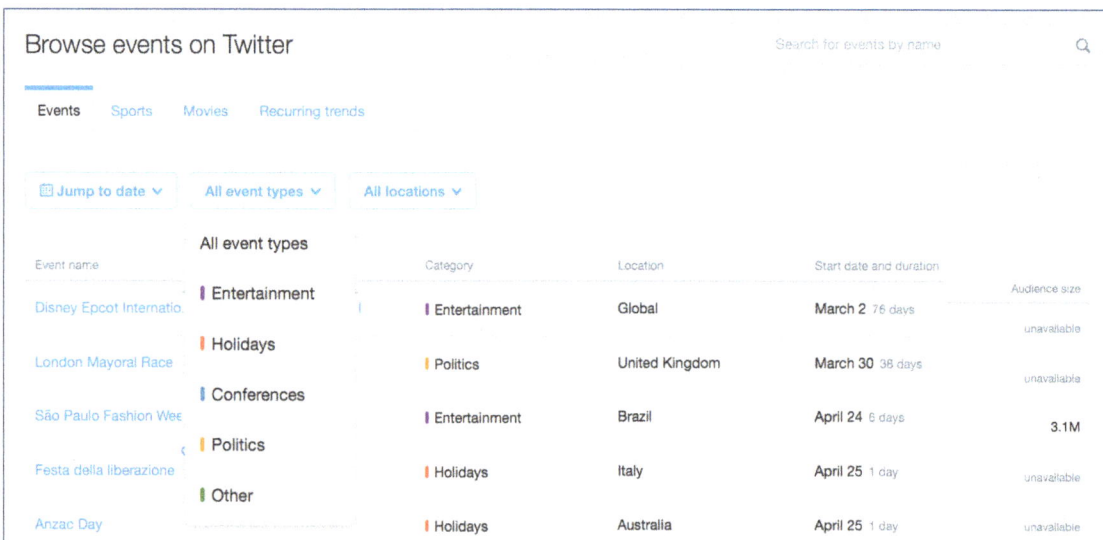

Figure 5.22: Twitter analytics—Events and other categories

The "Event" section of Twitter analytics also has a separate category for "Recurring Trends," which are status updates or conversations on Twitter that happen on a recurring basis, each of which has an associated hashtag. For example, #TBT refers to conversations about "Throw Back Thursday" where Twitter users share photos and videos or other kinds of posts that serve as a reminder of a past experience, every Thursday of the week (Figure 5.23). These recurring themes create a great opportunity to introduce your own unique way of participating in the larger conversation on Twitter. For example, a #TBT theme would make a great opportunity to draw attention to glaciers from the past, which are now shrinking and disappearing due to global warming.

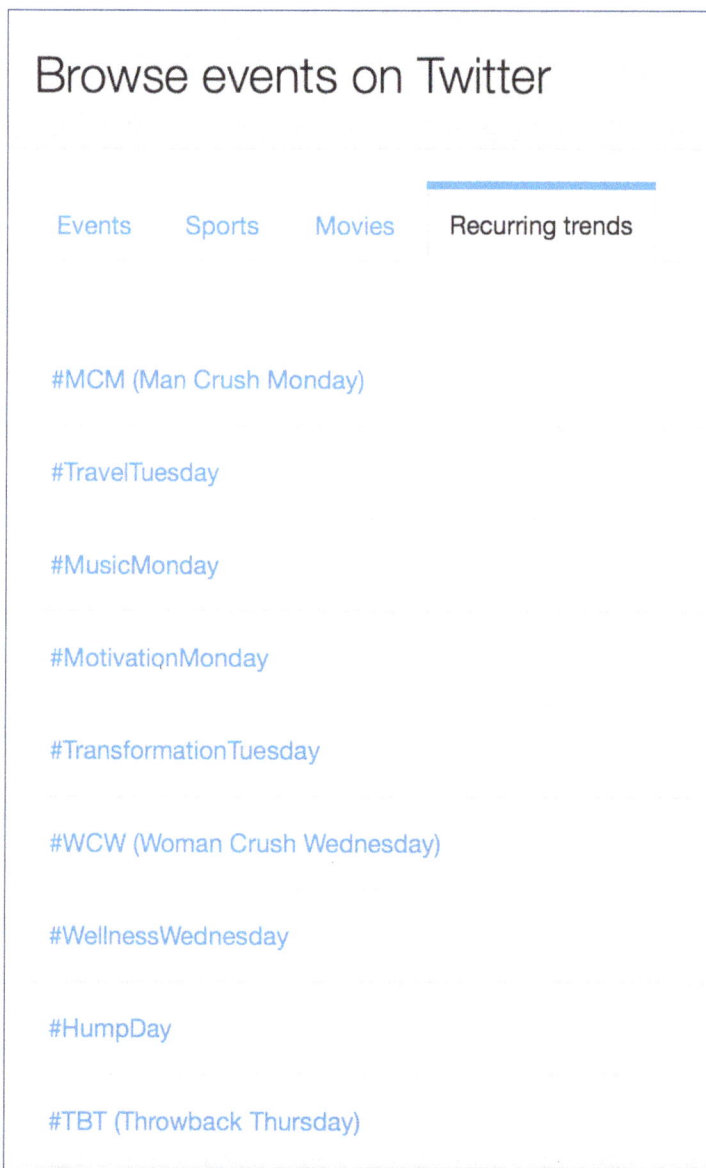

Browse events on Twitter

Events   Sports   Movies   **Recurring trends**

#MCM (Man Crush Monday)

#TravelTuesday

#MusicMonday

#MotivationMonday

#TransformationTuesday

#WCW (Woman Crush Wednesday)

#WellnessWednesday

#HumpDay

#TBT (Throwback Thursday)

Figure 5.23: Twitter analytics—Recurring trends

Outreach on Twitter

There is an unfortunate and common misconception about outreach via social media: "You just create a profile and start posting." This simplistic view of outreach via social media often leads to disappointment and frustration because it bears no fruit. To appreciate why this approach would not be effective, imagine yourself standing on your rooftop, reading a message and expecting all of the 322 million population of the United States (which is also similar in size to the entire Twitter user base) to notice, hear, understand, and act on your message. Pretty unrealistic, isn't it? Effective outreach is strategic and thoughtful, requiring patience and dedication to hard work, and a slow but steady progress. Another misconception is that the only worthwhile social media outreach comes in the form of viral campaigns, not realizing that much can be accomplished by focusing on making small-scale changes through our own small communities. Besides, successful campaigns require hours of hard work in first building a community. Viral campaigns don't become viral by accident, and people don't join social networking sites to join your campaigns. Therefore, if you are serious about using social media to entice action, prepare to get to work—patiently and consistently.

Using the foundation work we have already done for the Linda Goldberg sample profile, let's see how strategic and thoughtful outreach via Twitter would look like (Figure 5.24).

As a general rule, once you have built a foundation, a good workflow would start with research, followed by listening, then following relevant Twitter users, posting status updates, and participating in relevant conversations. The research you saw in this book with RiteTag allowed finding a number of hashtags that the Jewish community tends to use on Twitter (Table 5.1). Your first step in outreach is to "listen" to Twitter status updates that include these hashtags and to find the few, which present the best opportunity for you to participate and contribute to the conversation. Also, examine Twitter users who use these relevant hashtags, then follow those who match your audience profile, or those who would be considered their influencers.

Upon "listening" to Twitter status updates that include hashtags listed in Table 5.1, I decided that #beingjewish seemed to present the best opportunity for outreach as it is used by many Jewish moms and religious leaders. This hashtag also helps define the approach you would take in introducing climate action—drawing upon the fact that protecting the environment is an important Jewish value (5.4).

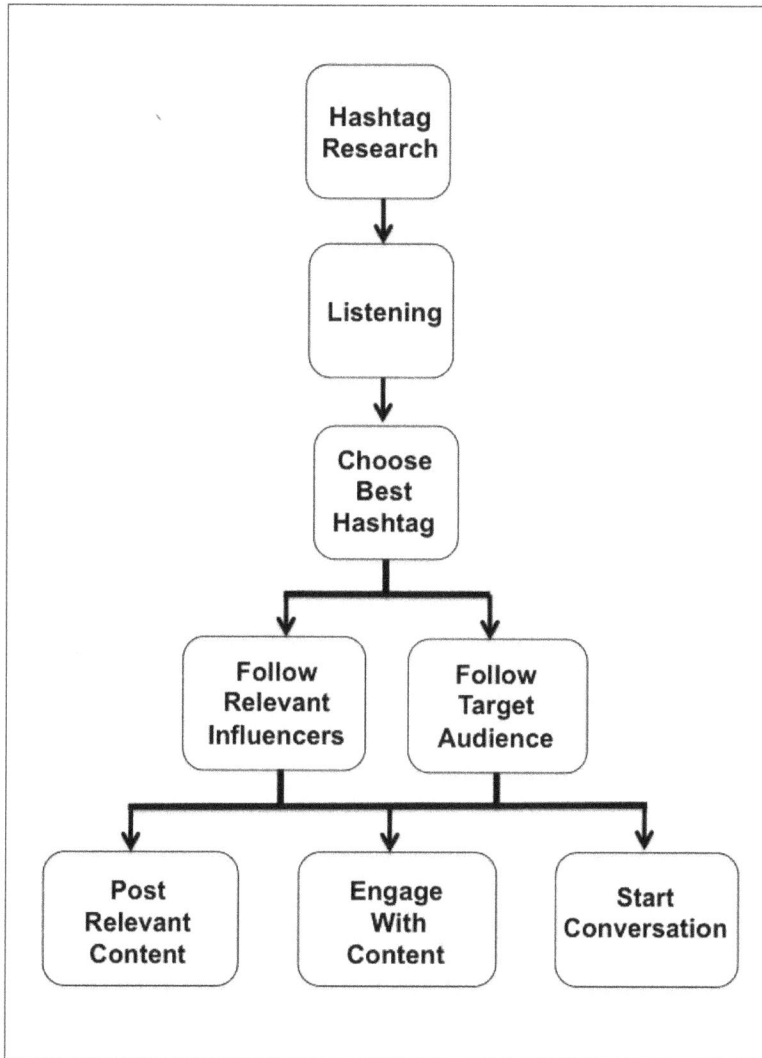

Figure 5.24: Workflow for outreach via Twitter

Now you are in the position to start posting status updates about Jewish values and their connection with being a good steward of the environment and other living creatures, making sure that you include #beingjewish. Note that you have to be economical when it comes to posting status updates on Twitter due to the 140-character limitation. Sometimes URLs can use up valuable space. To prevent long URLs from using up valuable characters, take advantage of link-shortening tools such as bit.ly (http://bit.ly) to save space. This tool not only shrinks the size of a URL, but will also provide analytics as to the source and number of clicks each link receives. You can even customize each link with a specific name so you can keep track of where the link was shared.

Figure 5.25 shows a sample Twitter update in which I included relevant hashtags, used Bit.ly to shorten a link and employed symbols instead of words in order to save space.

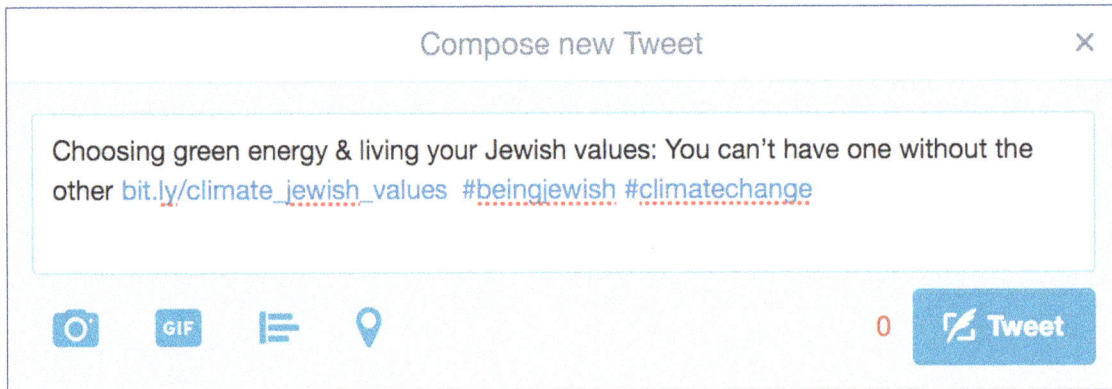

Figure 5.25: Sample Twitter status update

Set aside fifteen to twenty minutes per day to go through the workflow. Be patient and take the time to build a community of followers whom you meet every day on Twitter. Think of Twitter as the coffee shop you stop by each day to catch up with your community where you exchange news, ideas, and some laughs. It is only through this steady and consistent relationship-building routine you can ensure that when you are ready to deliver your climate action message, or start a campaign, there will be attentive, listening ears to hear your message.

The ideas you've seen with the sample Twitter updates in terms of reaching out and connecting is pretty much the same for any social network. The idea is to reach out, connect, and generate conversations, rather than being a billboard of your own ideas and opinions.

Practice what you learned

In this chapter you were introduced to Twitter and how its different features can be used to search for and identify communities matching a particular demographic. You have seen how this can be accomplished for the Linda Goldberg audience profile. Now, practice using Twitter by creating a profile (if you don't have one already); be sure to upload a photo and include a description of who you are. Then take Twitter for a test run with the other two example profiles in this book, or with the profile you have created for your own niche audience. Use the workflow presented in Figure 5.24 to start building your community and delivering your message. Be sure to join our Facebook group (http://facebook.com/groups/social4climate), share your results, and ask questions.

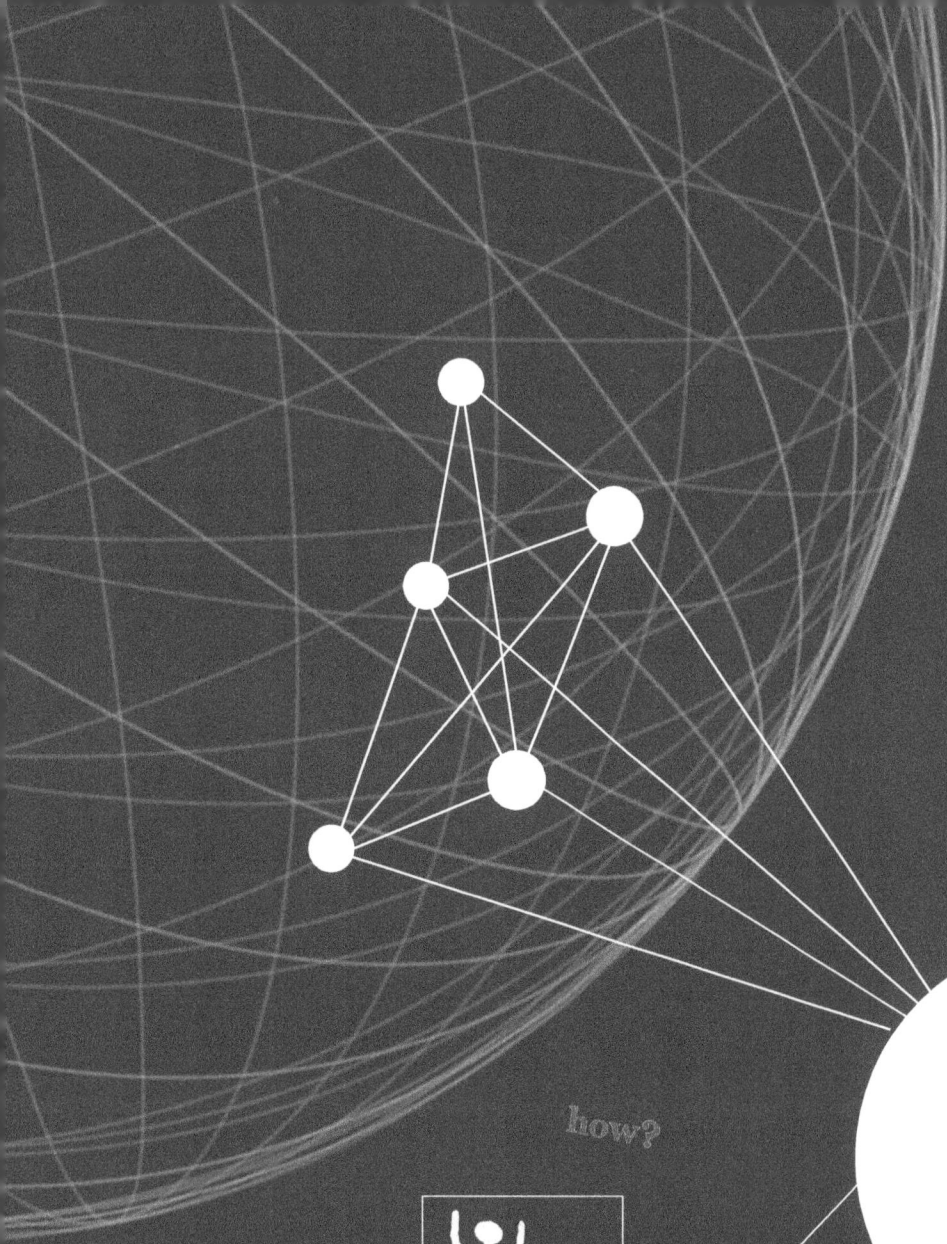

Chapter 6

Facebook

With more than one billion active users daily, Facebook continues to be the leader among the social networking platforms (6.1). Part of the appeal of this network for amplifying communication efforts is what happens when a Facebook user interacts with a specific piece of content on the platform. For example, if I comment on a particular Facebook page, my Facebook friends will see that activity in their own news feeds. Therefore, a page that might not have been on my friends' radar will be revealed to them because they can see my activities on Facebook due to their connection with me. This extended reach for a piece of content is unlike the "Retweet" feature of Twitter, where a user has to make a conscious decision to share something with his/her network. On Facebook, extended reach happens automatically as a user interacts with a piece of content.

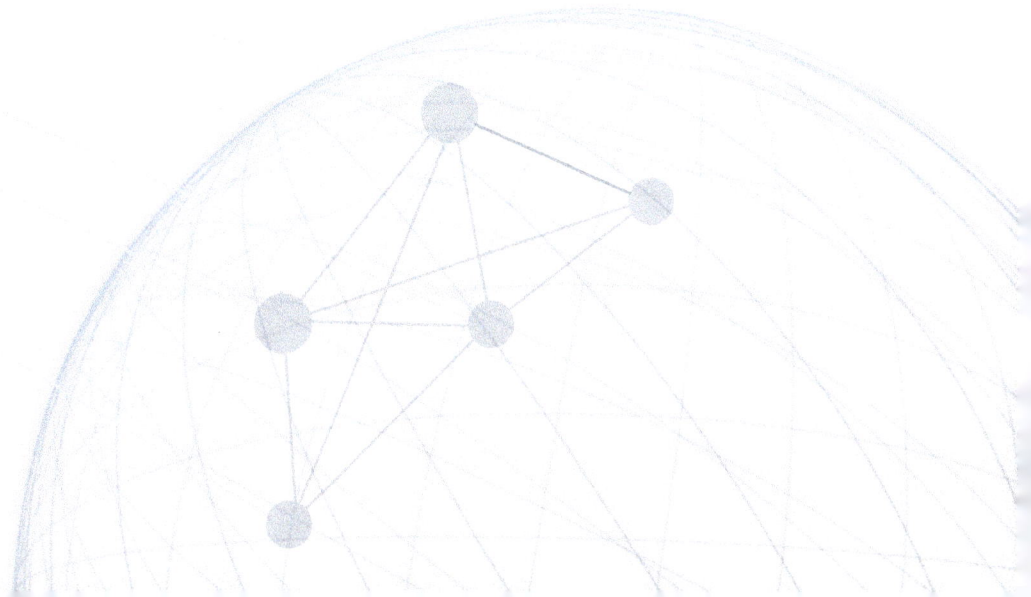

Another important feature of Facebook is the News Feed algorithm, which is the intelligent software that decides what we see in our news feeds, and how often. The basic premise behind the News Feed algorithm is to show a Facebook user content that is most relevant to him/her based on his/her history of interaction with the profile or page that is sharing the content. The algorithm has been put in place because it would be impossible to show a user content that was published by *all* of their friends at any given time when they happen to visit Facebook. In the early days of Facebook, the algorithm was less a factor in what users saw in their news feed, and it was less restrictive. However, as the size of the network grew, it became necessary to make the news feed more manageable. Note that you *still can* see all of your friends' activities, but you have to know where to look (6.2). The News Feed algorithm also allows Facebook to limit the reach for postings from Facebook pages to incentivize page owners to use its advertising platform to reach their communities. If you are going to use Facebook for your communication efforts, it is important to understand the factors that positively or negatively affect the News Feed algorithm. We will cover this topic in the "Status updates and terminology" section of this chapter.

Profile

Chances are you already have a personal Facebook profile, and are connected to your family and friends. You may also be following a few Facebook brand pages (or Facebook pages for short). When it comes to your climate change or science communication efforts, you may be wondering whether you should deliver your message through your personal profile, or start a Facebook page where you can publish your views to the public. I would suggest that you do both, and here is why:

1. Other than buying Facebook ads, it is quite challenging and time-consuming to build a community around your Facebook page. However, I would still encourage you to create a Facebook page as they offer features that personal profiles lack, such as analytics and ability to use the powerful Facebook ad platform when you have a budget. However, I feel that you will be more efficient at reaching your target audience and changing minds if you make personal connections.

2. Facebook pages offer analytics for measuring your success but this is not a feature available for personal profiles. In fact, you *can* get data about engagement for posts on your personal profile only if you publish them on your Facebook page

first and then share the post on your personal profile. We will cover Facebook analytics under the Analytics section of this chapter.

3. Joining Facebook groups, to which your target audience may belong, is a powerful way to make connections, similar to joining Twitter chats. However, there is no option for joining a group as a page; this is only available through your personal profile.

If you are worried about privacy and wonder what information on your personal profile is seen by the public (and not just your friends), you can actually view your profile as "Public"—as if someone not connected to you on Facebook is looking at your profile. Go to the "Settings" section of your profile (www.facebook.com/settings) and choose the "Timeline and Tagging" item on the left menu. Once you are there, find the "Who can see things on my timeline" section and click on "View As" (Figure 6.1). You can now view your profile either as how it appears to the public or as a friend. In this way, you can monitor what is viewable by those who are not your Facebook friend. You can also choose the audience for activities on your personal profile. (I will cover this topic later in this chapter.)

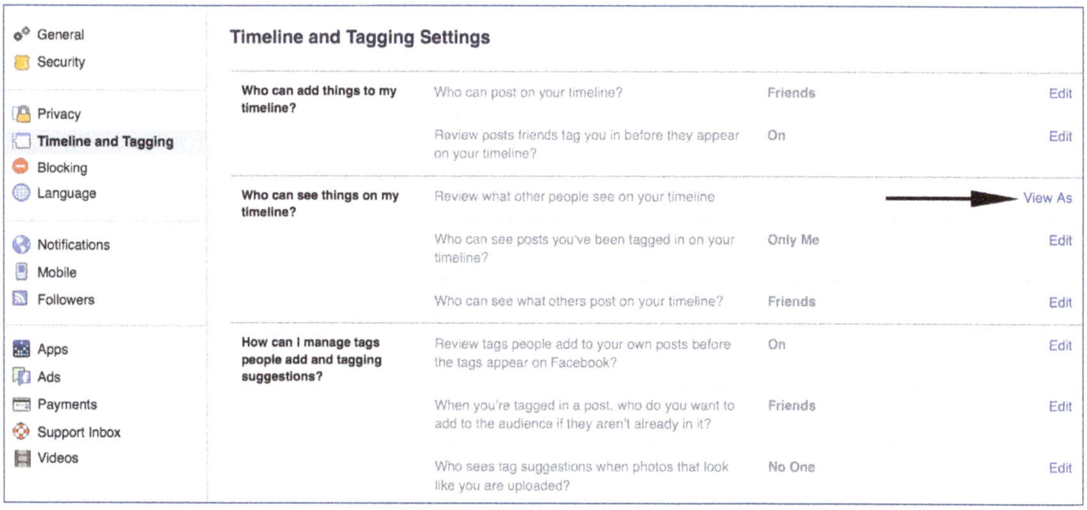

Figure 6.1: Viewing your profile as "Public"

If you decide you wanted to focus your outreach through your personal profile, you still have the option of sharing information with the public without having to connect with them as a Facebook friend. In other words, you can allow people to "follow" you without being your Facebook friend. To activate this feature for your personal profile, visit the "Followers" section of the "Settings" page to allow the public to follow you. Your followers will only be able to see your public posts, while your friends have access to all of your posts—public or private. We will cover choosing the audience for your posts in the "Status updates and terminology" section of this chapter.

Similar to Twitter, Facebook offers a number of features that are designed to provide you with control over your news feed. Let's review these options below.

Organization

We learned in the previous chapter about Twitter lists and how they can help organize your interactions. Facebook also offers lists:

1. Visit the "home page" of your Facebook profile (your news feed).
2. Look on the left column for a menu item called "Friends."
3. Hover your mouse over this menu item which reveals "More."
4. Click on "More," which will open a new page.
5. Click on the "Create List" button to create a list. You can either add friends to the list by typing their name (Figure 6.2) or do so from the "Friends" section of your Facebook profile (Figure 6.3).

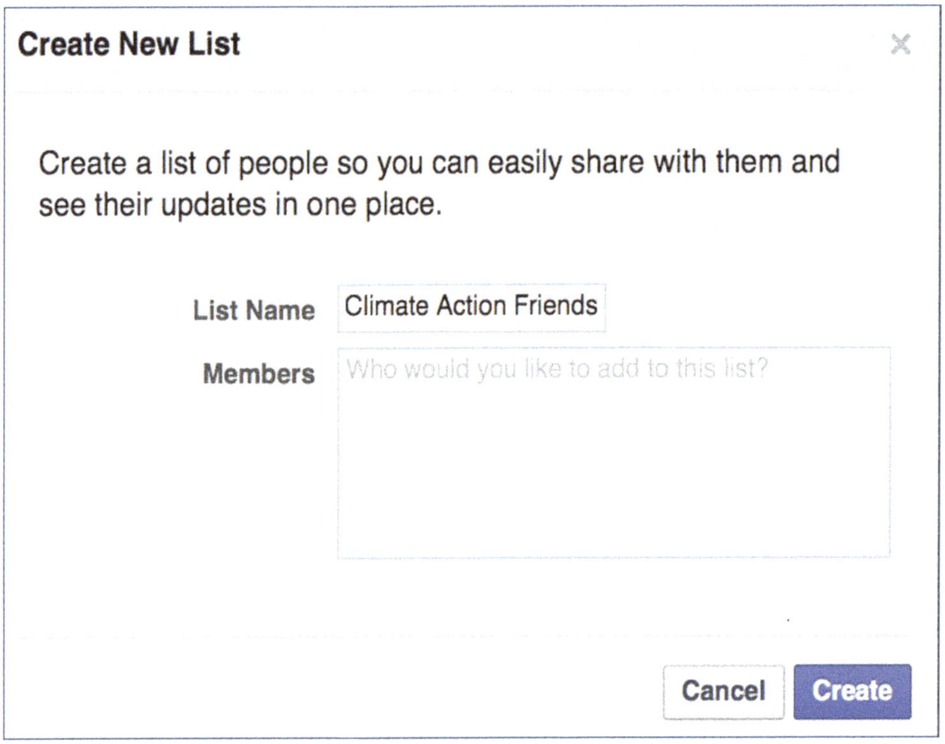

Create New List   ✕

Create a list of people so you can easily share with them and see their updates in one place.

List Name   Climate Action Friends

Members   Who would you like to add to this list?

                                              Cancel   Create

---

Figure 6.2: Creating lists on Facebook

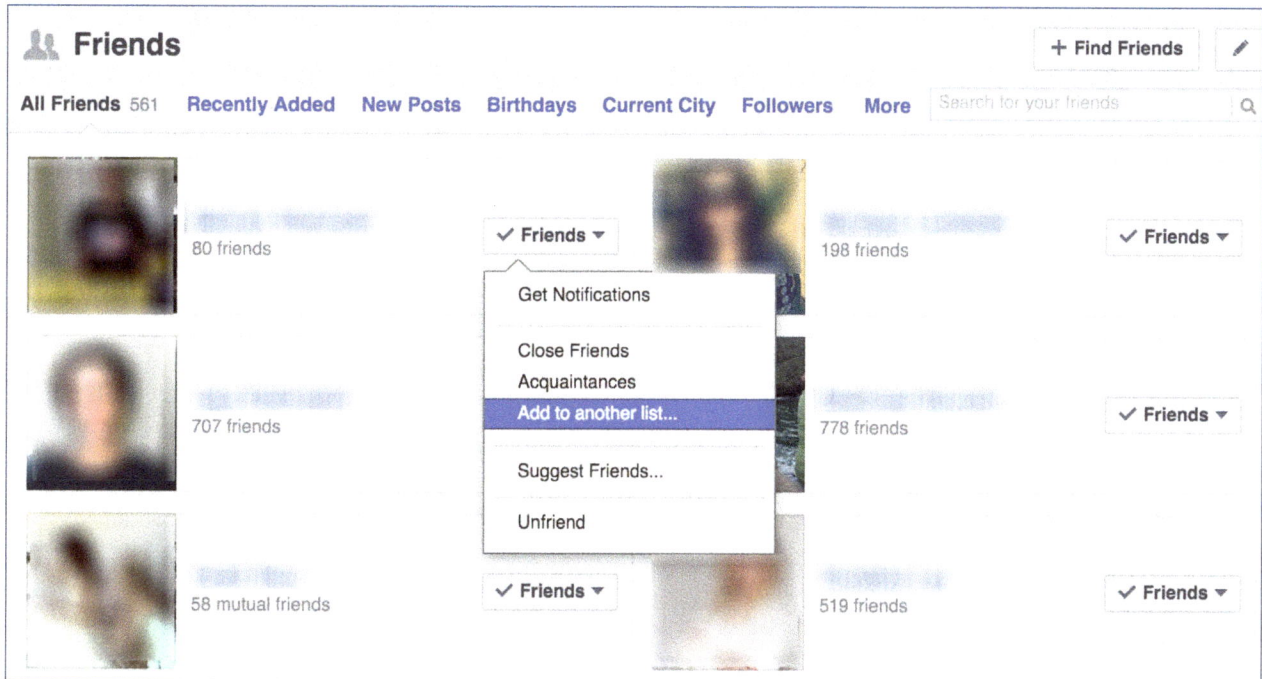

Figure 6.3: Adding friends to lists on Facebook

Similar to Twitter lists, you can choose to view only Facebook updates from a given group of friends. You can do something similar for the brand pages that you like by creating an "interest" list. The steps for creating interest lists are similar, except you begin by finding the "interest" menu item on the left-hand side of your news feed.

Search

When finding answers for questions or looking for information, most people tend to search Google. Google's robots crawl the Web and index information in blogs, websites, forums, and even social networking sites. Google is a search engine for information. Facebook does something similar with the information its users share with the platform about their interests and their lives. Facebook is a people search engine, and the depth of search results it provides is mindboggling, which can be valuable.

If you were on a quest to find your target audience, what would you search for? For one thing, you could search for Facebook profiles of the relevant people you already discovered on Twitter. Alternatively, you could start from scratch and use the keywords that we used in Chapter 5 to search Facebook. Let's use both of these methods to search Facebook as each one has its own merit. The former approach can help you develop a better understanding of and build deeper relationships with those you've already found on Twitter, and the latter can help you find new people and communities, which you otherwise would not have discovered.

Starting from scratch

Let's continue with our example profile Linda Goldberg. Searching for "Jewish mom" on Facebook reveals dozens of communities of Jewish mothers networking and exchanging parenting ideas (Figure 6.4).

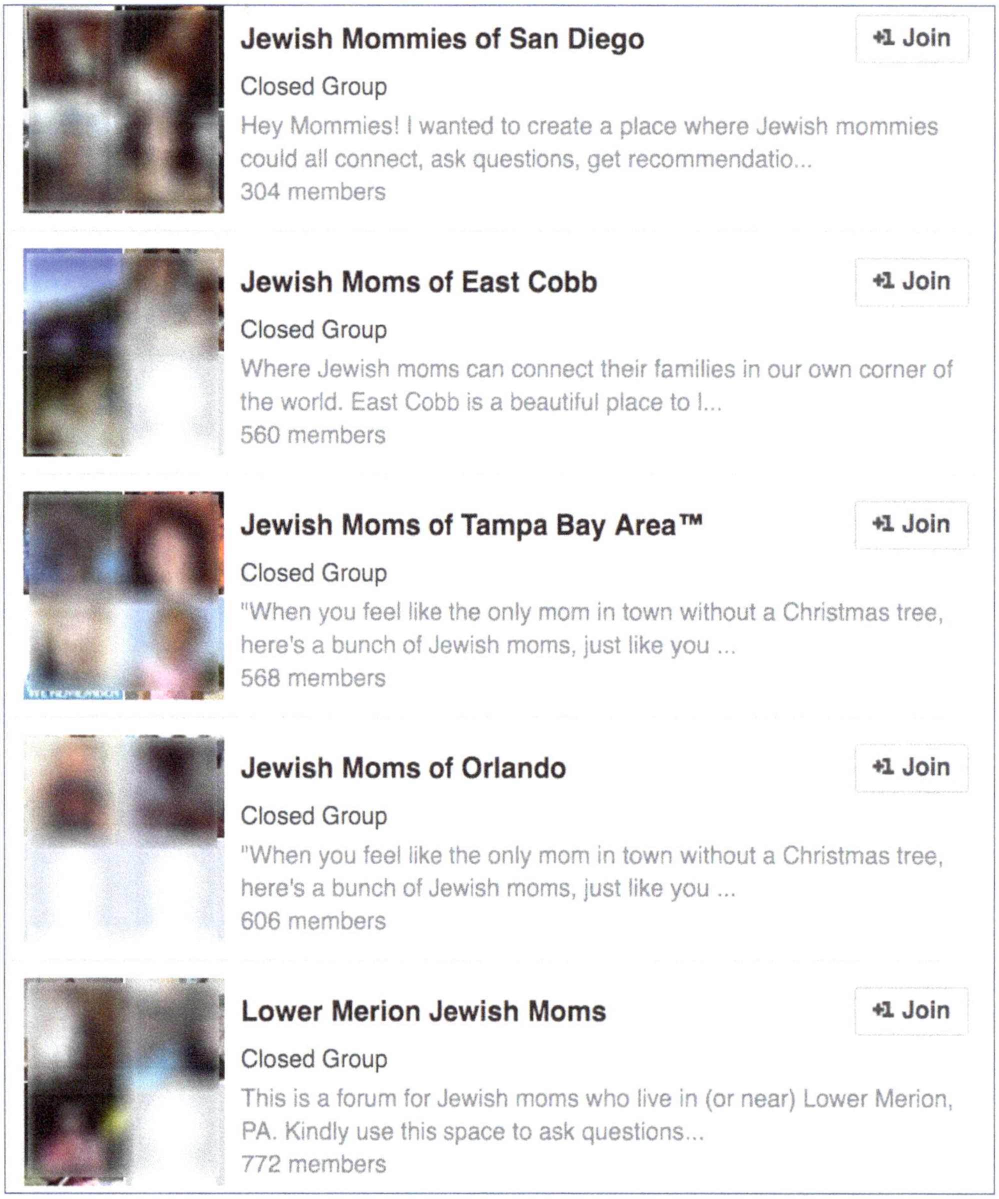

Jewish Mommies of San Diego   +1 Join

Closed Group

Hey Mommies! I wanted to create a place where Jewish mommies could all connect, ask questions, get recommendatio...
304 members

Jewish Moms of East Cobb   +1 Join

Closed Group

Where Jewish moms can connect their families in our own corner of the world. East Cobb is a beautiful place to l...
560 members

Jewish Moms of Tampa Bay Area™   +1 Join

Closed Group

"When you feel like the only mom in town without a Christmas tree, here's a bunch of Jewish moms, just like you ...
568 members

Jewish Moms of Orlando   +1 Join

Closed Group

"When you feel like the only mom in town without a Christmas tree, here's a bunch of Jewish moms, just like you ...
606 members

Lower Merion Jewish Moms   +1 Join

Closed Group

This is a forum for Jewish moms who live in (or near) Lower Merion, PA. Kindly use this space to ask questions...
772 members

Figure 6.4: Jewish mom Facebook groups

This search also reveals popular Facebook pages serving the Jewish mom community (Figure 6.5), physical locations (places) that serve Jewish moms (Figure 6.6), events that Jewish moms have attended (not shown), and a whole host of content relevant to this demographic. With this information in hand, you can reach out to Jewish moms either by joining their communities online, building relationships with influencer page owners who serve them, or visit the physical "places" or the events they tend to frequent. The sky is the limit!

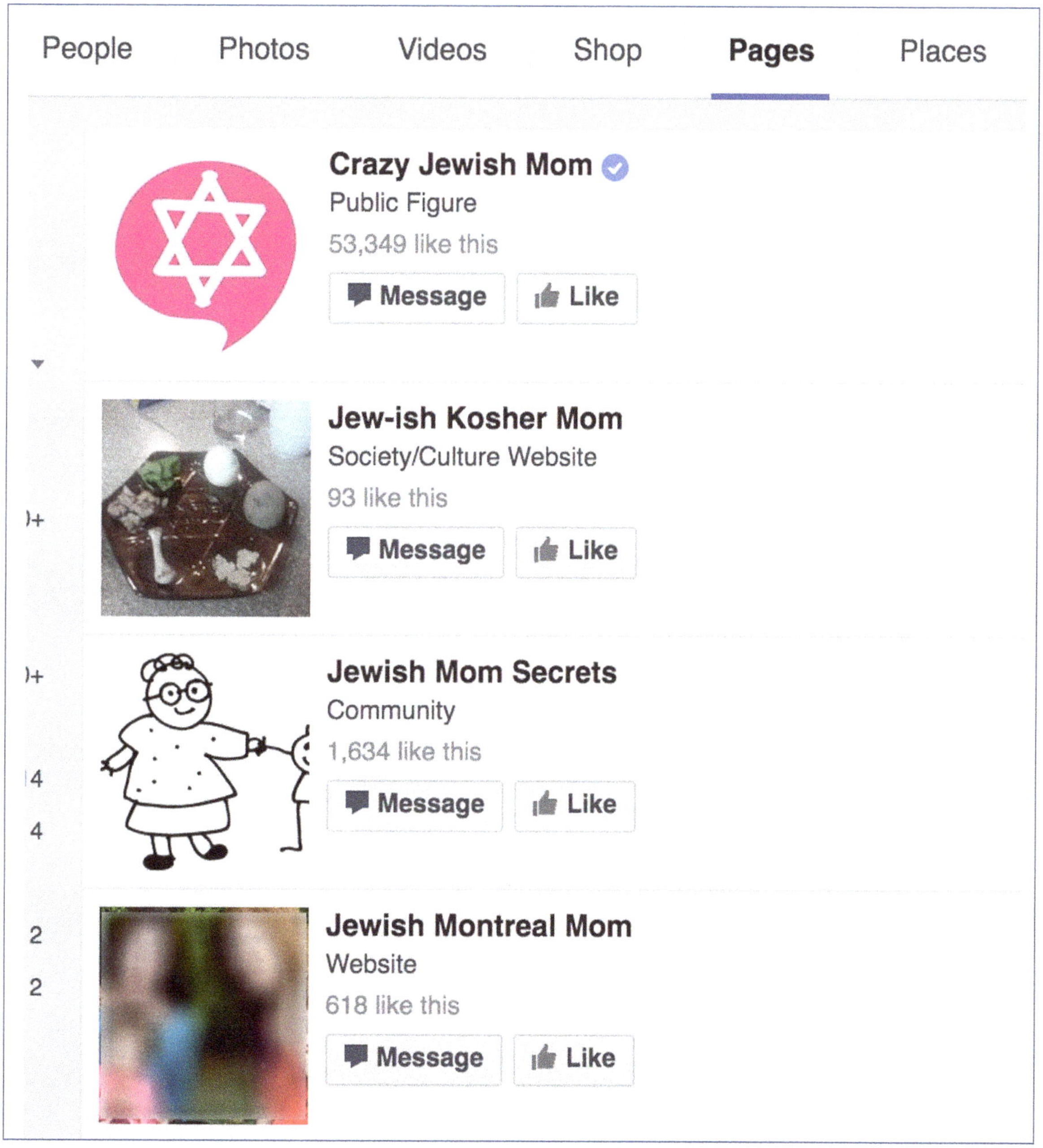

Figure 6.5: Facebook Pages serving the Jewish mom community

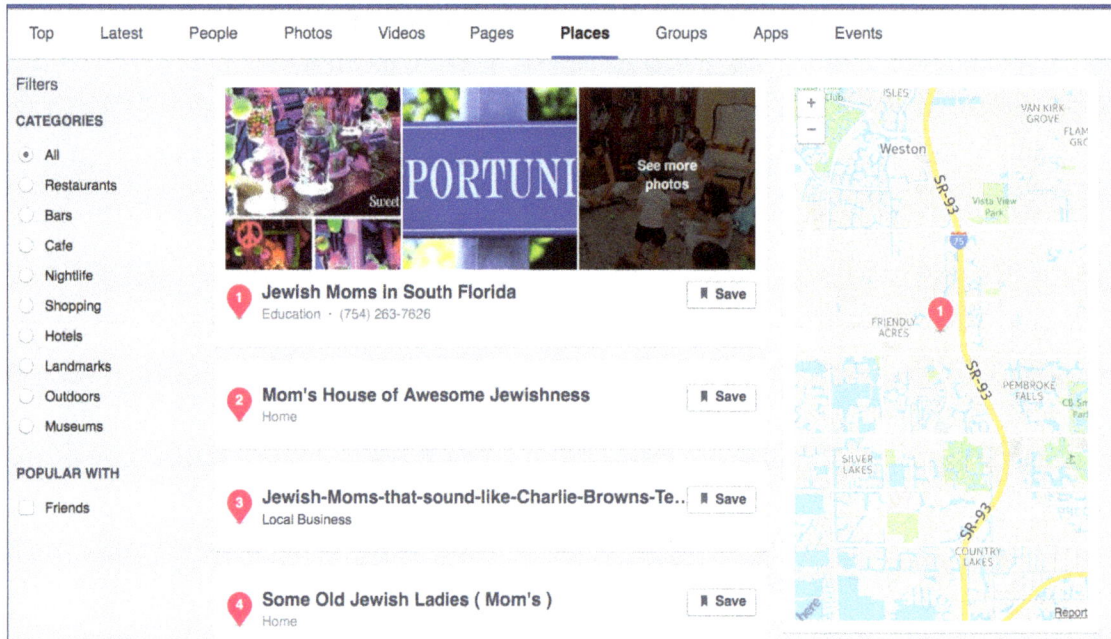

Figure 6.6: Places visited by Jewish moms

Here is a great social media learning moment: Searching through Twitter, we discovered that using the keyword "Jewish mom" or the hashtag "#jewishmom" was not fruitful for finding communities of Jewish mothers. This hashtag was not fruitful in our search for communities of Jewish moms because #jewishmom on Twitter is used as a satirical reference to stereotypical overprotectiveness ascribed to Jewish mothers, rather than an invitation to join a community. With Facebook however, this is a totally different story; a search with the phrase "Jewish mom" did in fact reveal communities of Jewish mothers. The lesson here is that every social network has its own culture and language, similar to when you enter a new country. If one tactic is not fruitful in one social network, don't necessarily give up on it, as it may be more productive in another.

Connecting Twitter profiles to Facebook communities

Let's refer back to Table 5.1 where we listed the people we found by searching Twitter, and see if we can find Facebook communities to which they may belong. To illustrate how this would work, I will use one of these profiles as an example, a blogger mom belonging to the Jewish faith (to be respectful of her privacy, I will not reveal which one I chose). Searching Facebook using her full name, I discovered her profile and learned more about her. However, we are only interested in the communities to which she belongs, so we need to focus on the "Groups" section of her profile. This investigation revealed a few public groups, one of which seems relevant to our target audience: Jewish Life in West Hartford (Figure 6.7). Exploring the group description and membership revealed that this is a community mostly composed of Jewish families. For a Jewish climate change activist in the Connecticut area who is trying to "be the Jewish voice" for climate change, joining this group would be a great way to begin forming connections.

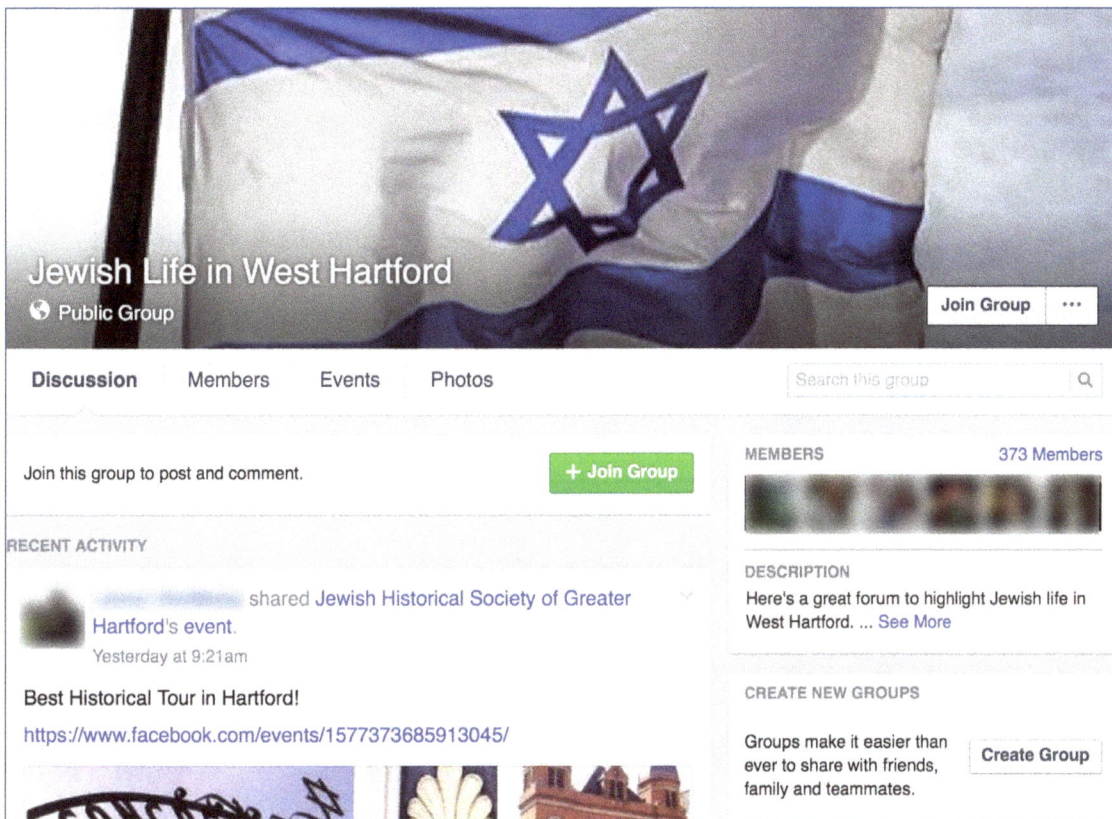

Figure 6.7: Jewish Life in West Hartford

Using pages to find groups

Once we find Facebook pages that serve our target audience, we can discover more communities that we have missed in our previous search. Asking the Facebook search engine to show us groups for users who "like" a particular page will help accomplish this task. An example is shown in Figure 6.8, starting with the page "Jewish Mom Secrets." Type the following phrase in the Facebook search box— "Groups joined by women who like Jewish Mom Secrets"—and you will discover a world of new communities (Figure 6.8).

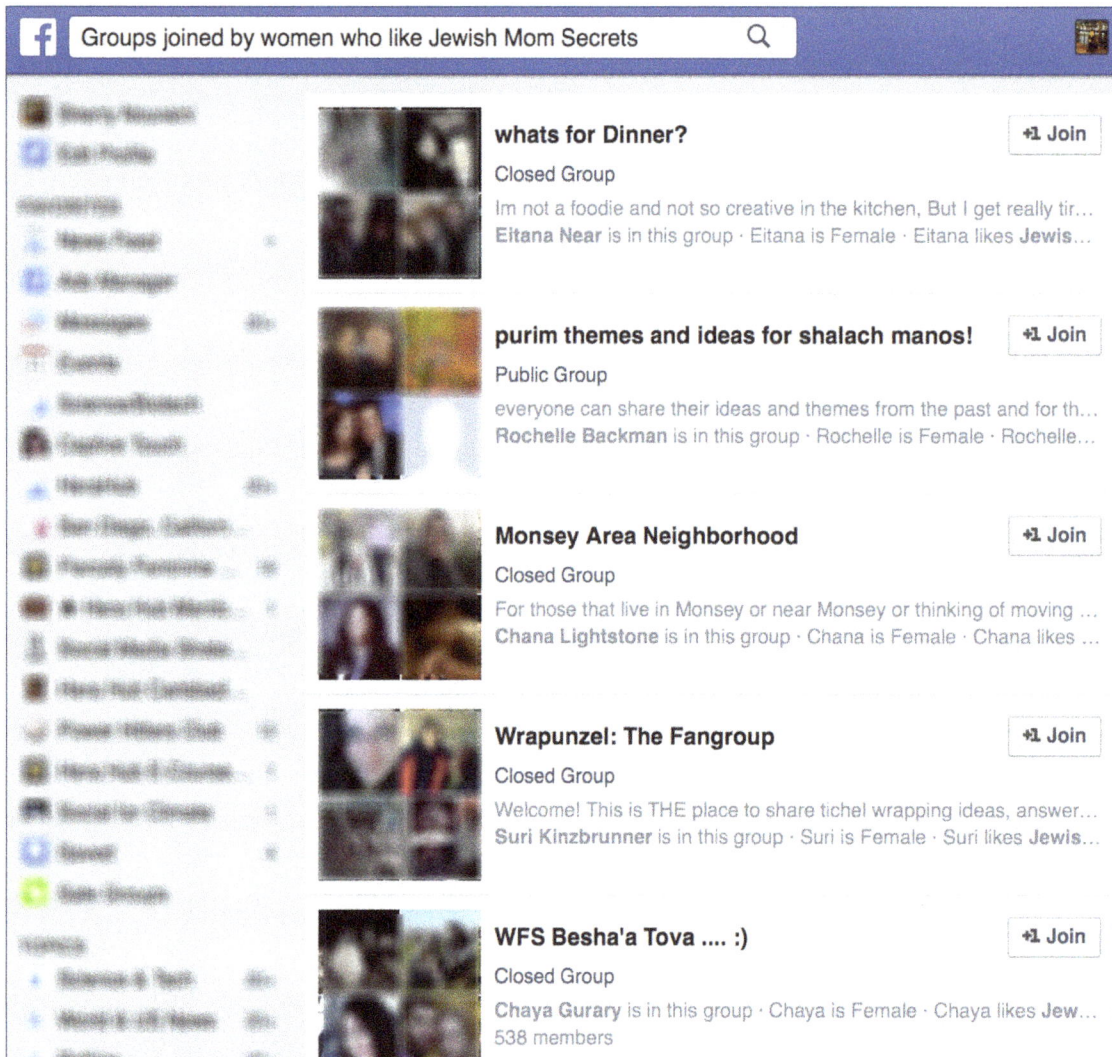

Figure 6.8: Using Facebook searching to discover specific groups

Note that finding what you want with the Facebook search engine will not always be so straightforward. More often than not, you will have to explore different versions of a particular search phrase to find what you are looking for.

A few words of clarification and caution

Reading these passages, you may think I am encouraging you to start stalking people's profiles on social media channels. I want to make it clear that this is not what I am advocating. I truly believe that making a meaningful difference in people's attitudes and tendency toward taking action against climate change will require invitations to do so by their peers and members of their communities. So if you find a community with which you do not share common values, interests, or identities, don't try to use the information you find on social media to appeal to its members. On the other hand, if you happen to be well-positioned to make connections in communities that are a match with you along with the values you hold, then by all means join these groups—but don't be in a hurry to try to force them into conversation about climate change. When you join a community, first listen. Get to know your community members, participate in discussions that *they* care about, before bringing up the subject of climate change. In other words, you need to be patient or your communication efforts will backfire. Let me give you a personal example to illustrate:

I am not a climate scientist, but wanted to expand my knowledge about the science behind climate change and about how the science is being misrepresented by deniers. To that end, I enrolled in and completed the Denial 101x massive open online course put together by John Cook (3.21). Shortly after I completed the course, I accepted a "friend request" from a woman who was also a Denial101x student, as indicated in her profile. Immediately after I accepted the friendship request, she sent me a private message with a link asking me to sign a petition! You might think that I signed it right away, but actually I did not, as I found it offensive that she would not take the time to build a relationship with me before asking me for something, even though it may have been toward making a difference on an issue about which both of us cared. Imagine the reception she would have received from a person who was disengaged from climate change issues! What was even more disappointing about this development was that, about a month into our Facebook friendship, I received a message from her indicating that she would be "unfriending" me as she had reached her "friends" limit on Facebook! She needed to unload her connections with me so she could make more friends and submit more requests for signing her petition. If you approach your climate change activism efforts on social media channels in this way, you will only build walls instead of breaking down communication barriers.

What if you are an organization, and joining communities and making personal connections is not an option? What are you going to do with all these communities you would discover on Facebook? As I mentioned at the end of Chapter 4, an alternative option is to reach out to influencers of your target audience, who tend to be moderators of Facebook groups, Facebook page owners, or opinion leaders within their communities. To identify these opinion leaders you may have to spend some time to listen to conversations within these communities. As a general rule, just taking the time to listen is the best place to start. Once you gain an understanding of the community, along with the leaders of these communities, you are in a better position to approach them in a thoughtful and strategic manner. You will understand the types of projects, activities, or campaigns that would be the best fit for a particular community. Climate change activism can take many forms, and as an organization, you have the responsibility to match your campaigns and outreach efforts with the needs and preferences of the community you approach, as opposed to expecting them to adapt to yours.

Updating the Linda Goldberg persona using our search results

In Chapter 2, I used Facebook Audience Insights to gain a deeper understanding of our target audience using data. I listed what I learned from this exercise in Table 2.3. Now an audience profile in Table 2.3 can be updated again using the information that Facebook search provided, particularly the "Interests" and "Facebook pages" section of this table. The data revealed by analyzing Audience Insights pointed to three pages—Israel Defense Forces, United With Israel, and GodVine—but these topics are too broad. Now we can make this section of the audience persona more focused by adding Facebook pages and groups that we discovered through Facebook search. The "Location" category can also be updated as we found groups of Jewish moms in additional locations (new information is show in italic, Table 6.1 on opposite page).

Status updates and terminology

You may already be familiar with updating your personal profile on Facebook. However, what you may not know is that you can choose the audience of your Facebook status updates. For example, if you want to limit your climate change discussions to a limited group of your friends, you can allow access to only *them*. What you need to do first, however, is generate a list of those friends, as you learned in the "Organization" section of this chapter.

To limit the audience of your posts, click on the white arrow next to the "Publish" button. This will reveal an array of viewing choices for your audience by selecting a particular list. You can even exclude or share with only a few specific friends without choosing a list by selecting the "Custom" option (Figure 6.9).

Table 6.1: Updated profile of Linda Goldberg

Name	Linda Goldberg
Job title/profession	Homemaker/mother
Age	56
Gender	Female
Family income	~~150,000~~ $75K–more than $500K
Location (top 5)	Greater NYC area> LA> Miami> Chicago> Philadelphia> *West Hartford> South Florida> Portland> Baltimore> Orlando*
Identity	Wife, mother
Attitude toward climate change	Disengaged
Influencers	Meteorologists, charismatic news anchors, friends, *Facebook group moderators, page owners*
Political affiliation	Democrat
Barriers to taking personal action on climate change	Lack of knowledge, low interest level
Religion	Judaism
Online activity	Facebook, email with friends, home pages as news source (i.e. Yahoo, MSN)
Home ownership	Home owner $200K and above
How does she shop	High-end/upscale, primarily uses credit cards
Interests	Israel Defense Forces, United With Israel, GodVine
Facebook pages they like	United With Israel, Israel Defense Fund, GodVine, *Crazy Jewish Mom, Jewish moms, Jewish Mom Secrets*
Education level	College graduate and above
Mobile device used	More likely to use iPhone & iPad rather than Android. Uses both computer & a mobile device.
Activity on Facebook	Much more likely than other Facebook audiences to like, share, comment, and click on ads. Not likely to redeem promotions.

Table 6.1: Updated profile of Linda Goldberg

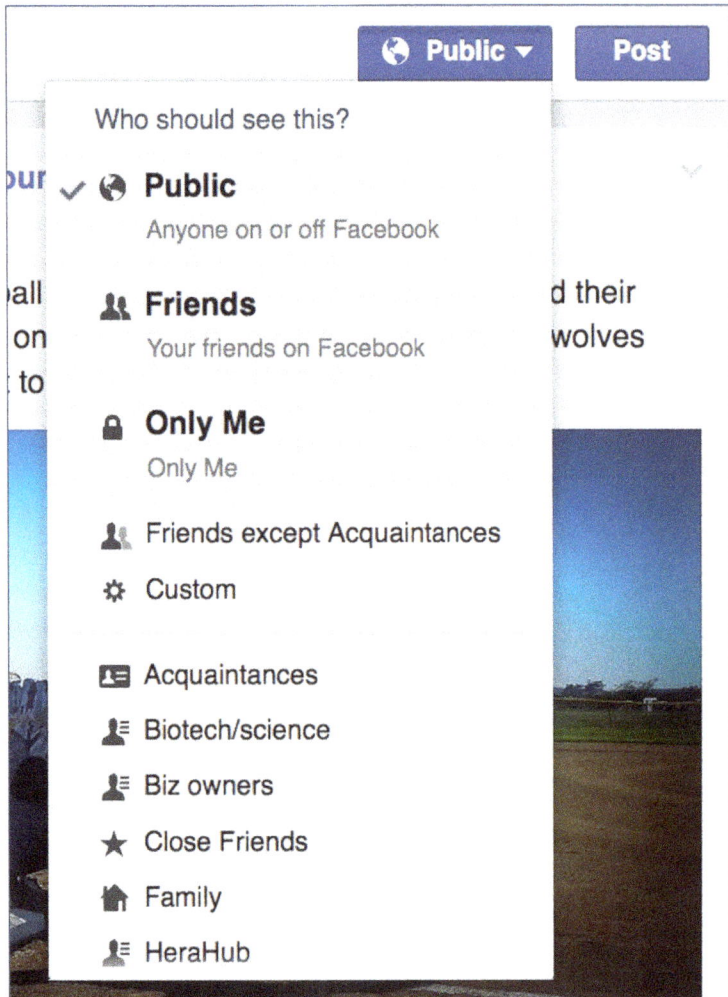

Figure 6.9: Selecting an audience for your posts

Facebook page status box

A page status box allows a number of different types of content (Figure 6.10).

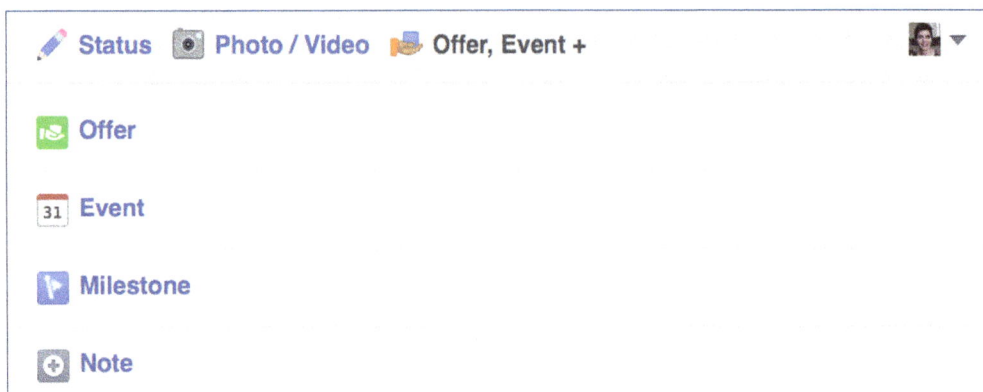

Figure 6.10: Status box on a Facebook page.

1. Status (Text)

2. Image or video

3. Offer: an equivalent of a coupon

4. Event: page owners can set up an event page and invite their followers to attend

5. Milestone: page owners can post updates about their accomplishments

6. Note: equivalent of a blogging platform but integrated within Facebook. The advantage of blogging on Facebook is that your page followers do not have to leave the platform to read your article. An added advantage is when they engage with your articles (likes, reactions, and comments), their friends will see that activity thus increasing the potential reach of your articles.

Publishing now, in the future, or the past

Social media power users usually use third-party tools to schedule their status updates on social networking sites. The only social network that includes scheduling of posts is Facebook, and it even allows page owners to post content dated in the past. Once a status update has been created, click on the arrow to the right of the "Publish" button to either schedule the post for the future, back date, or save the post as a draft (Figure 6.11).

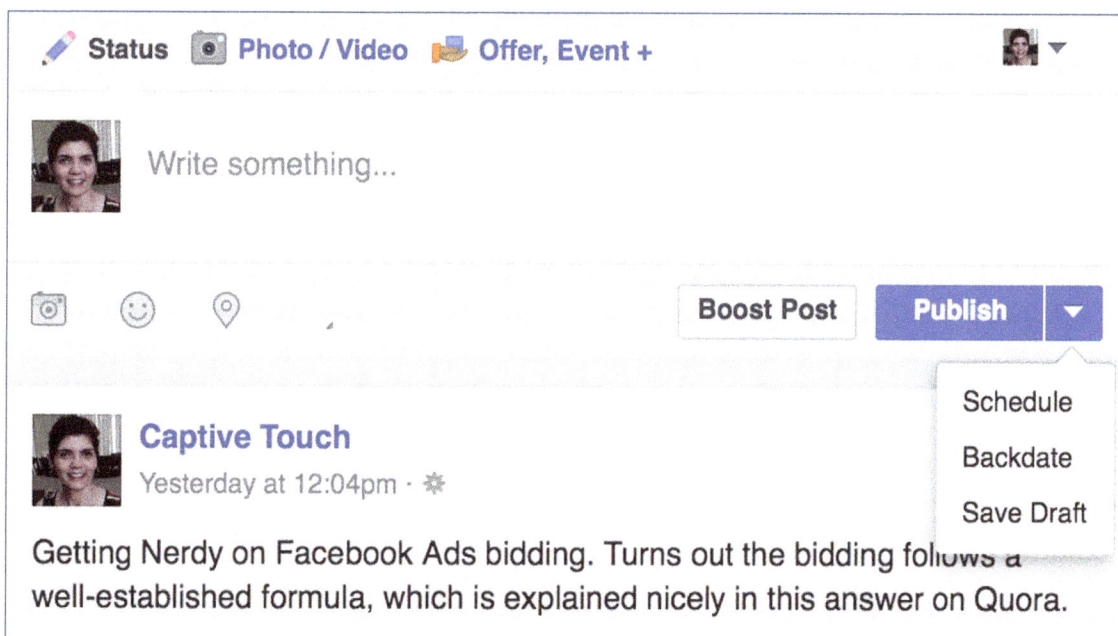

Figure 6.11: Facebook allows scheduling of posts

The Facebook News Feed algorithm

Facebook status updates are in competition for a user's attention, and which one wins the battle is decided by Facebook's algorithm. The criteria for choosing winners are partly determined by how often a user interacts with the content posted by a friend or a Facebook page, and how he/she interacts with the content. Note that I said "partly determined," that's because no one really knows the full set of criteria that the software uses. We only know what Facebook reveals, and *that* even changes from time to time. Basically, the News Feed algorithm team on Facebook continually tweaks the software based on tests they perform, and research they conduct on how page owners try to game the algorithm, and how Facebook users react with content shared on Facebook (6.3). However, one thing is for sure: Facebook aims to show its users the highest-quality, most relevant, and timely content. Ultimately, the best way to be algorithm friendly is to post high-quality, timely content that resonates with your audience. Here is a list of a few set of criteria based on which Facebook judges and assigns scores to status updates:

1. Positive Score:
   a. Posts that entice a reaction quickly: Topic of your posts should be timely. For example, posting about topics that are trending on Facebook would have a better chance of gaining traction immediately. To see trending topics, look on the right-hand side of your Facebook news feed and find "Trending."
   b. The more time that a Facebook user spends on a post, the higher the ranking of that post. One way to keep your audience's attention for a longer time is to upload an interesting video.

2. Negative Score:
   a. Overly promotional posts (buy this or that, sign up, etc.)
   b. Click bait: These are posts that entice a reader with an interesting title only to show them a piece of content that is not relevant to the title. Facebook detects these types of posts by measuring the length of time a reader spends on the website once they click through the link. If they click right back to Facebook, this is a signal that they did not get what they expected.
   c. Posts that entice users to label them as spam, to "hide" them, or to "hide all posts" from a particular source (see Facebook Insights below).

From "like" button to "reactions"

It used to be that users could only "like" updates; however, in 2016, Facebook introduced "reactions" on both personal profiles and pages (Figure 6.12).

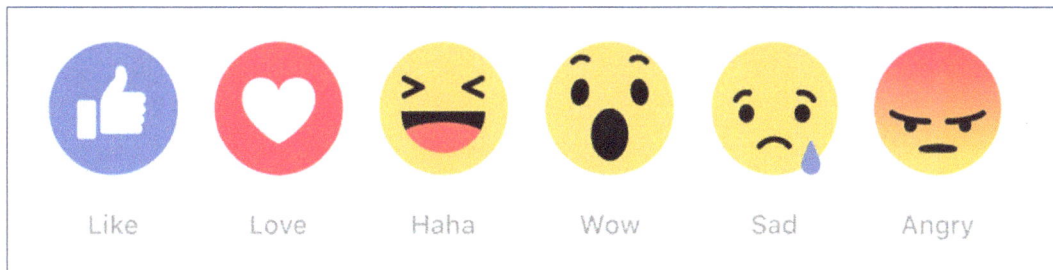

Figure 6.12: Facebook reactions

As of this writing, Facebook is experimenting and learning how the community uses this new feature. In addition, Facebook page owners do not get data about the different types of reactions to the posts in the analytics of their pages; these are simply reported collectively as "reactions." However, if and when Facebook decides to parse this data based on types of reactions, you can imagine how powerful it could be to learn how your target audience reacts to different types of posts so you can gauge their emotional reactions!

Communities

We already saw communities on Facebook in the "Search" section of this chapter: Facebook groups and pages. What I want to cover in this section are types and features of Facebook groups. Groups come in three flavors; all have the same features, but each comes with a different level of privacy. Table 6.2 lists different types of groups and a comparison of their privacy settings.

Group features

As of the third quarter of 2015, Facebook reported that 1 billion people join groups on this platform! Groups offer a number of features that facilitate seamless community building. Figure 6.13 is a screenshot of a Facebook group, which shows a number of these features.

Table 6.2: Facebook groups

	Public	Closed	Secret
Who can join?	Anyone can join or be added or invited by a member	Anyone can ask to join or be added or invited by a member	Anyone can be added or invited by a member
Group name visibility	Anyone	Anyone	Current and former members
Group description visibility	Anyone	Anyone	Current and former members
Group posts visibility	Anyone	Only current members	Only current members
Group visibility in search	Anyone	Anyone	Current and former members
Group membership visibility	Anyone	Anyone	Current and former members

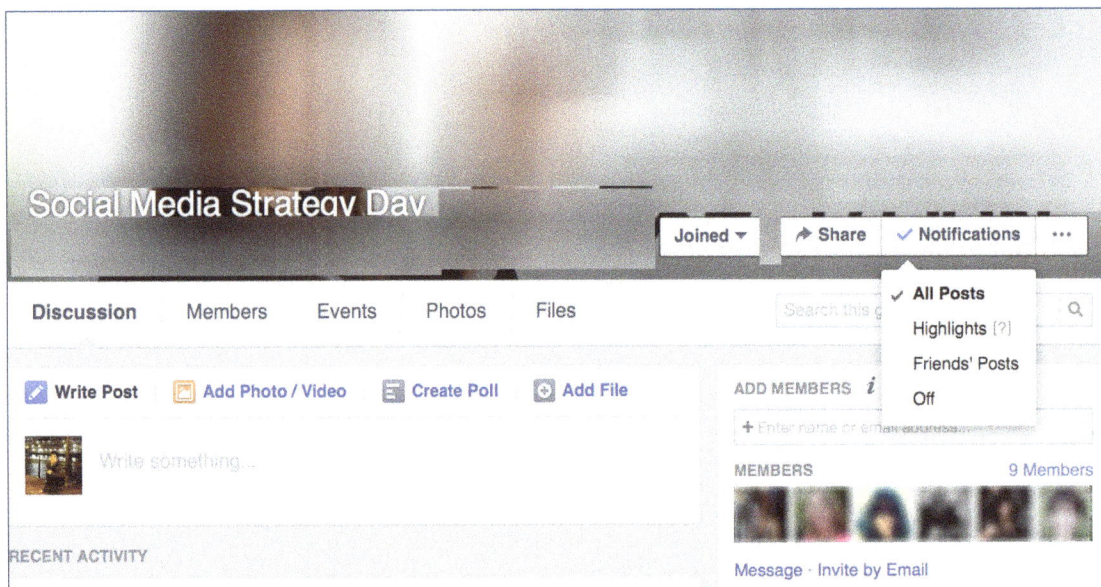

Figure 6.13: Facebook groups

When you join a Facebook group, it is important to keep up with conversations in the group and participate in them. To ensure you are up-to-date about these conversations, be sure to turn on the notifications settings for your favorite groups (shown in Figure 6.13).

Other features in Facebook groups are creating events, uploading files, sending private messages to group members, and broadcasting a live video to group members (available only on the mobile Facebook app). You can also survey members of a group using the poll feature. Joining Facebook groups is a great way to connect with like-minded people without having to go through the process of "friending" them. When you join a group, remember to spend some time to listen, and participate in conversations first, before initiating climate action outreach.

Analytics

Facebook offers page owners a whole host of analytics on the performance of their pages. When a page is newly established, Facebook does not provide analytics data until at least thirty people have liked the page. In the beginning, all that a new page gets is notifications of activities that others perform on the page (like, comment, and share). As a Facebook page gets more likes and engagement, the "Insights" tab appears on the page, which is where the analytics are reported. Let's look at the different types of data that Facebook Insights provides, and why they matter. Figure 6.14 shows a screenshot of the Insights home page where you get an overview of a page's performance. On the Insights home page, data is presented for the present day, the past day, the past week, or the past twenty-eight days. In other sections of Facebook page Insights, there is more flexibility for choosing the range of dates that one can view data about a page. The date range can be set by a calendar that appears at the top of each section (not shown).

Promotions

Promotions are an easy way to advertise on Facebook without the need to visit the advertising platform. You can promote the call-to-action feature of your Facebook page, your website, your local business to nearby Facebook users, or your page right from this section of Facebook Insights. This is also where you can keep track of all of the promotions you have placed and their performance (not shown). If you have the budget to pay for ads, I recommend against using this option, as setting up Facebook ads from the advertising platform, Ads Manager, is superior. Advertising on Facebook is not in the scope of this book so I will not elaborate further.

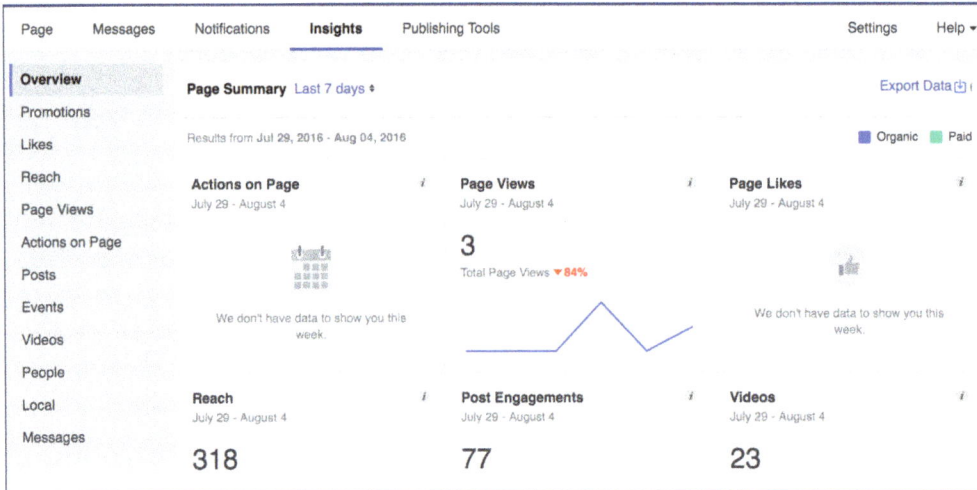

Figure 6.14: Facebook Insights home page

Likes

The Likes section of Facebook Insights is composed of three parts, each of which provides a different aspect of the performance of a page in terms of its number of followers (number of page likes):

1. Total page likes provides an overall count of the total number of Facebook users who have "liked" a page. It provides a visual representation of how a page is growing over time (Figure 6.15).

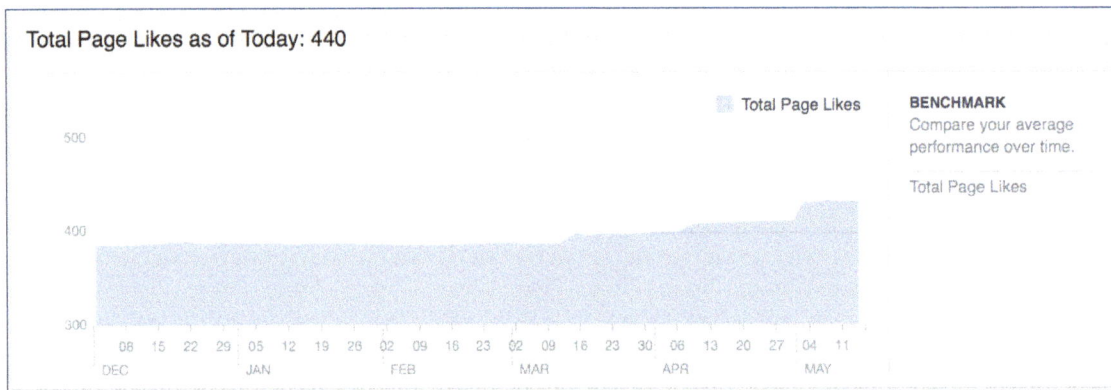

Figure 6.15: Total Page Likes

2. Net likes shows individual "like" and "unlike" events, as well as "like" events that occurred as a result of advertising (paid likes) as opposed to those that occurred without paying for ads (organic likes) (Figure 6.16). It is important

to pay special attention to the "unlike" events and correlate them with what you posted on the page. As you will see later in this section, you can actually determine whether or not a particular post led to someone unliking your page (in the Posts section of Insights).

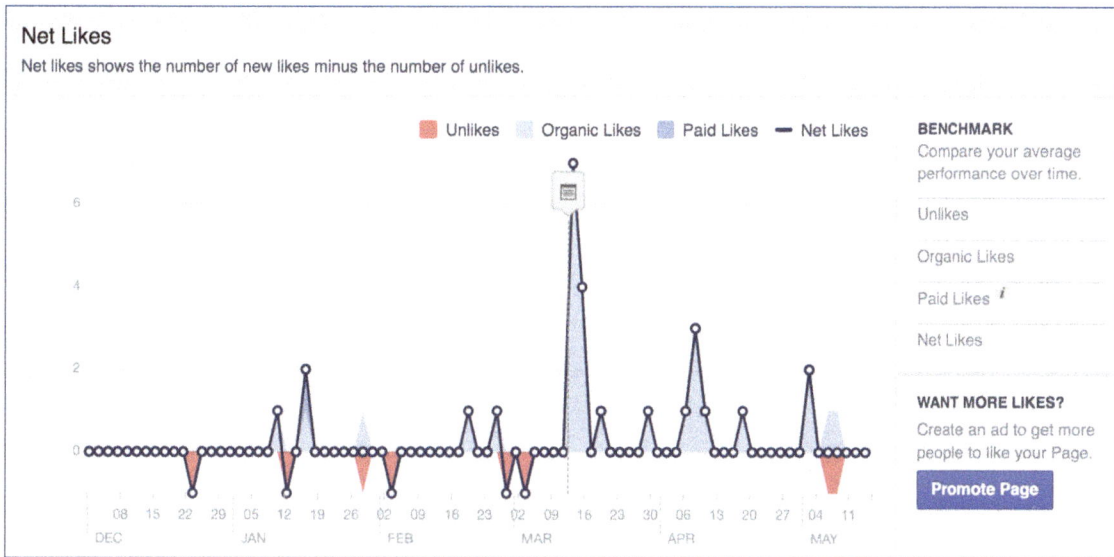

Figure 6.16: Facebook page Insights—Net Likes

3. "Where your page likes happened" provides information about where the "like" event occurred (Figure 6.17). Facebook users can "like" a page from a variety of different locations on the Web. It is important for a page owner to keep track of where page likes are coming from to find out which space is more effective in attracting a following.

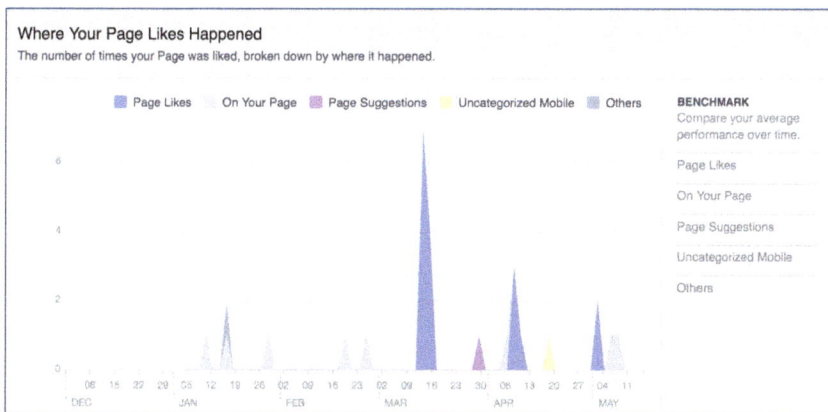

Figure 6.17: Facebook page Insights—Where Your Page Likes Happened

Reach

This section of Facebook Insights provides information about the performance of a page's posts. Reach is divided into four sections:

1. "Post Reach" shows the number of Facebook users who could potentially see a page's post, both organically or as are result of an ad. Keeping an eye on this data will offer visual cues as to which posts have reached the most people (not shown).

2. "Likes," "Comments," and "Shares" show line graphs, which indicate the number of each of these individual events on a page's posts (Figure 6.18). Clicking on data points for each event will reveal a summary of the page's posts that occurred on that day. This data can be used to identify best-performing posts and learn what tends to resonate with a page's followers (not shown).

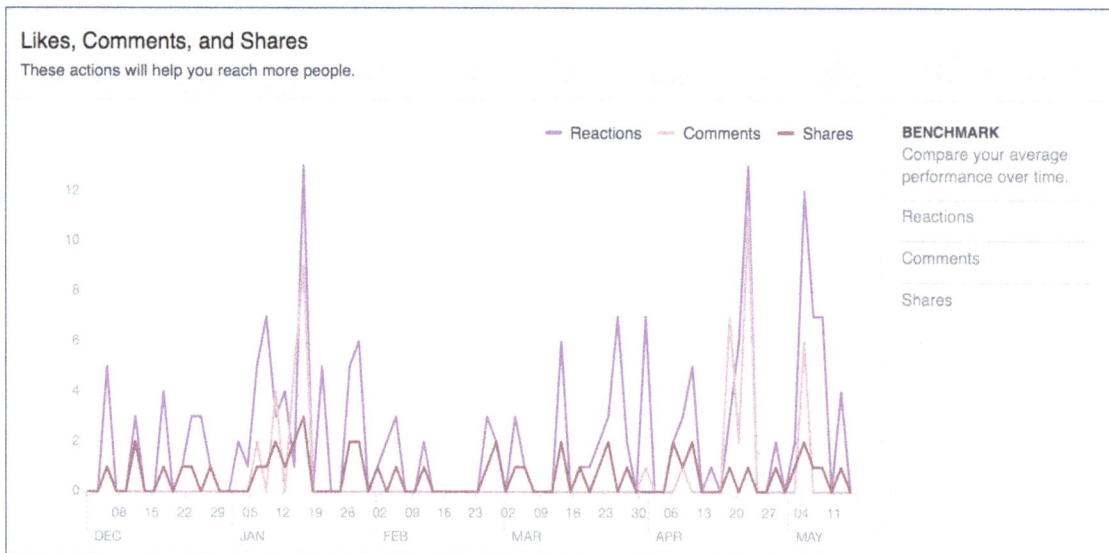

Figure 6.18: Likes, Comments, and Shares of page posts

3. "Hide," "Report as Spam," and "Unlikes" is one of the most important sections of Facebook Insights because it allows page owners to identify page content that tends to drive their followers away (Figure 6.19). This data is also used by the Facebook algorithm to penalize posts and pages that get reported as spam too often (see Facebook News Feed algorithm above).

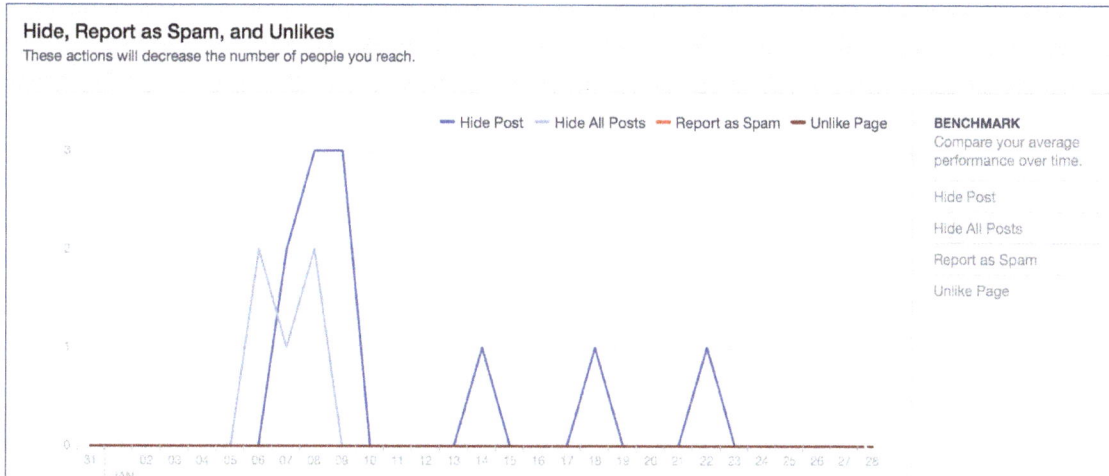

Hide, Report as Spam, and Unlikes
These actions will decrease the number of people you reach.

Figure 6.19: Negative Feedback on page posts

4. "Total Reach" provides data about the number of people who were served any activity from a page—including page posts, posts to a page by other people, page like ads, mentions of the page, and check-ins by people into an establishment that owns the page (not shown).

Page Views

"Page Views" provides data about visits to the page and is divided into three sections.

1. "Total Views" provides data about frequency of visits to the Facebook page. It also breaks down this data into visits to the different sections of a page (Figure 6.20). This data can be valuable as one can determine the section of the page that attracts more visitors.

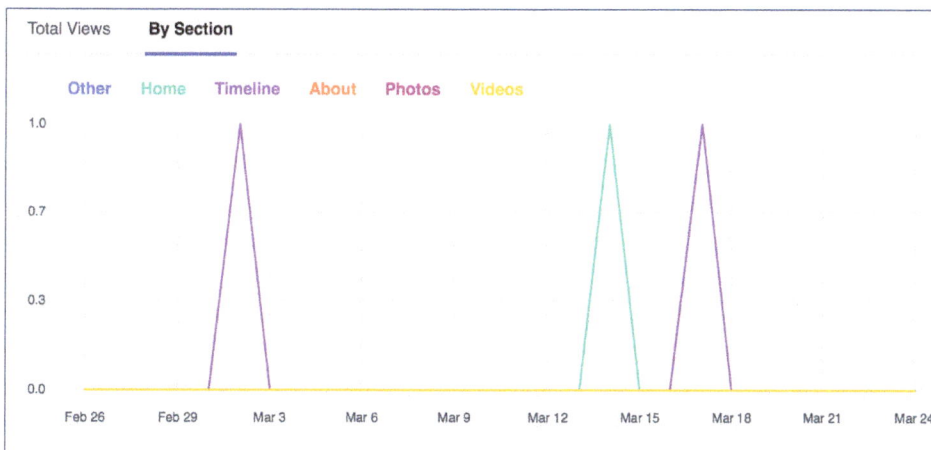

Figure 6.20: Total Page Views—By Section

2. "Total people who viewed" shows the number of people who viewed a page, the demographics of the viewers, and the devices they used to visit the page (Figure 6.21). This data can help a page owner understand more about their page visitors. For example, the page owner can ask if the people visiting the page match the demographic of the target audience they are hoping to reach.

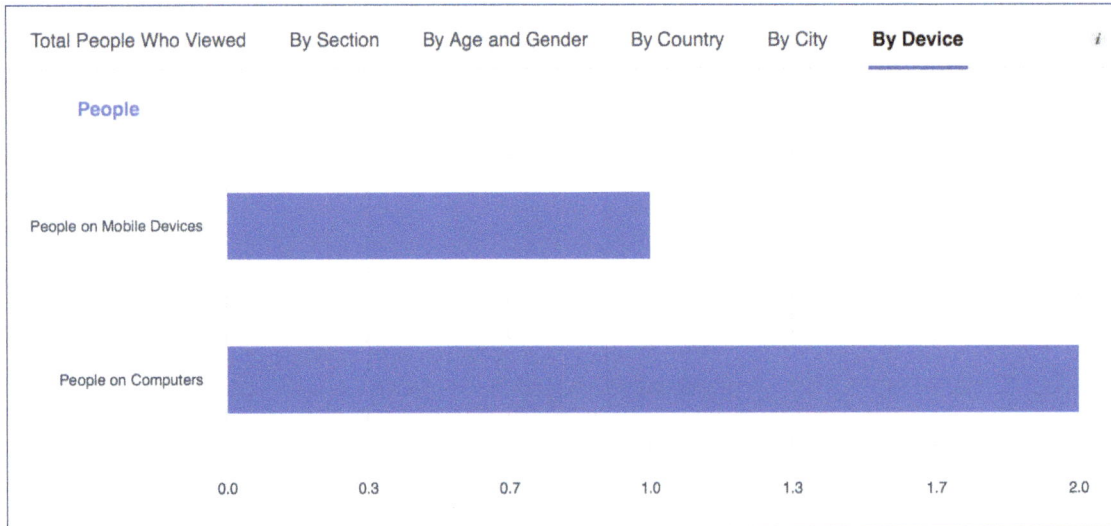

Figure 6.21: Total People Who Viewed—By Device

3. "Top Sources" displays where the viewers of a page are coming from, which helps determine sites that send the most traffic to their pages (not shown). Growing a page's audience is hard work, and page owners should provide every opportunity to their audiences online and offline to visit the page. This section of Insights will help determine which of these efforts are paying off.

Actions on page

Facebook owners should provide as much information about the entity behind the page as possible in the "About" section, where they can include links to a website, include an address, and list a phone number. Page owners can also include a "Call to Action" on the page to entice page visitors to make an inquiry with the page owner (Figure 6.22).

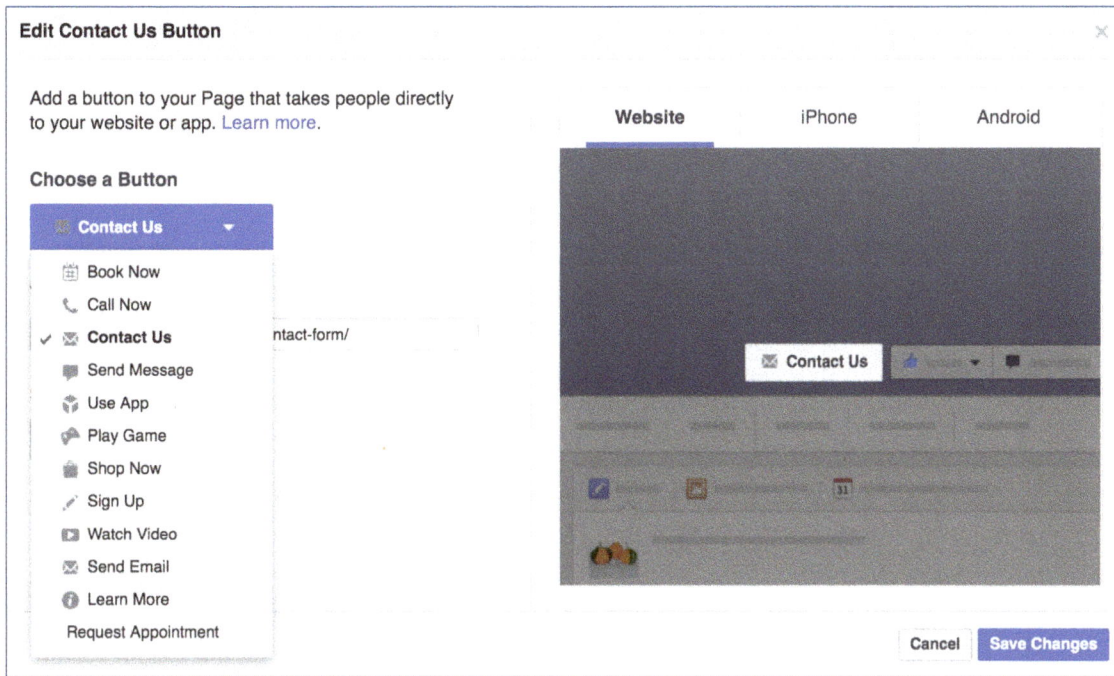

Figure 6.22: Types of "Call to Action" on Facebook pages

"Actions on Page" offers data about how often page visitors took any action on the following information: clicks on address, phone number, website URL, or Call to Action. It also displays information about the demographics of those who took these actions (not shown).

Posts

The "Posts" section of Facebook Insights is rich with valuable information, which is divided into four sections:

1. When Your Fans Are Online. One frequently asked question in the social media industry is "What time should I be posting?" There is no universal answer to this question, and it depends on the Facebook habits of followers of a particular page. Of course, when the page is new, there is no choice but to do some guesswork and learn the answer to best timing by trial and error. However, the "Posts" section of Insights takes the guesswork out once the page accumulates enough followers. "When Your Fans Are Online" shares days of the week, and times of the day that followers of a page are active on Facebook (Figure 6.23).

When viewing this information, page owners should seek out days and times that their follower base is most active on Facebook, and that is when page owners should be posting.

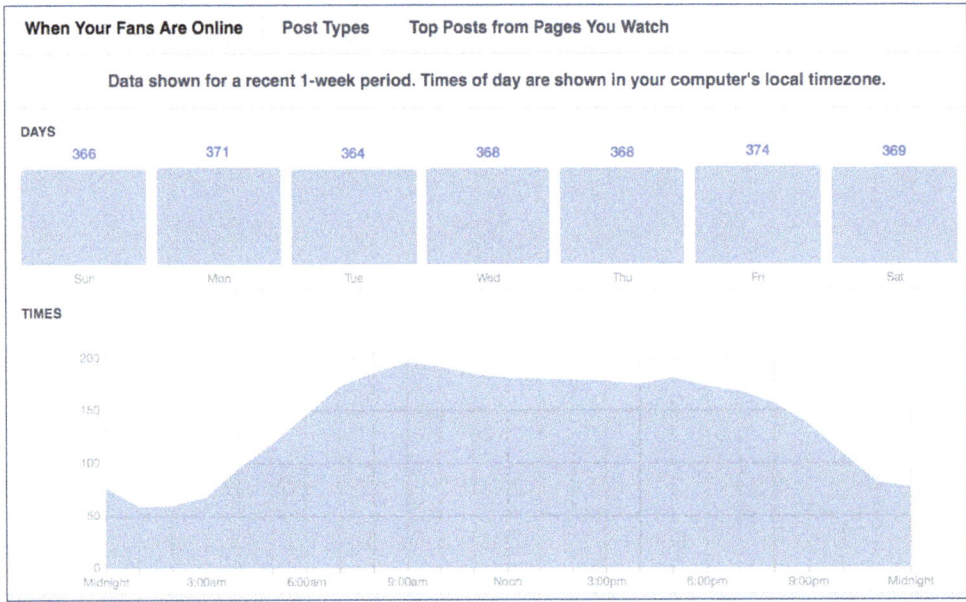

Figure 6.23: When Your Fans Are Online

2. "Post Types" compares different types of content shared on a page with respect to reach and engagement. This data will help determine the type of content that most resonates with a page's audience (Figure 6.24).

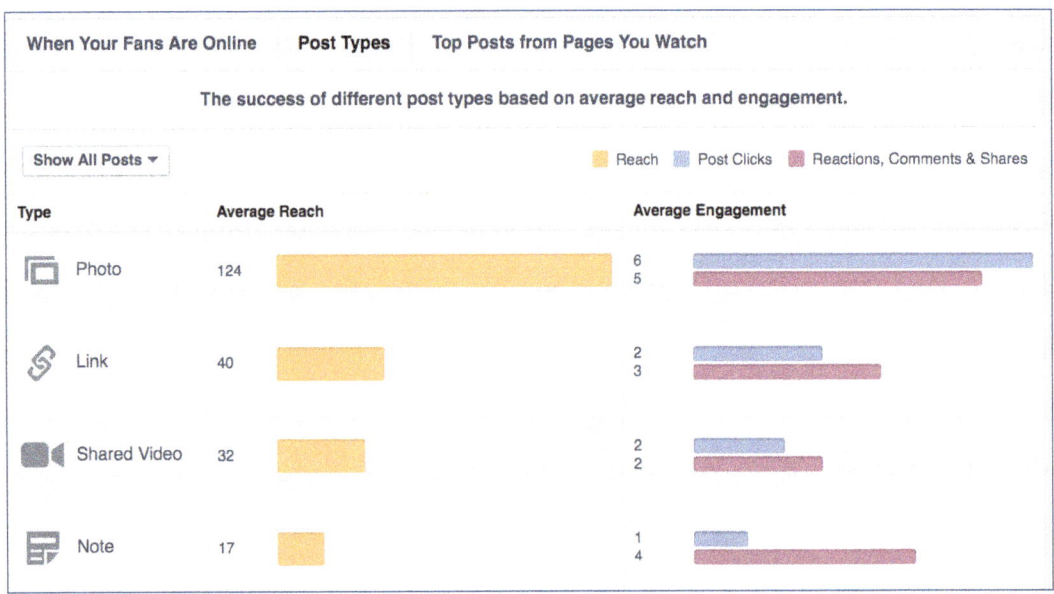

Figure 6.24: Post Types

3. Top Posts from Pages You Watch. A page owner is able to track the activities of up to five other pages. This section of Insights lists posts from these pages that enjoyed the highest amount of engagement. Page owners can set up this monitoring in the "Overview" section of Page Insights (not shown). This data provides a great learning opportunity, in case there are successful pages that a page owner wishes to emulate. Conversely, this data could also be a good source of information about a competition. For example, a climate scientist activist could monitor pages owned by climate change deniers to gauge the level of response to their efforts in spreading misinformation.

4. "All Posts Published" gives reach and engagement information about each and every post published on the page (Figure 6.25). Reach and Engagement columns can be displayed in four different ways, which can be selected from a drop-down menu. (Figure 6.25). My personal favorite is the Fans/Non-Fans version of Reach, because it displays a chart for reach among those who already follow a page (Fans) and those who don't (Non-Fans). Posts that are viewed by a large number of non-fan audiences tend to be those that enjoy a high amount of engagement by page followers, because when followers of a page engage with a piece of content, their friends will see that activity and may respond to it.

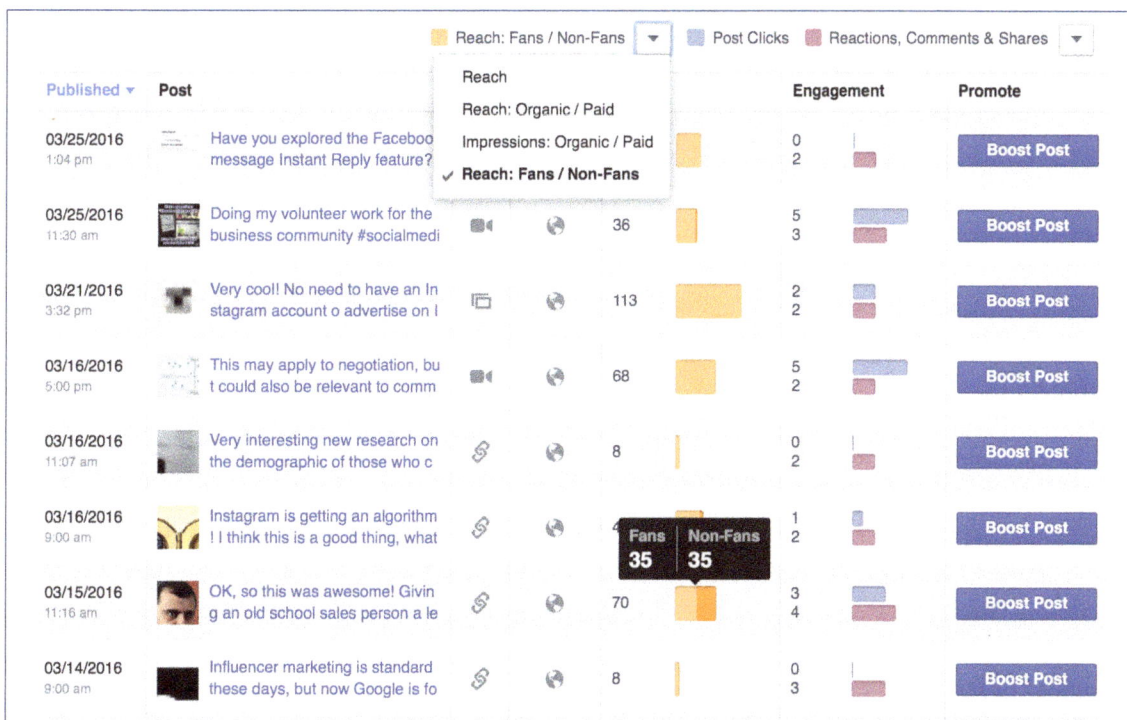

Figure 6.25: All Posts Published

In the "All Posts Published" section of Insights you not only get an overview of performance for each post, you can also view more details about each post if you click on the post. This reveals a new window with a whole host of important data (Figure 6.26). If you notice on Figure 6.26, you will see "Likes," "Comments," and "Shares" metrics associated not only for the actions taken on the content when posted on a page itself, but also when that post is shared from the page to elsewhere on Facebook. This data is listed under "On Shares." If you remember in the beginning of this chapter, I mentioned that Facebook does not provide analytics for content posted on personal profiles. However, if you post content on a Facebook page first, then share it on your personal profile, you can get metrics for engagement with that content on your personal profile in the "On Shares" section of Facebook Insights.

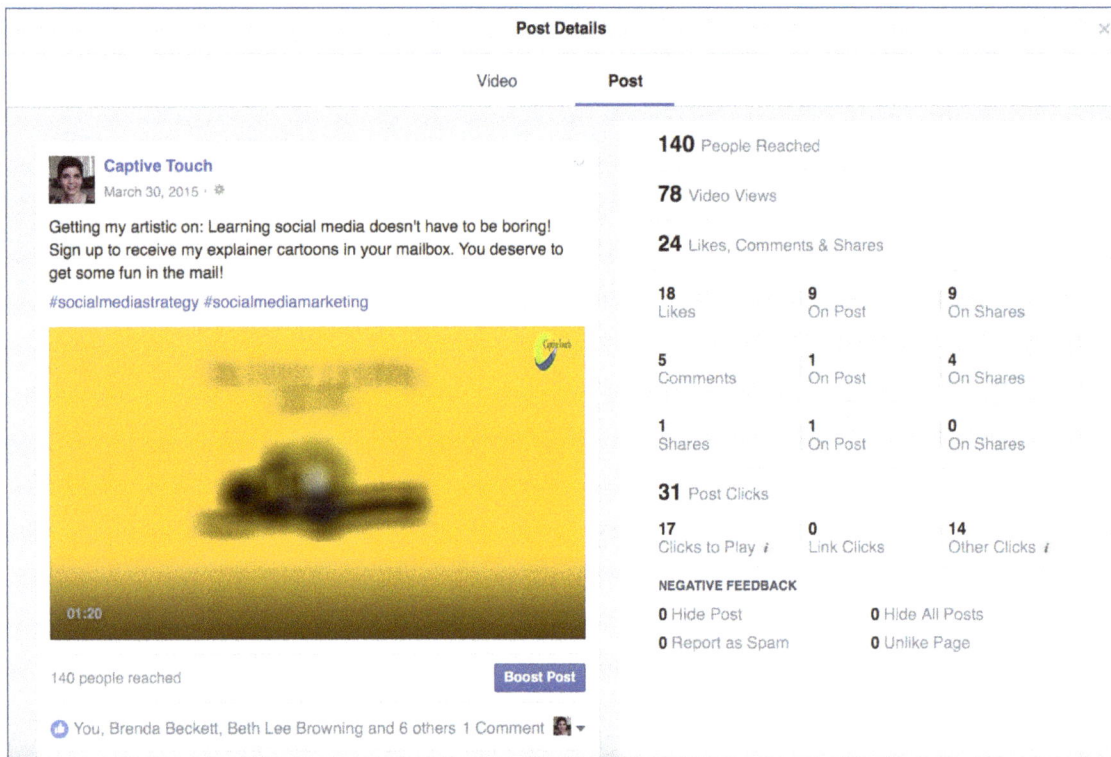

Figure 6.26: Facebook Insights—Individual post metric

Events

If you use the Events feature of your Facebook page to host a gathering, this section of Insights will allow you to keep track of the performance of your event in terms of raising awareness, engagement, ticket sales and the audience (not shown).

Videos

Video content is exploding on Facebook and will only continue growing. As a general rule, videos that are most successful are less than two minutes long, include captions, immediately catch the attention of viewers, and compel them to watch to the end in the first few seconds. That is no easy task, but the data presented in the video section of Insights will help you optimize your videos toward this goal.

For most people on Facebook, videos automatically play with no sound, because most do not know how to turn off this feature. Auto-play is a good thing as it allows viewers to get a preview as the video passes through their timeline, which helps them decide if they want to watch to the end or click to hear the sound. That is why videos need to be somehow compelling in the first few seconds. When you view the analytics for videos, be sure to review the "Auto-played vs. Clicked-to-Play" option (Figure 6.27). The aim should be to have the percentage of Clicked/Auto as close to 100 percent as possible, because this means your audience became interested enough in the content to take an action on the video.

The "Video" section of Facebook Insights is divided into three parts and the data is provided only for videos directly posted on Facebook, not videos shared on the platform from another website or video hosing site (like YouTube).

1. "Video Views" shows the total number of times a video has been viewed for at least three seconds (Figure 6.27).

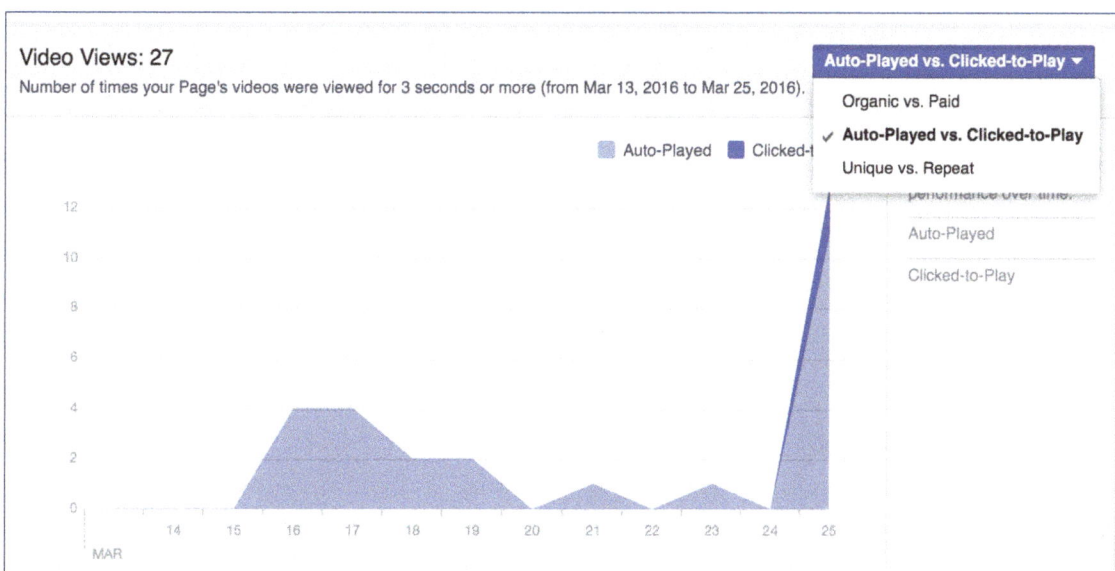

Figure 6.27: Facebook Video Metrics—Video Views

2. "Ten-Second Views" reports on the number of times a video has been watched for ten seconds or more, which is always lower than the number reported in Video Views; not everyone is going to watch videos for longer than ten seconds (Figure 6.28).

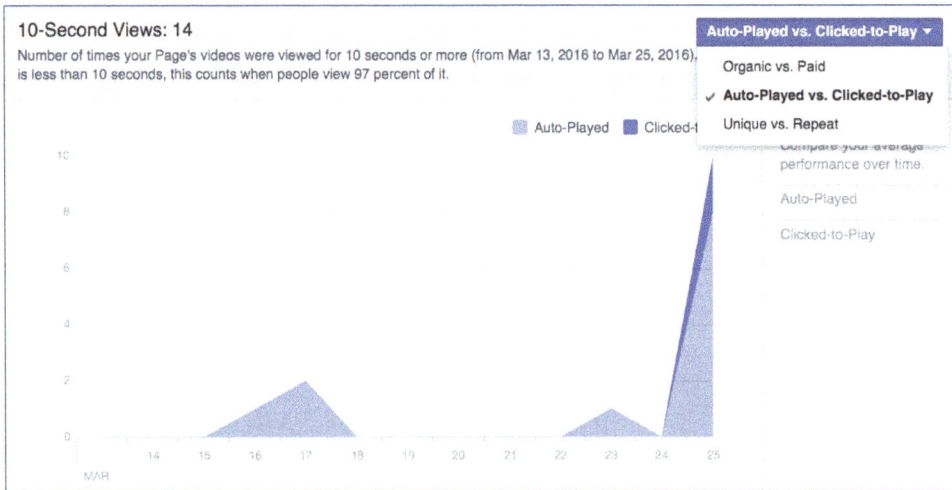

Figure 6.28: Facebook video metrics—10-Second Views

3."Top Videos" features those videos with the best performance (not shown). Clicking on the "Video Library" button will reveal the entire set of videos that a page has posted and its relevant metrics. This is where you will find a whole host of new data about the effectiveness of your videos but I want to highlight one thing that is extremely important for understanding video performance: Percentage Completion (Figure 6.29).

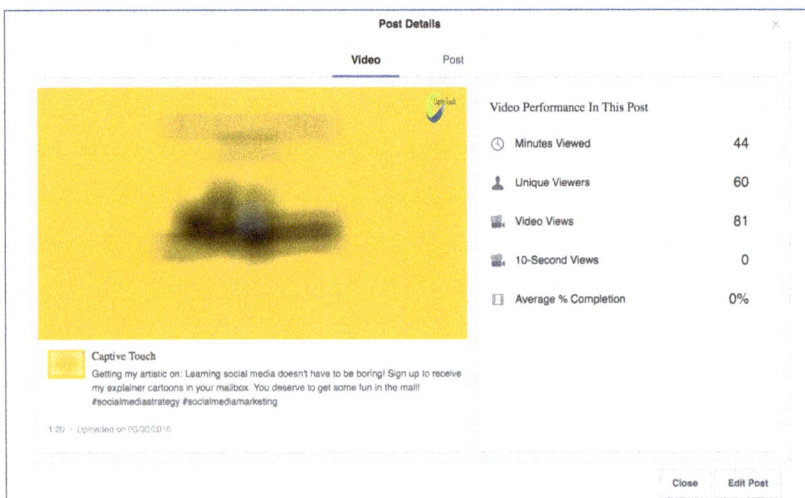

Figure 6.29: Analytics for individual videos-Percentage Completion

"Percent Completion" measures how many viewers watched a video until the end, which is a very rare occurrence. If you click the arrow next to "Average % Completion," a new window opens that reveals "Audience Retention" (Figure 6.30), a metric that will help you pinpoint where viewers lose interest and stop watching the video. Longer videos tend to have a lower retention rate, which is why I mentioned in the beginning that you should aim for creating very short videos. It may sound easier said than done, but it *is* possible.

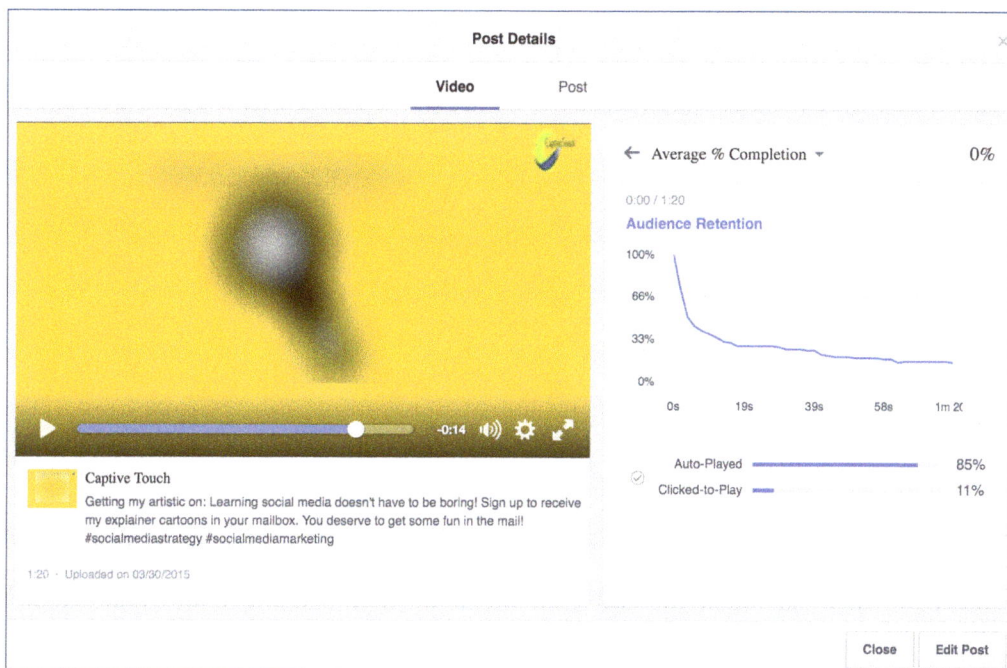

Figure 6.30: Facebook Video—Audience Retention

People

This is where you can find demographic information about followers of your page as well as those to whom your page content has been served by Facebook (they potentially saw the content in their news feed). The demographic data include gender, age range, and location (not shown). You can use this information to judge how the audience actually following your page and seeing your content matches with the audience profile you created, and with whom you are hoping to engage. If there is very little match between the two, either you need to work smarter to attract the audience you want, or use this data to consider an audience who is interested in your content whom you did not originally consider.

Local

If you have indicated a physical address in the About section of your page, Facebook Insights provides demographic data about Facebook users who are in the vicinity of that address (not shown). You can compare this demographic data to your target audience and decide if you want to advertise to gain their attention and invite them to visit your establishment.

Messages

Followers of a Facebook page have the option of sending private messages to page owners. This section of your Facebook page keeps an inventory of the private conversations you have had with your page followers (not shown). Followers who take the time to send you private messages can be considered the most engaged of your audience, so the data provided in "Messages" will help you keep a record of them.

Outreach on Facebook

Outreach on Facebook can take many forms, and it depends on whether you want to focus on using your personal profile or a Facebook page. As mentioned before, I suggest that you do both, but for the purposes of having a structure, I present a separate workflow for each. Although the two workflows are different, both begin with audience research (which you have already seen earlier), and require slow and steady work in the art of patiently building relationships.

Outreach through a personal profile: I have illustrated the steps in this workflow in Figure 6.31. If you recall, the audience research described in the beginning of this chapter revealed groups joined by—and Facebook pages liked by—the sample profile, Linda Goldberg. Your outreach efforts would begin by joining these communities. When you join a group, it is important to introduce yourself to the community so they notice and get to know you. Spend the first few weeks after you join a group to "listen" to conversations and engage with the topics being discussed. This is important, as you want to better understand the relationship dynamics in the community, and their level of knowledge and involvement in taking action on climate change. Taking the time to build this understanding will help you decide how to approach the community about climate change. Try to participate in ongoing discussions in the community and offer your perspective from a global warming point of view. For example, if there is a discussion in the group about finding summer camps for kids, and you may know of a science camp that offers a climate change education component, suggest the camp and state why you think it would be a good choice. This approach will ensure that you introduce the topic of climate change around the needs of the community, as opposed to what *you* think is the most relevant.

Building relationships on a Facebook page liked by your target audience will center on the topics posted by the owner of the page. Listen and engage with the posts on the page, be a "regular," and try to build relationships with other "regulars," especially those who best fit with your audience profile. Sometimes these relationships lead to offline gatherings, which can lead to friendships on Facebook. You can even start your own Facebook group and invite those who fit your target audience to your group. Once people become part of your tribe, they will be more open to your viewpoints and suggestions about climate change and taking action.

No matter which path you choose, ultimately the best way to drive climate action through your personal profile is living by example. If you are going to share news about climate change, be sure that your sources are credible, take on a positive tone, and share how *you* are taking action against global warming.

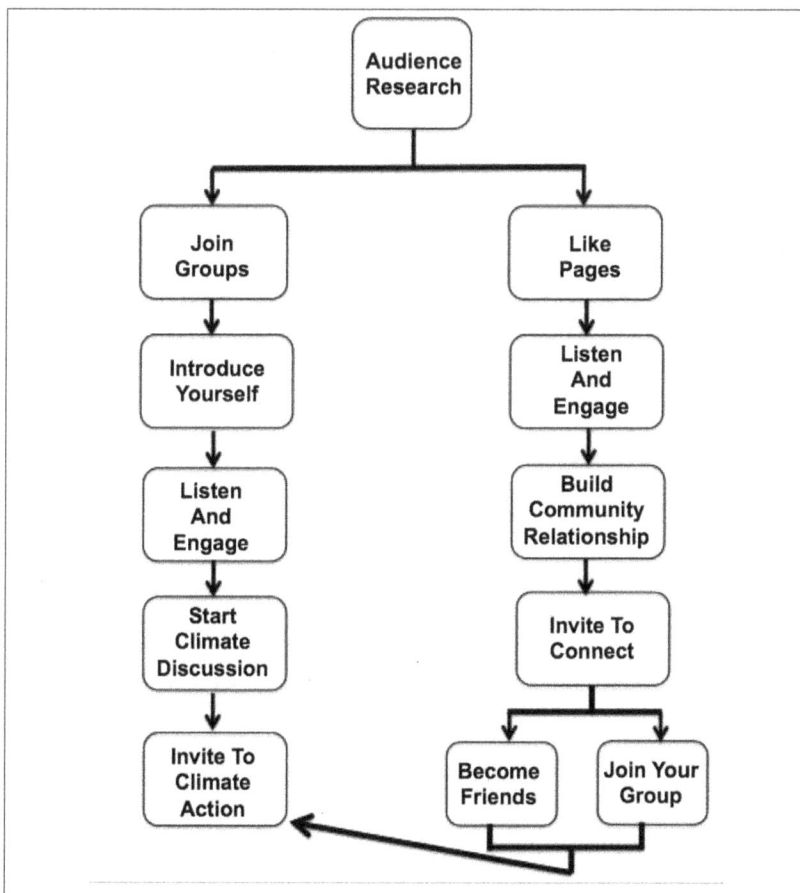

Figure 6.31: Workflow for outreach through a personal Facebook profile

Outreach through a Facebook page: Please view Figure 6.32 for this workflow. The steps you would take for your outreach through a Facebook page depends on whether you have a budget for serving "Ads," in which case you should make sure to "Save" your audience profile when you use Facebook Audience Insights (see Chapter 2). This custom-made audience will allow you to focus your targeting when using Facebook Ad products. I feature a "page like" Ad in Figure 6.32 (which invites people to like your page), but note that this is just *one way* you can target your audience using Ads. Details about Facebook Ads products are outside the scope of this book, but you can teach yourself the basics of Facebook advertising by taking the free online Facebook Blueprint course at https://www.facebook.com/blueprint.

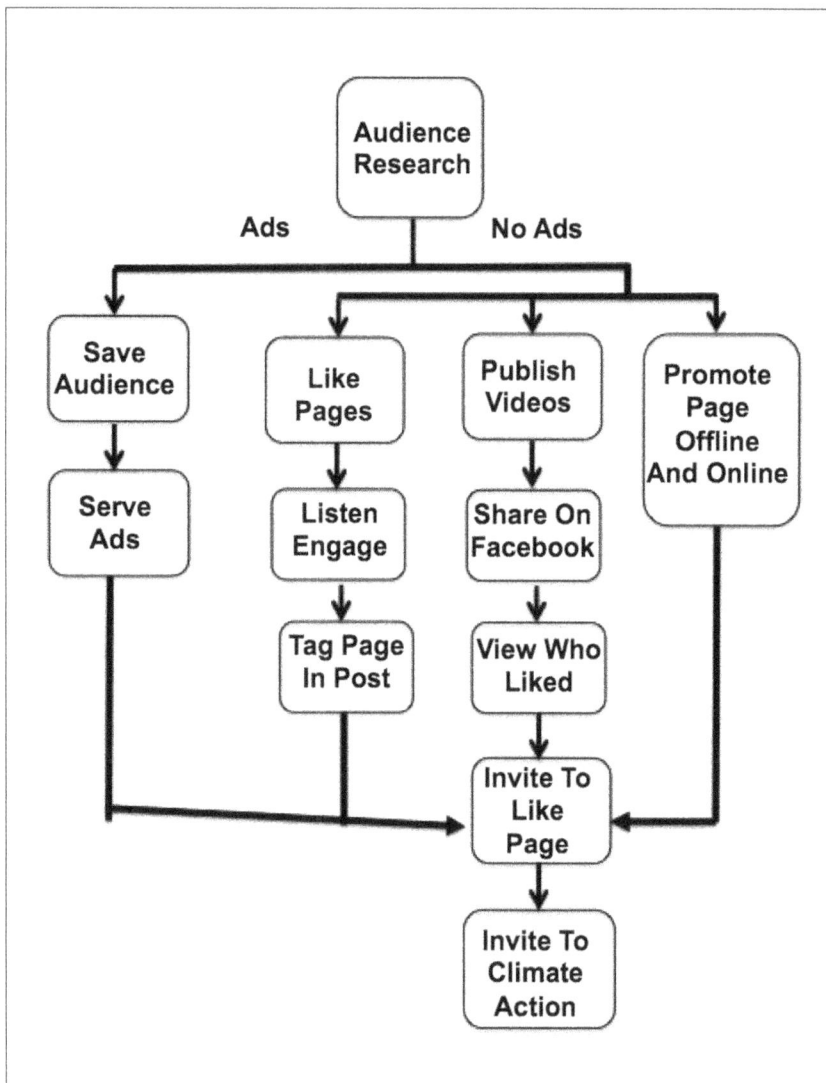

Figure 6.32: Workflow for outreach through a Facebook page

If you do not have a budget for advertising, there are a few things you can do to build an audience for your Facebook page:

- At the most basic level, you should promote your page at every opportunity: on your business card or other printed material, in your email signature, or by putting a "Page Plugin" on your blog (6.4), which allows your blog visitors to "like" your page without having to visit Facebook.

- As a page owner, you have the option of using Facebook under two different identities: as your personal profile, or as your page. You should take advantage of this option and "like" pages followed by your target audience as your page. In this way, when you build relationships with a community, your page will get noticed. Furthermore, liking other pages as your page has another advantage. When you tag a Facebook page in a status update, the followers of that page will see that activity on their timeline, even if they do not follow your page. Now, this does not happen every time you mention a page (not clear how Facebook decides when), but the chances of your target audience noticing your page will definitely increase. To illustrate how this works, let's look at an example. "Jewish Mom Secrets" is a page the Linda Goldberg sample profile likes. A status update on this page mentions the merits of eating leftovers (Figure 6.33).

Jewish Mom Secrets
June 25, 2014 ·

#362 "I eat leftovers from shabbos throughout the week. My husband doesn't eat leftovers though. I just can't let good food go to waste. Still tastes great and sometimes it tastes better than when I first cooked it!"

Like   Comment   Share

26   Top Comments

Figure 6.33: Status update on a Facebook page that the Linda Goldberg sample audience profile follows

Now, on your page, you can comment on the energy-saving value of eating leftovers, tag the "Jewish Mom Secrets" page, and commend them for this climate-friendly practice (Figure 6.34).

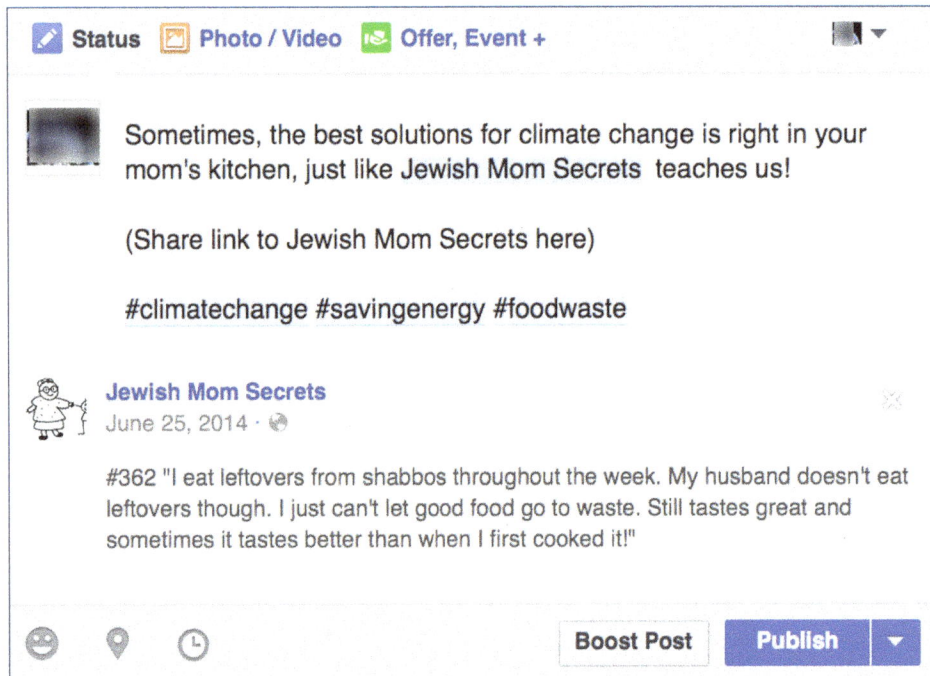

Figure 6.34: Tagging "Jewish Mom Secrets" in post

To follow a Facebook page, as your page, visit the "See Pages Feed" section of your Facebook page (Figure 6.35), and click on the green "Like Other Pages" button. This action will allow you to type in the name and follow other Facebook pages (not shown). The "Pages" feed is also where you can view updates from the pages you follow, and interact with them either as your page or as your personal Facebook profile. If you own more than one Facebook page, you have the option of choosing any of them when you engage with other pages (see drop-down menu in Figure 6.36).

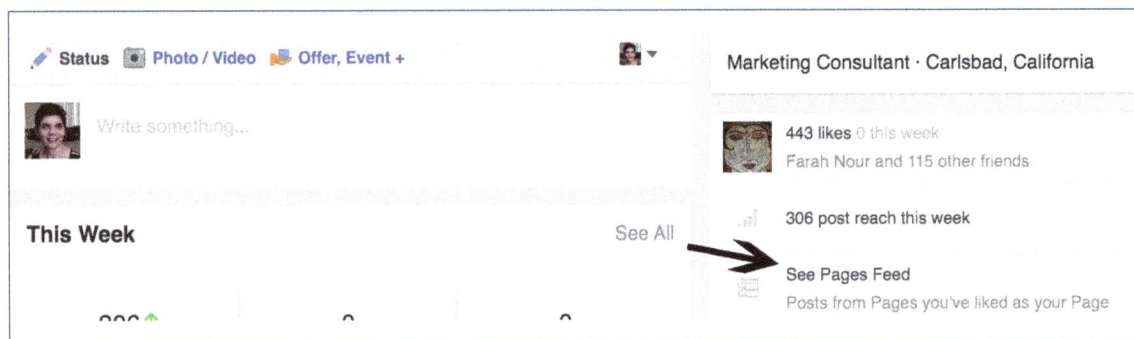

Figure 6.35: Finding the "See Pages Feed" on your Facebook page

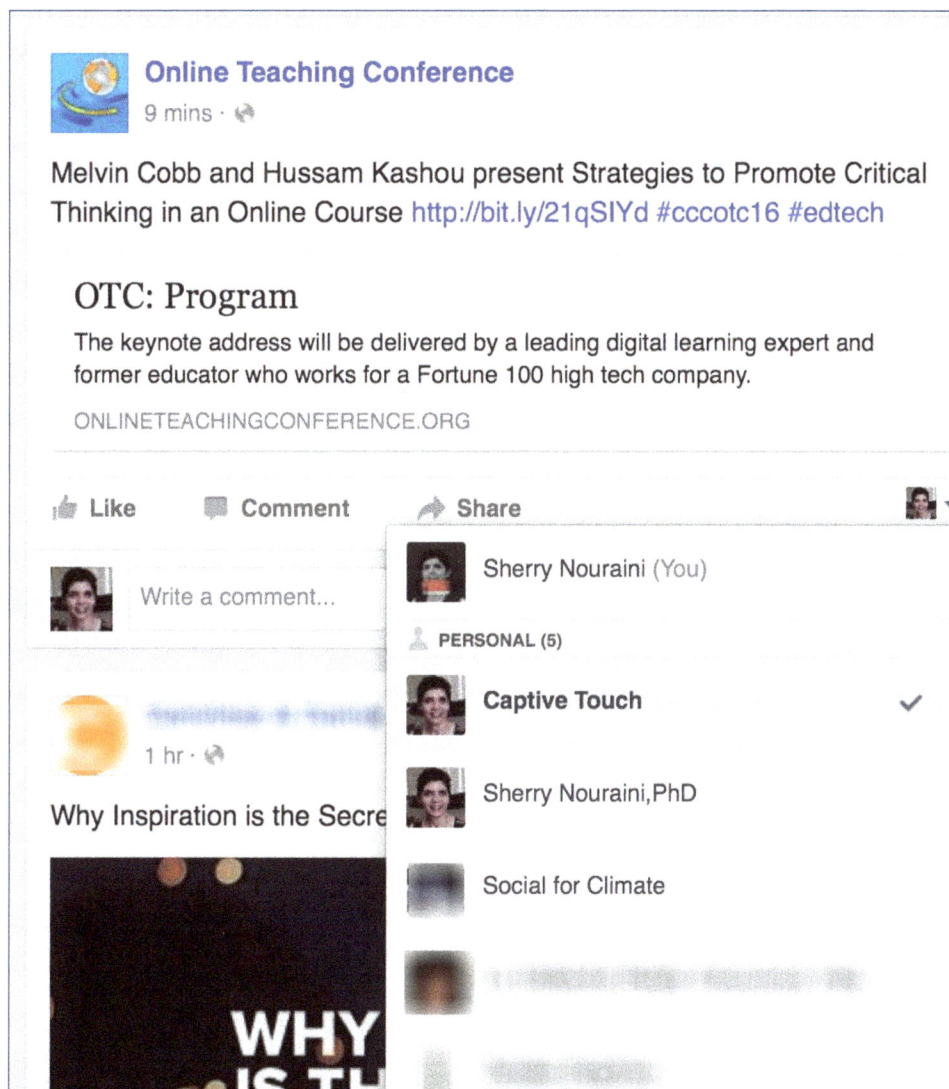

Figure 6.36: Engaging with updates from other pages on the "See Pages Feed" section of your page

- Publishing videos on Facebook offers an added opportunity to invite users to follow your page without paying for ads. Facebook analytics shows who has liked a particular video, and whether or not they are following your page. For those who liked your video but do not follow your page, you are given an opportunity to invite them to do so. To take advantage of this feature, visit the "Video Library" on your page, then choose any video you have published and click on the "Like" counter icon. This action will open a window where you can view who has liked your video and whether or not they are following your page (Figure 6.37).

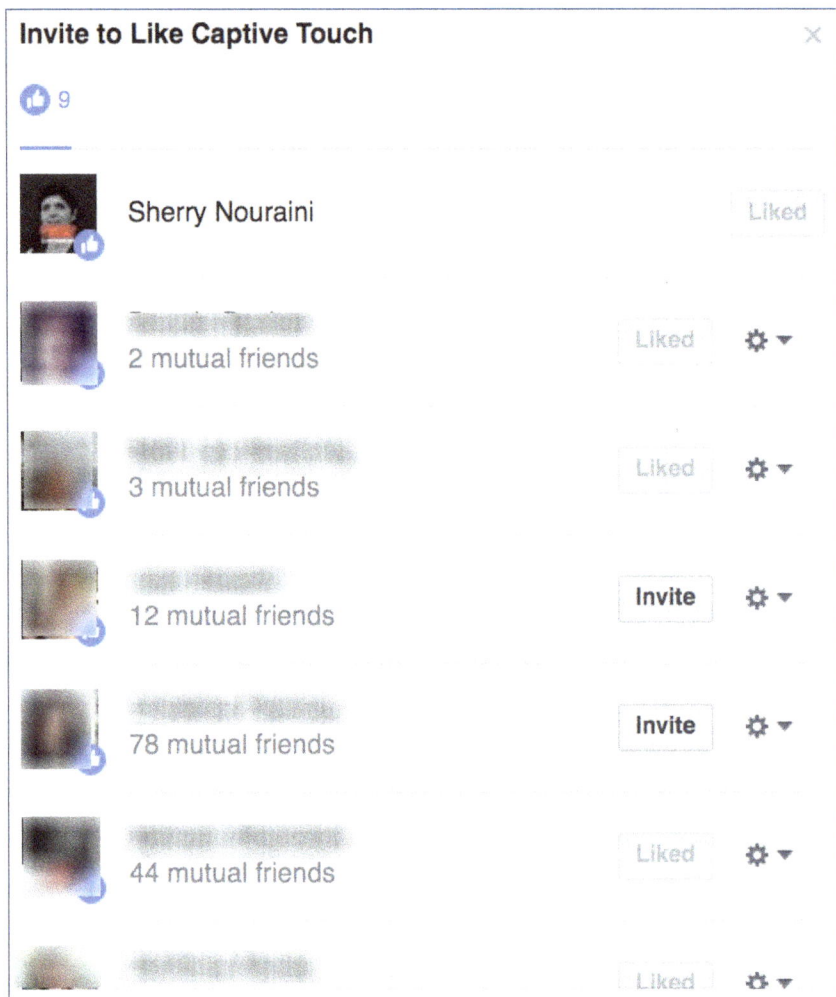

Figure 6.37: Inviting viewers of your videos to like your page

Practice what you learned

Facebook is a *huge* topic, and not everything offered by this platform can be covered in one book. I have covered what I consider to be the most enduring and important features of Facebook for the purposes of raising awareness. You saw an example of using the power of the Facebook search engine to find communities of an example audience profile, Linda Goldberg, and learned tactics that you can use to grow your community on Facebook. Now it is time to get your feet wet, and take the other audience profiles in this book—or your own defined audience—through the research and discovery steps outlined here. Don't hesitate to join our Facebook group (http://facebook.com/groups/social4climate), come share your findings, and ask any questions you may have.

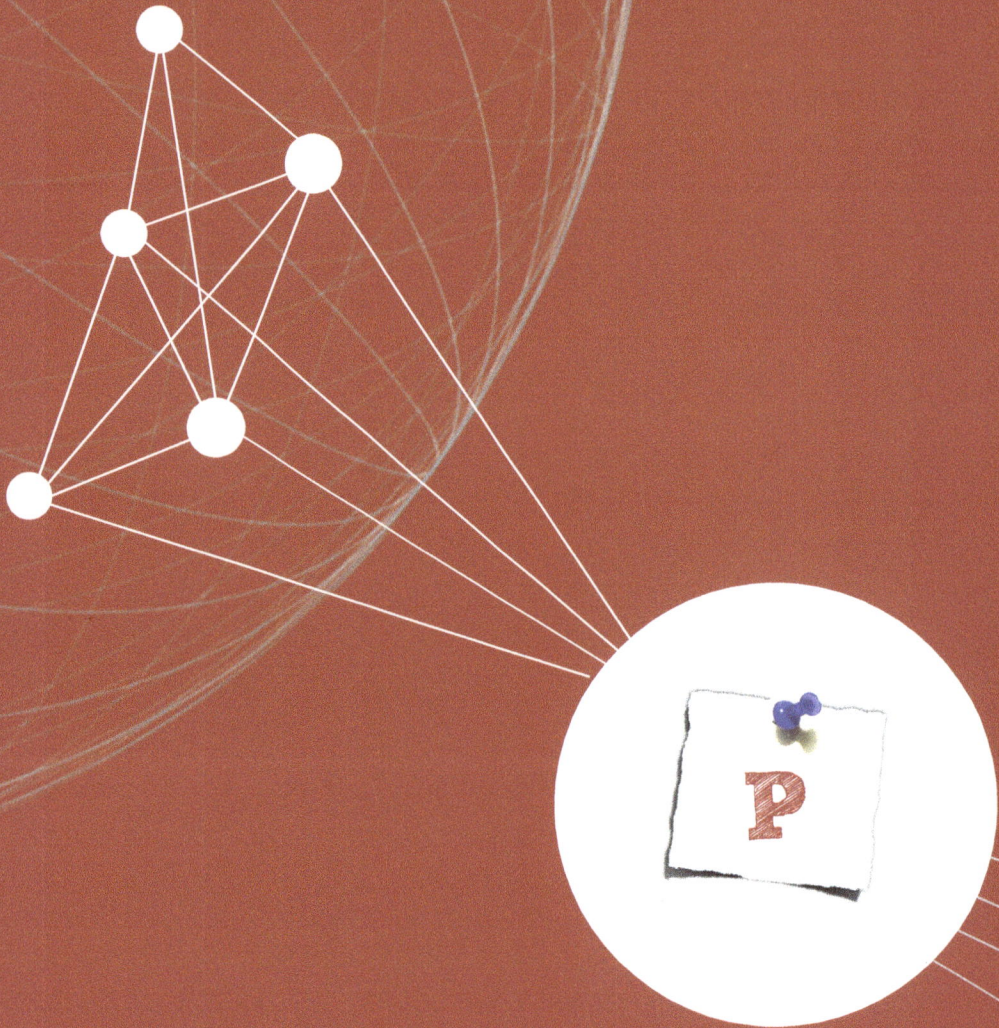

Chapter 7

Pinterest

Pinterest has 100 million monthly active users (7.1)—the majority of whom are women (7.2). This social network is simple to use, addictive, and very popular with people interested in fashion, food, decorating, DIY projects, and education (7.1). Pinterest is a destination for idea discovery, which combined with its mostly female demographic makes Pinterest a perfect place for communicating climate action by sharing sustainable alternatives to clothing, food, and home décor and improvement. It is also a great place to reach out to teachers for climate change education. Among the three sample demographic profiles we are covering, Linda Goldberg may be the best match; although, I don't want to give the impression that searching for other demographics is a waste of time on Pinterest.

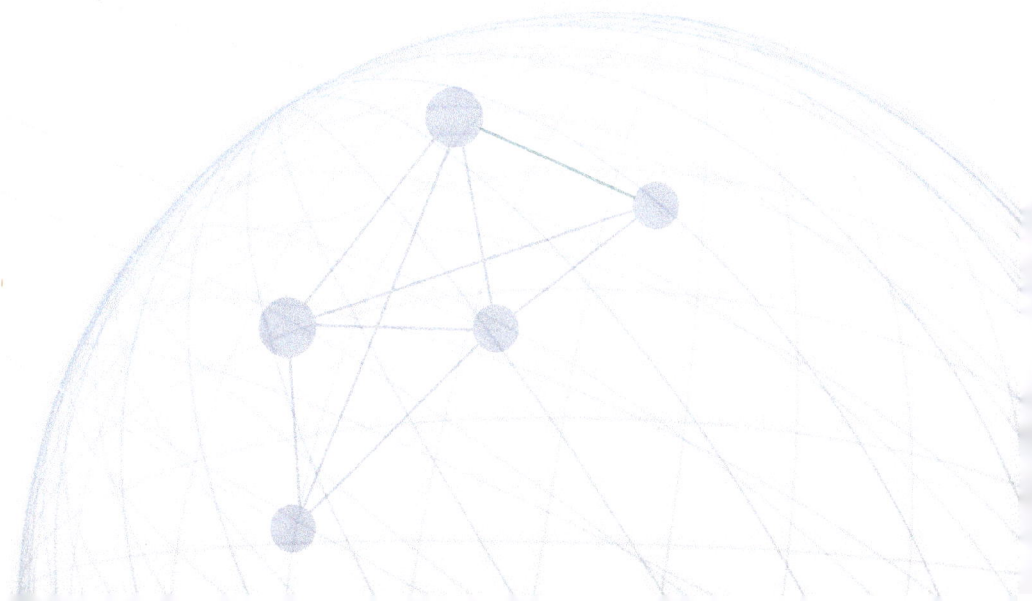

Let's first learn some basic features of Pinterest, and then use this site to search for the Linda Goldbergs of the world.

Profile

Pinterest offers both personal and business profiles, but business profiles have the added advantage of having access to analytics and advertising. You can begin your Pinterest adventure with a personal profile and then convert to a business entity. If you are going to set up a blog (which I hope you will) and wish to develop an audience, then start with a business profile so you can leverage Pinterest analytics on website traffic.

Just like other social networking sites, your profile includes a photo and a description (Figure 7.1).

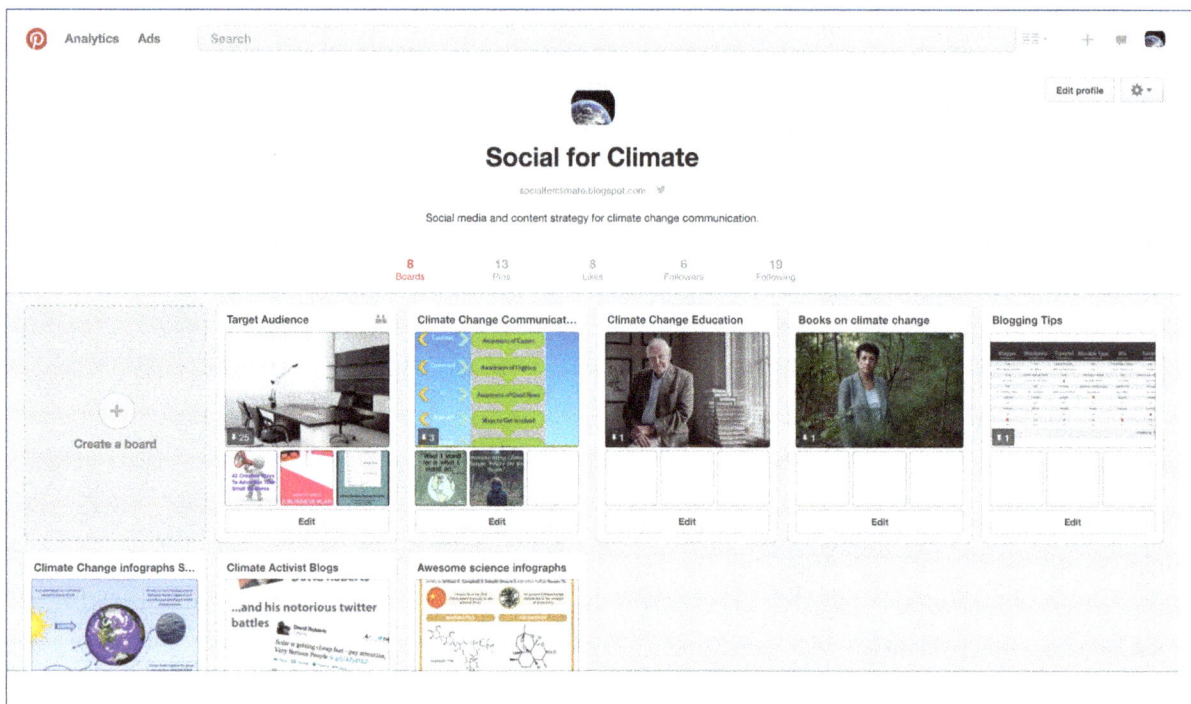

Figure 7.1: Pinterest profile

When you create the description, be sure to include relevant keywords, just like we discussed in Chapter 5 for a Twitter profile. Like Twitter, you need to create your profile first, then add the description by using the editing option (Figure 7.2).

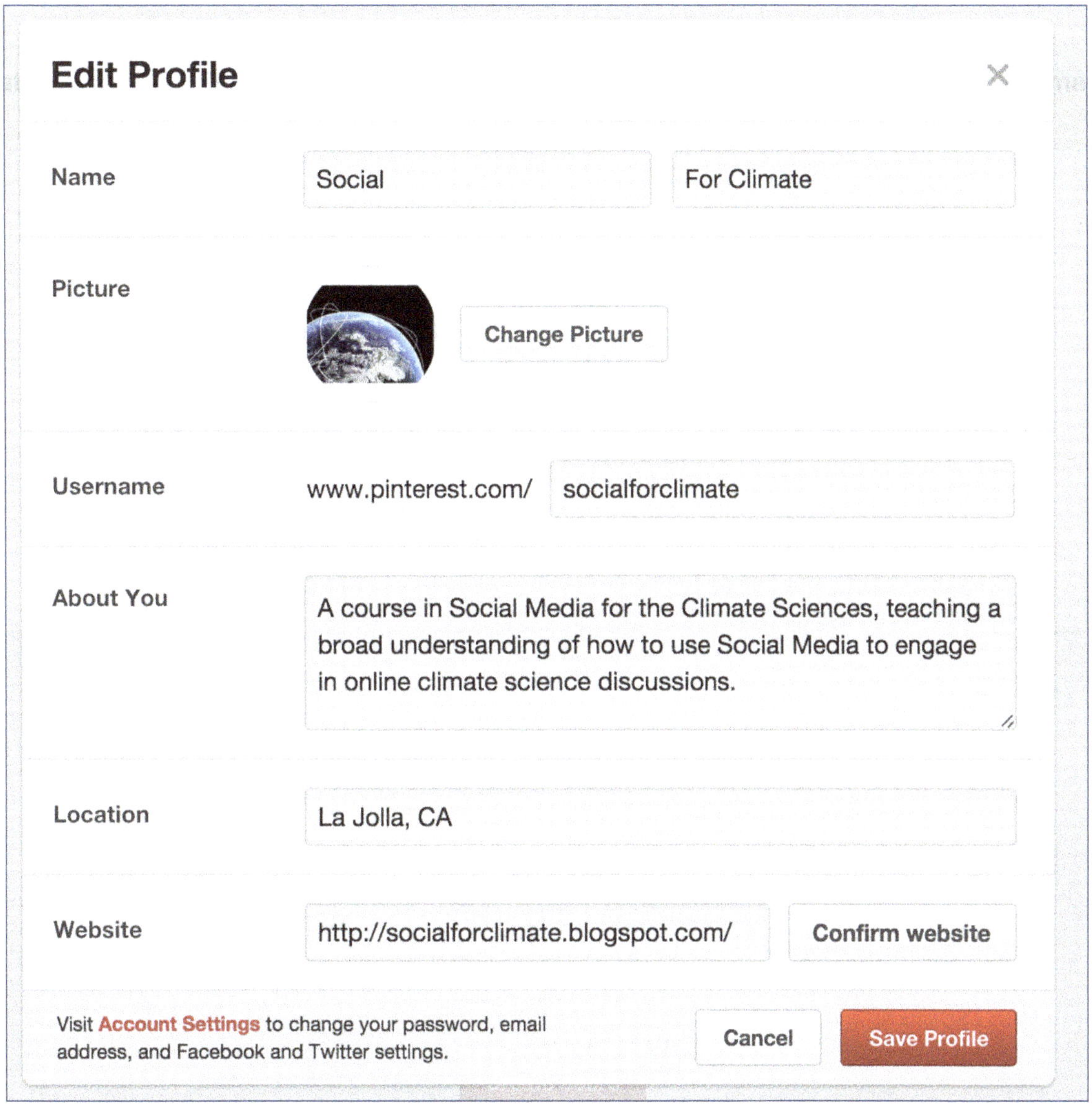

Edit Profile ✕

Name | Social | For Climate

Picture | Change Picture

Username | www.pinterest.com/ socialforclimate

About You | A course in Social Media for the Climate Sciences, teaching a broad understanding of how to use Social Media to engage in online climate science discussions.

Location | La Jolla, CA

Website | http://socialforclimate.blogspot.com/ Confirm website

Visit **Account Settings** to change your password, email address, and Facebook and Twitter settings. | Cancel | **Save Profile**

Figure 7.2: Editing your Pinterest profile

Confirming your website

I mentioned that creating a business profile on Pinterest is superior because of the associated analytics, which provides information on your effectiveness on the platform. If you wanted to get insight on traffic to your website, or learn about your website visitors who share images from your website to Pinterest, you will have to verify your website:

1. Click on the "Confirm your website" button (see bottom of Figure 7.2).

2. Copy the code in the gray-colored window (Figure 7.3).

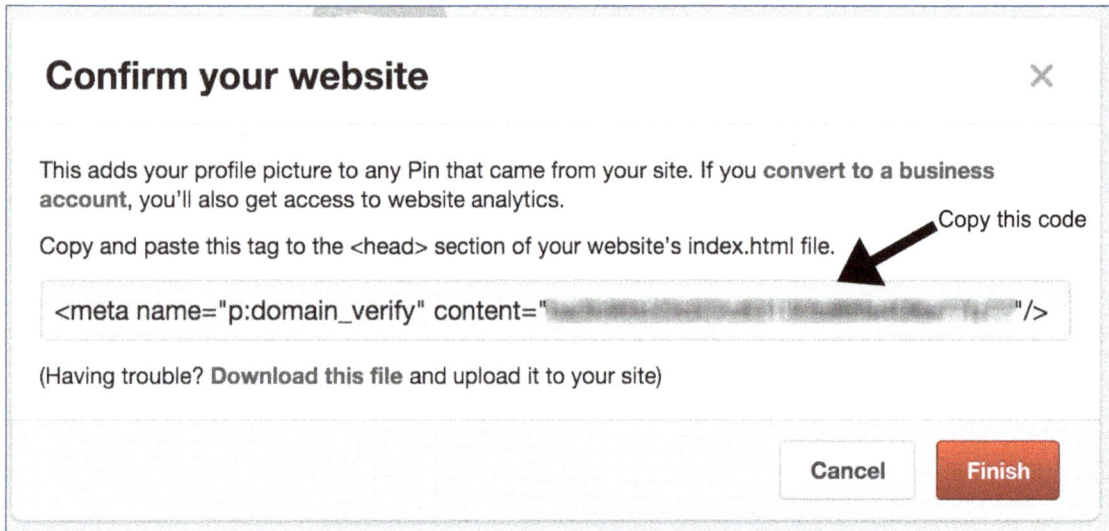

Figure 7.3: Code needed to confirm your website with Pinterest

3. Log in to your blogger site and choose "Template" from the left menu. Click on the "Edit HTML" button (Figure 7.4).

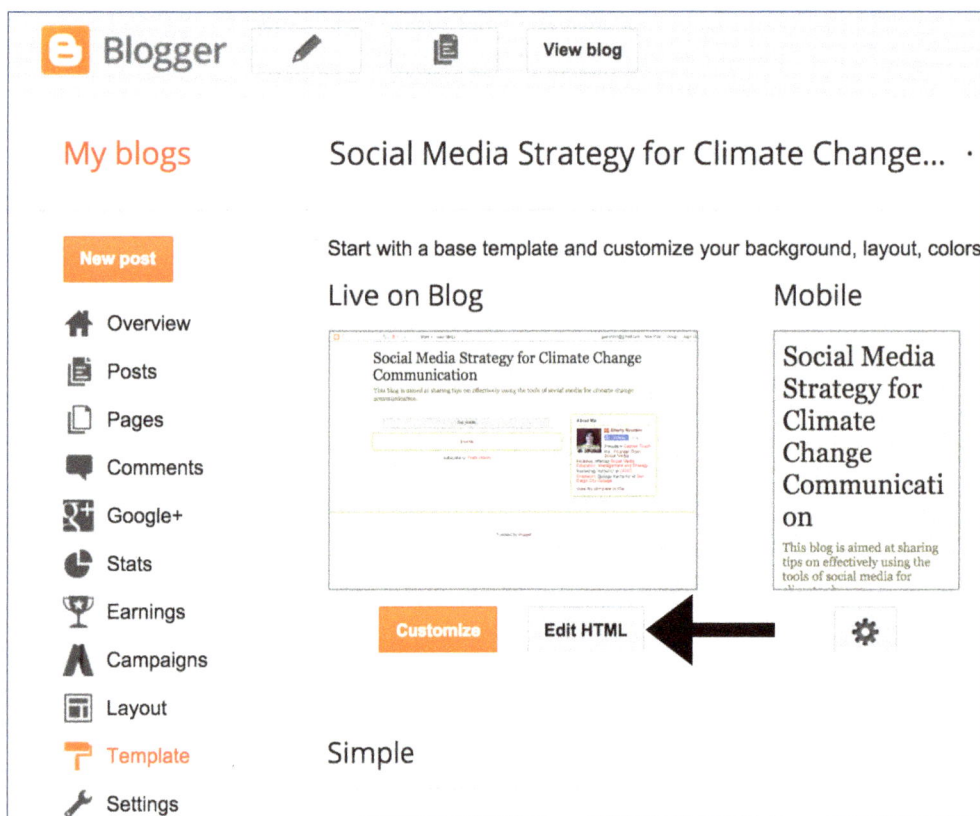

Figure 7.4: The "Template" section of your blog

4. Insert the code just underneath the <head> section of your website's code and save the changes (Figure 7.5).

Figure 7.5: Adding the Pinterest code to the "<head>" section of your blog.

To check if the confirmation process has worked, check the status of your website in the "Edit your profile" section of Pinterest for a "website verified" message. Note: it may take a while for this change to happen.

Organization

When you use Pinterest, you have no choice but to be organized, as organization is a major part of this platform's design, and an important reason why people use it. However, this organization does not center on people, per se, but around the *content* that people share on the platform. Once you set up your Pinterest profile, consult the list of categories that you created when you were getting organized for a Twitter account. If you recall, for the Linda Goldberg example profile, these categories were "Jewishmoms," "Jewishorganizations," "Jewishschools," and "Jewishspiritualleaders." Using each of these categories, create a "board" on Pinterest for each of these topics. You can think of a board as a shoebox or a bulletin board where you organize all of your photos, collections, or things to remember. To create a board, visit your profile and click on the plus sign labeled "Create a board" (see Figure 7.1). When you create a board, give it a title and a description, making sure to include keywords to increase the likelihood of being discovered when your target audience is searching for information on Pinterest. You can make a board private or public and invite others to contribute.

Search

Similar to other social networking platforms, Pinterest is rich with data compiled, based on user activities. Pinterest "Search" is a powerful discovery tool as it allows users to find exactly what they are looking for in three different ways:

1. Searching based on categories allows one to find content and people on Pinterest based on broad topics. These topics have been created based on how Pinterest users categorize information, so it is not arbitrary. To search Pinterest by categories, click on the icon to the right of the search bar (Figure 7.6).

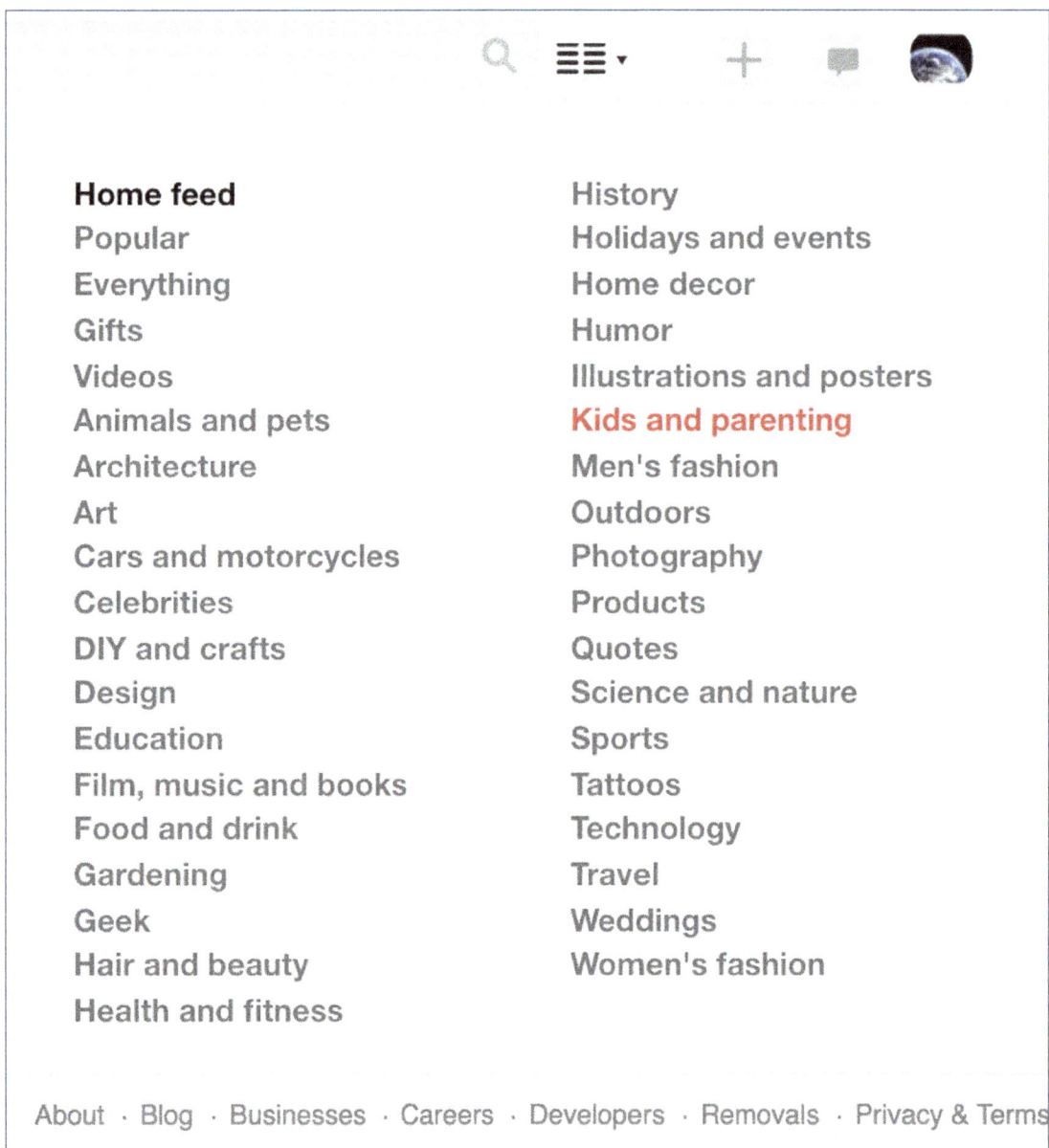

Home feed	History
Popular	Holidays and events
Everything	Home decor
Gifts	Humor
Videos	Illustrations and posters
Animals and pets	Kids and parenting
Architecture	Men's fashion
Art	Outdoors
Cars and motorcycles	Photography
Celebrities	Products
DIY and crafts	Quotes
Design	Science and nature
Education	Sports
Film, music and books	Tattoos
Food and drink	Technology
Gardening	Travel
Geek	Weddings
Hair and beauty	Women's fashion
Health and fitness	

About · Blog · Businesses · Careers · Developers · Removals · Privacy & Terms

Figure 7.6: Searching Pinterest based on categories

Let's say we want to find content about "Kids and parenting." When one chooses this option, Pinterest not only discovers content and users (pinners) who post media about this category, it also provides a menu of subtopics to choose from (Figure 7.7).

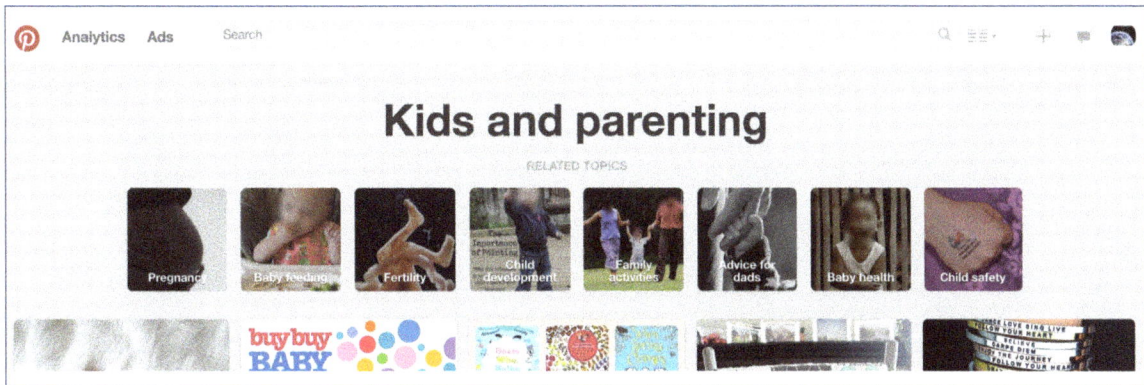

Figure 7.7: Search results with "Kids and parenting" category

2. "Guided Search" (an image-based search, see below) sets Pinterest apart from other search engines, because it helps one begin with a broad keyword and then zeroes in on targeted content. The best way to show how this works is with an example. If you'll remember, one word common to all keywords for the Linda Goldberg target audience is "Jewish" (review the "Organization" section of this chapter). However, this is a broad keyword because there are so many different aspects to a Jewish identity. We are mainly interested in the Jewish mom community and the relationship between Jewish values and protecting the environment. We can look to guided search to help us find relevant content and users in this demographic.

Start typing the word "Jewish" in the search box, and you will see that Pinterest provides not only content relevant to this word, but it also presents a horizontal menu of all sorts of words that relate to the word Jewish (Figure 7.8).

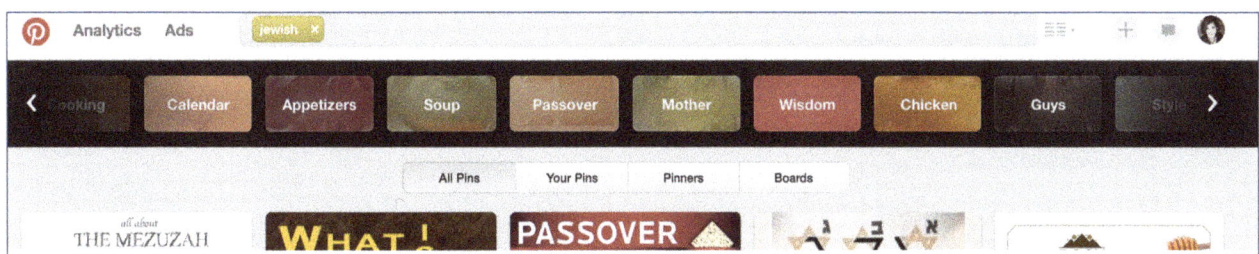

Figure 7.8: Pinterest Guided Search for the keyword "Jewish"

You can scroll through this menu and try to narrow the search to something closer to what you are looking for. In this case, we can choose the item "Mom" from this menu. Reviewing the "Boards" section of this more granular search result will reveal dozens of pins and the names of those who have created them—who tend to be mothers in the Jewish community (not shown). In this case, we were able to find our targeted demographic with just a few clicks. *That* is the magic of Pinterest Guided Search.

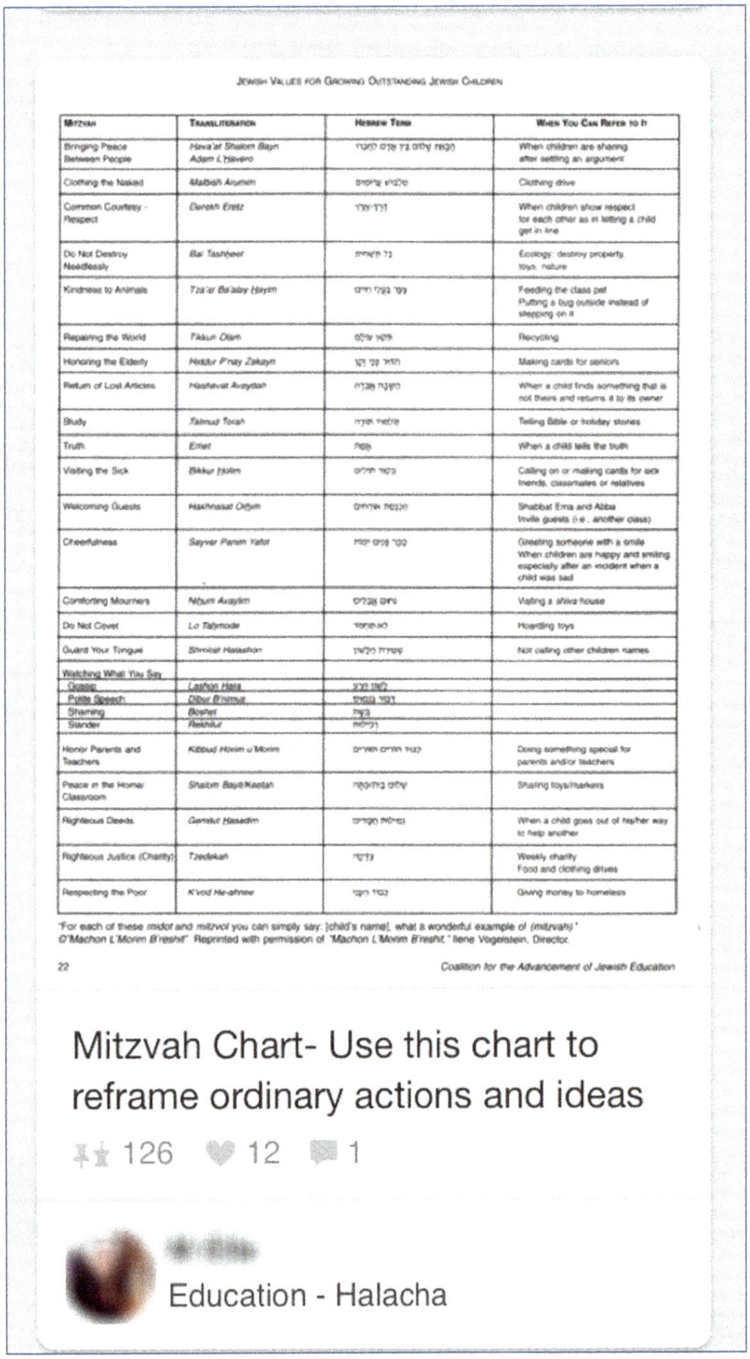

JEWISH VALUES FOR GROWING OUTSTANDING JEWISH CHILDREN

Mitzvah	Transliteration	Hebrew Term	When You Can Refer to It
Bringing Peace Between People	Hava'at Shalom Bayn Adam L'Havero	הבאת שלום בין אדם לחברו	When children are sharing after settling an argument
Clothing the Naked	Malbish Arumim	מלביש ערומים	Clothing drive
Common Courtesy - Respect	Derekh Eretz	דרך ארץ	When children show respect for each other as in letting a child get in line
Do Not Destroy Needlessly	Bal Tashheet	בל תשחית	Ecology, destroy property, toys, nature
Kindness to Animals	Tza'ar Ba'alay Hayim	צער בעלי חיים	Feeding the class pet. Putting a bug outside instead of stepping on it
Repairing the World	Tikkun Olam	תיקון עולם	Recycling
Honoring the Elderly	Hiddur P'nay Zakayn	הידור פני זקן	Making cards for seniors
Return of Lost Articles	Hashavat Avaydah	השבת אבידה	When a child finds something that is not theirs and returns it to its owner
Study	Talmud Torah	תלמוד תורה	Telling Bible or holiday stories
Truth	Emet	אמת	When a child tells the truth
Visiting the Sick	Bikkur Holim	ביקור חולים	Calling on or making cards for sick friends, classmates or relatives
Welcoming Guests	Hachnasat Orhim	הכנסת אורחים	Shabbat Ema and Abba Invite guests (i.e., another class)
Cheerfulness	Sayver Panim Yafot	סבר פנים יפות	Greeting someone with a smile. When children are happy and smiling especially after an incident when a child was sad
Comforting Mourners	Nihum Avaylim	ניחום אבלים	Visiting a shiva house
Do Not Covet	Lo Tahmode	לא תחמוד	Hoarding toys
Guard Your Tongue	Shmirat Halashon	שמירת הלשון	Not calling other children names
Watching What You Say:			
Gossip	Lashon Hara	לשון הרע	
Polite Speech	Dibur B'nimus	דיבור בנימוס	
Shaming	Boshet	בושת	
Slander	Rekhilut	רכילות	
Honor Parents and Teachers	Kibbud Horim u'Morim	כבוד הורים ומורים	Doing something special for parents and/or teachers
Peace in the Home/ Classroom	Shalom Bayit/Keetah	שלום בית/כיתה	Sharing toys/markers
Righteous Deeds	Gemilut Hasadim	גמילות חסדים	When a child goes out of his/her way to help another
Righteous Justice (Charity)	Tzedakah	צדקה	Weekly charity Food and clothing drives
Respecting the Poor	K'vod He-ahnee	כבוד העני	Giving money to homeless

'For each of these midot and mitzvot you can simply say: [child's name], what a wonderful example of (mitzvah).'
©'Machon L'Morim B'reshit' Reprinted with permission of 'Machon L'Morim B'reshit,' Ilene Vogelstein, Director.

22   Coalition for the Advancement of Jewish Education

Mitzvah Chart- Use this chart to reframe ordinary actions and ideas

⚲ 126 ♥ 12 💬 1

Education - Halacha

Figure 7.9: Pinterest Guided Search with "Jewish+Life+Torah"

Let's look at another example that illustrates how Pinterest users are connecting Jewish values with actions toward protecting the environment—a great example of the power of framing your climate action message. Starting with the word "Jewish" and using Guided Search, we can select "Life" and "Torah" from among the items that appear in the horizontal menu. This Guided Search helps discover a piece of content that lists Jewish values that should be taught to children (Figure 7.9). In this list, each Jewish value is connected to a specific real-life topic.

Careful examination of the list reveals Jewish values that are connected to environmentally friendly behaviors: "Repairing the World" and "Do Not Destroy Necessarily," which have been connected to teaching topics "Recycling, and "Ecology," respectively (Figure 7.10).

Jewish Values for Growing Outstanding Jewish Children
compiled by Ilene Vogelstein (posted with permission at BibleBeltBalabusta.com)

MITZVAH	TRANSLITERATION	HEBREW TERM	WHEN YOU CAN REFER TO IT
Bringing Peace Between People	Hava'at Shalom Bayn Adam L'Havero	הֲבָאַת שָׁלוֹם בֵּין אָדָם לְחַבֵרוֹ	When children are sharing after settling an argument
Clothing the Naked	Malbish Arumim	מַלְבִּישׁ עֲרוּמִים	Clothing drive
Common Courtesy - Respect	Derekh Eretz	דֶרֶךְ-אֶרֶץ	When children show respect for each other as in letting a child get in line
Do Not Destroy Needlessly	Bal Tashheet	בַל תַּשְׁחִית	Ecology; destroy property, toys, nature
Kindness to Animals	Tza'ar Ba'alay Hayim	צַעַר בַּעֲלֵי חַיִים	Feeding the class pet Putting a bug outside instead of stepping on it
Repairing the World	Tikkun Olam	תִּקוּן עוֹלָם	Recycling
Honoring the Elderly	Hiddur P'nay Zakayn	הִדוּר פְּנֵי זָקֵן	Making cards for seniors
Return of Lost Articles	Hashavat Avaydah	הַשָׁבַת אֲבֵדָה	When a child finds something that is not theirs and returns it to its owner
Study	Talmud Torah	תַּלְמוּד תּוֹרָה	Telling Bible or holiday stories
Truth	Emet	אֱמֶת	When a child tells the truth
Visiting the Sick	Bikkur Holim	בִּקוּר חוֹלִים	Calling on or making cards for sick friends, classmates or relatives
Welcoming Guests	Hakhnasat Orhim	הַכְנָסַת אוֹרְחִים	Shabbat Ema and Abba Invite guests (i.e., another class)
Cheerfulness	Sayver Panim Yafot	סֵבֶר פָּנִים יָפוֹת	Greeting someone with a smile When children are happy and smiling, especially after an incident when a child was sad
Comforting Mourners	Nihum Avaylim	נָחוּם אֲבֵלִים	Visting a shiva house
Do Not Covet	Lo Tahmode	לֹא תַחְמֹד	Hoarding toys

Figure 7.10: Table of Jewish Values discovered on Pinterest. Reprinted with permission of "Machon L'Morim B'reshit," Ilene Vogelstein, Director

If one wants to be the Jewish voice for climate change, connecting with the Pinterest user who created this table—along with those who liked it and shared—would be a good start.

3. "Image-based search" is a powerful tool in Pinterest, which allows one to search for content using images rather than text. Let's look at an example. Some of the items revealed by searching Pinterest with the keyword "Jewish" are pieces of jewelry with Jewish symbolism. You can imagine that people who would be interested in this content are of the Jewish faith. We can use the image from one of the pieces of jewelry to search Pinterest and find similar images, and then by extension to find people of Jewish faith. To carry out image-based search, find a pin, click to enlarge it, and then hover your mouse to the top right of the image to see a magnifying glass encased in a bracket (Figure 7.11).

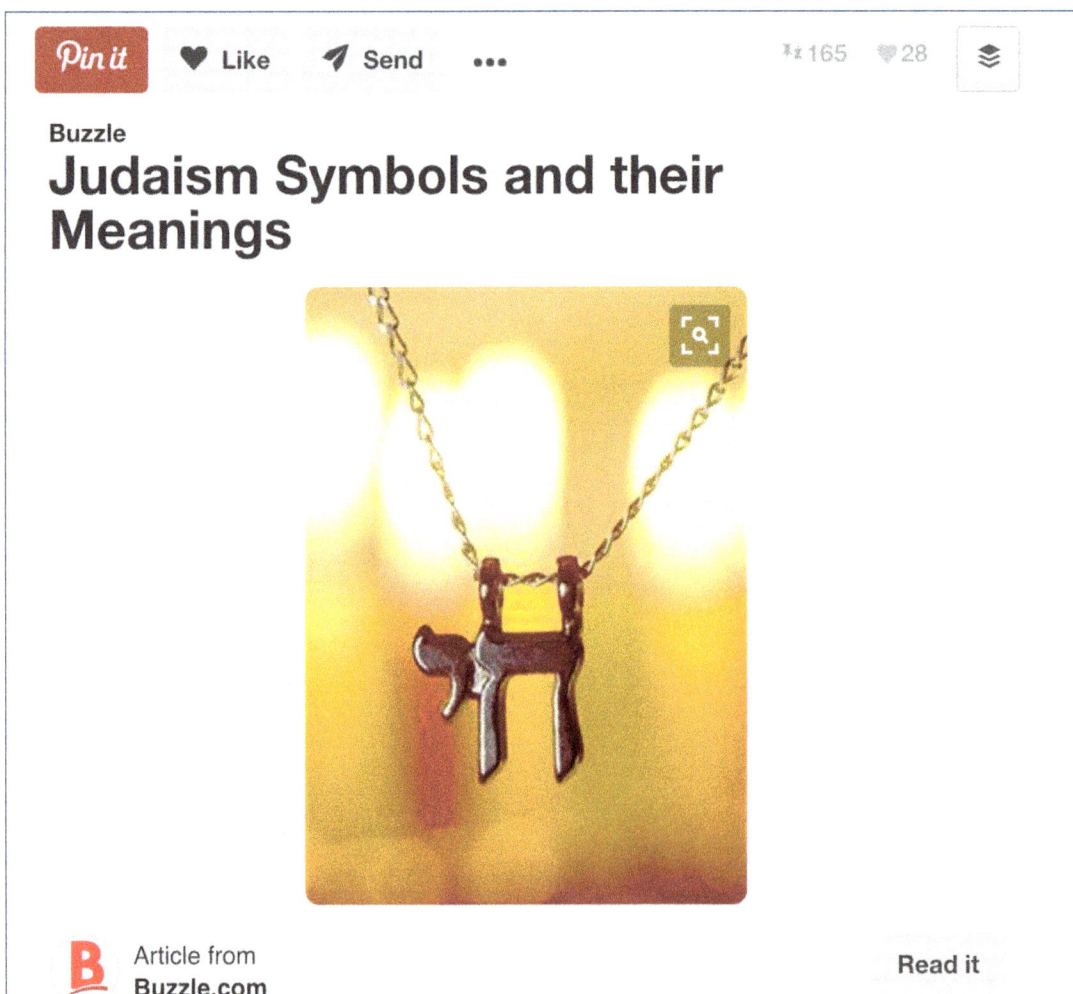

Figure 7.11: "Image-based search" on Pinterest

Clicking on this icon will reveal a new set of content containing similar images pinned by those who are passionate about the Jewish faith (Figure 7.12).

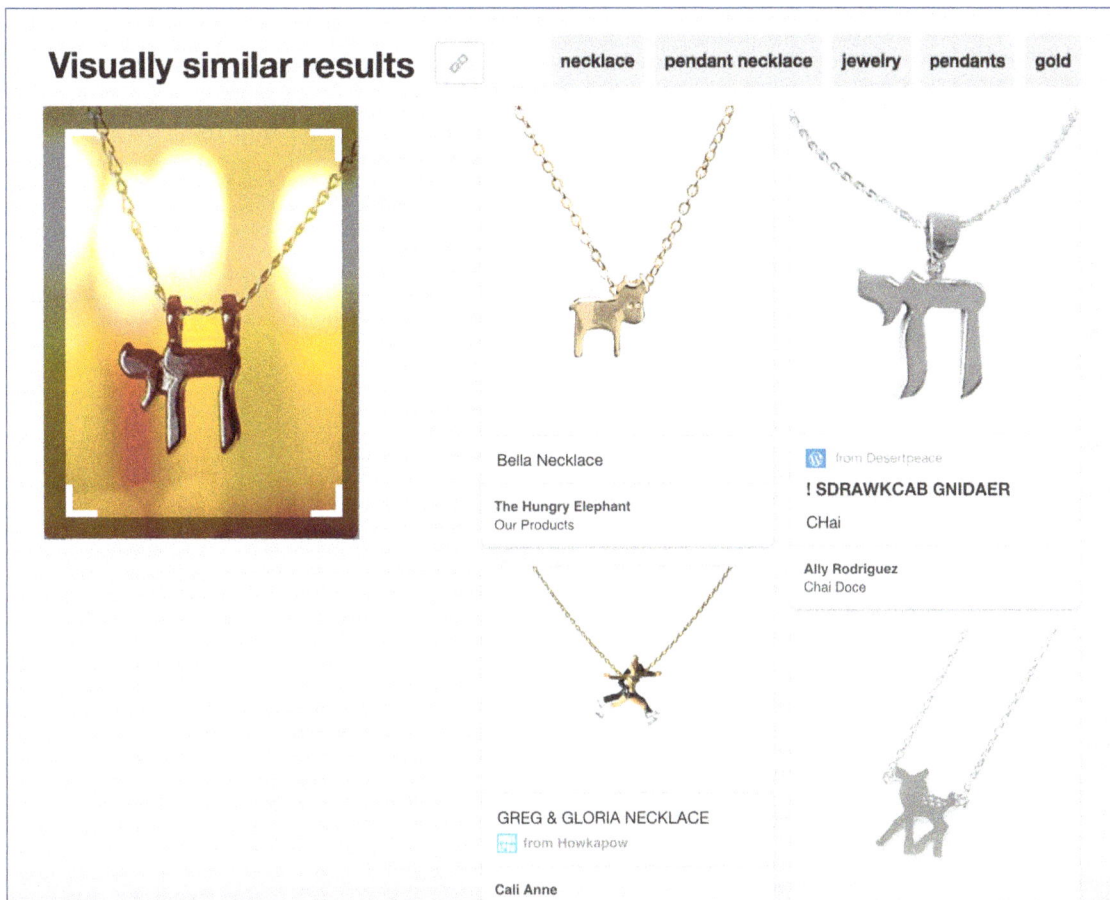

Figure 7.12: Results of "Image-based search" on Pinterest

Pinterest is rich with content, and it provides powerful ways to find what and who you are looking for. There are millions of people creating content on Pinterest every day that are available for you to explore. What you can discover on this amazing platform is only limited by your imagination. Now let's discuss the different types of content that can be created on Pinterest and how it is shared.

Status updates and terminology

Pinterest is a visual medium, so content on this platform comes in the form of images, videos, and slide presentations. Images can be directly uploaded to Pinterest or shared to Pinterest from a website. Videos and slideshows must be hosted on external websites and shared onto Pinterest by choosing the "Pin from a Website" option. In addition, most websites have included a "Pin it" button (recently renamed to the "Save" button by Pinterest, see citation 7.3), which allows adding images, videos, or slide presentations directly from these websites onto a Pinterest account. If you want to increase the odds of your blog gaining exposure to the Pinterest community, it is important that you include this functionality in your blog. Fortunately, this option is already included in a blog setup on Blogger. You can see it along with other social media "Share" buttons at the bottom of each of your blog posts (Figure 7.13).

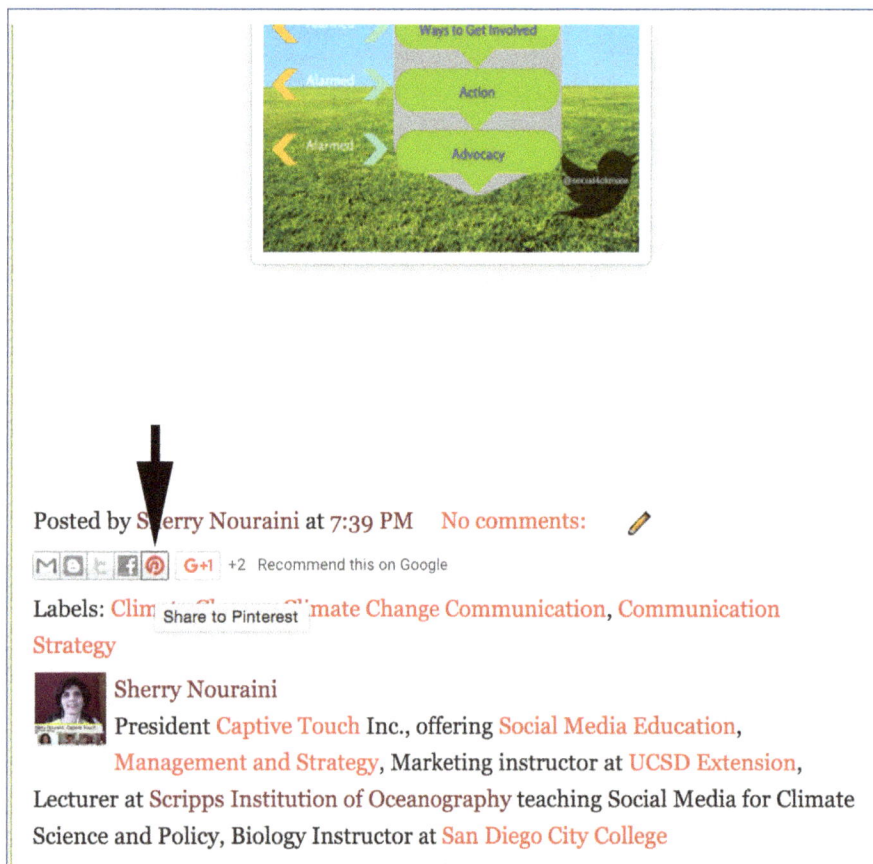

Figure 7.13: The "Pin it" and other social sharing buttons on your blog

An alternative means of reducing the barrier for sharing your content to Pinterest is to add a few lines of code to your website/blog, which makes a "Pin it" (recently renamed to "Save") button appear when a viewer of your content hovers their mouse over an image. Changing code may sound scary, but it is pretty straightforward. Follow these steps to enable this powerful feature for your blog (Figure 7.14):

1. Visit the "Widget builder" section of Pinterest using this URL https://business.pinterest.com/en/widget-builder.
2. Choose the "Save" button option.

3. Set button type to "Image Hover."

4. Choose shape, color, and language for the button.

5. Copy the code in the gray box below the image.

6. Paste the code just above the \</body\> tag in the HTML code for your blog (Figure 7.15).

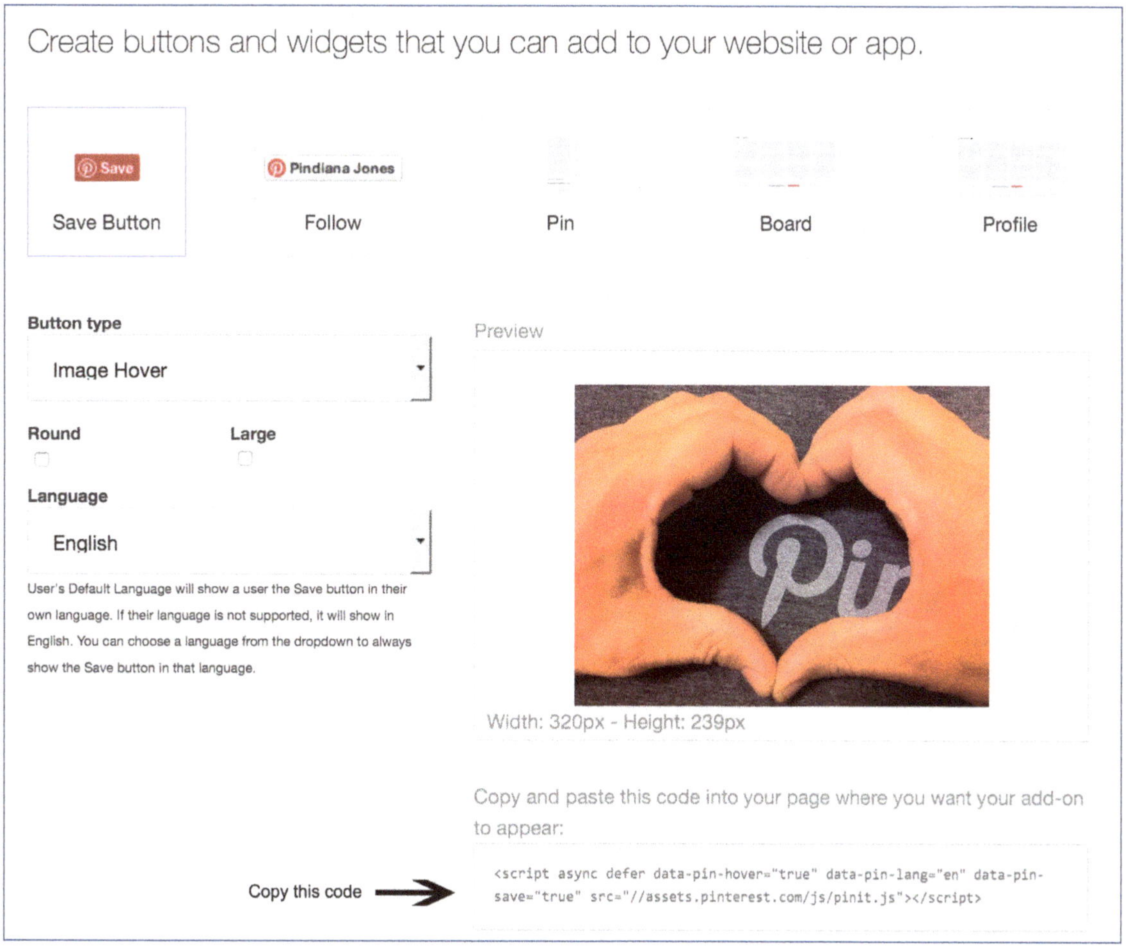

Figure 7.14: Creating a "Pin it" (now called "Save") button

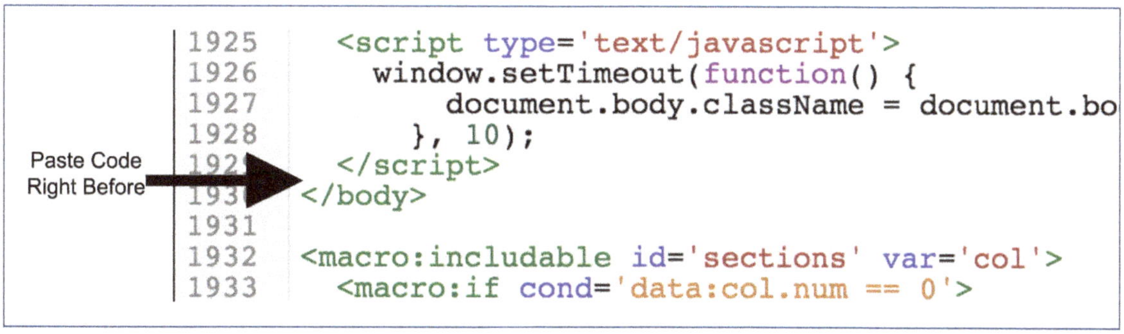

Figure 7.15: Pasting the "Pin it" code into your blog

Now, when someone visits your blog and finds an image appealing, they can hover their mouse on that image and pin or save the image to a board on their profile. The image that is posted on Pinterest is linked to your website. When a Pinterest user clicks on the image on Pinterest they will be led to your website. Just like that, you will gain a new website visitor.

The Pinterest status box and terminology

On Twitter and Facebook, we saw that content is created within a status box. On Pinterest, the equivalent of a status box is a "board" and the equivalent of posting content is a "pin." To pin an item, use the plus sign at the top right corner of your profile (Figure 7.16), or within a board (not shown).

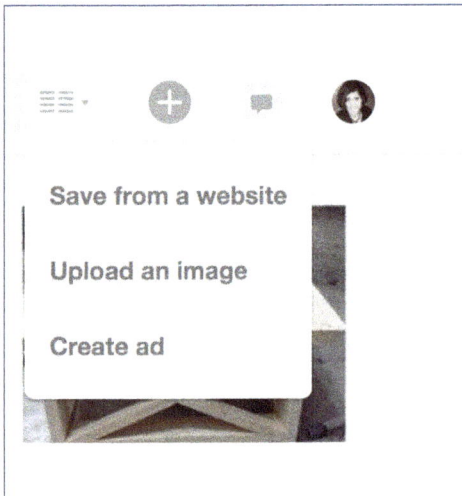

Figure 7.16: Uploading content to Pinterest

When you directly upload an image from your computer into one of your boards on Pinterest, be sure to add a description to the bottom of the image before pinning it to a board (Figure 7.17).

You can later edit this pin to add a URL or a location. When you pin an image, video, or slide presentation from an external website, Pinterest automatically populates the description box with the name of the media file and adds the URL associated with it. Be thoughtful about how you name your media files: include keywords in the titles when you first create and save them. Once an image has been pinned from a website, Pinterest adds a "Visit" button to the pin to entice viewers to visit the original website where the content originated (Figure 7.18).

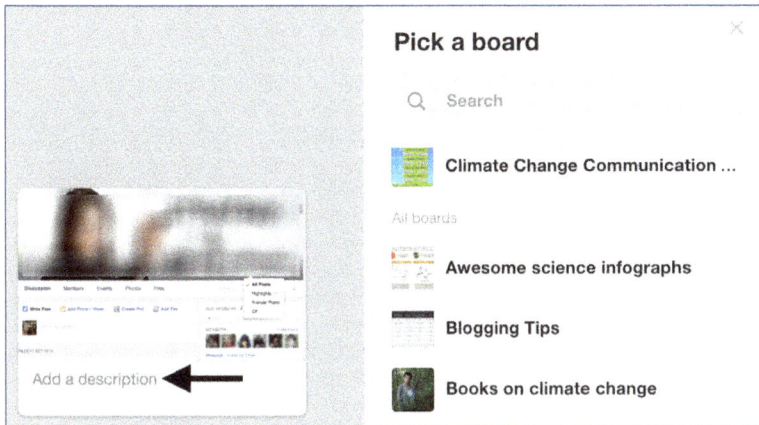

Figure 7.17: Adding a description to uploaded images

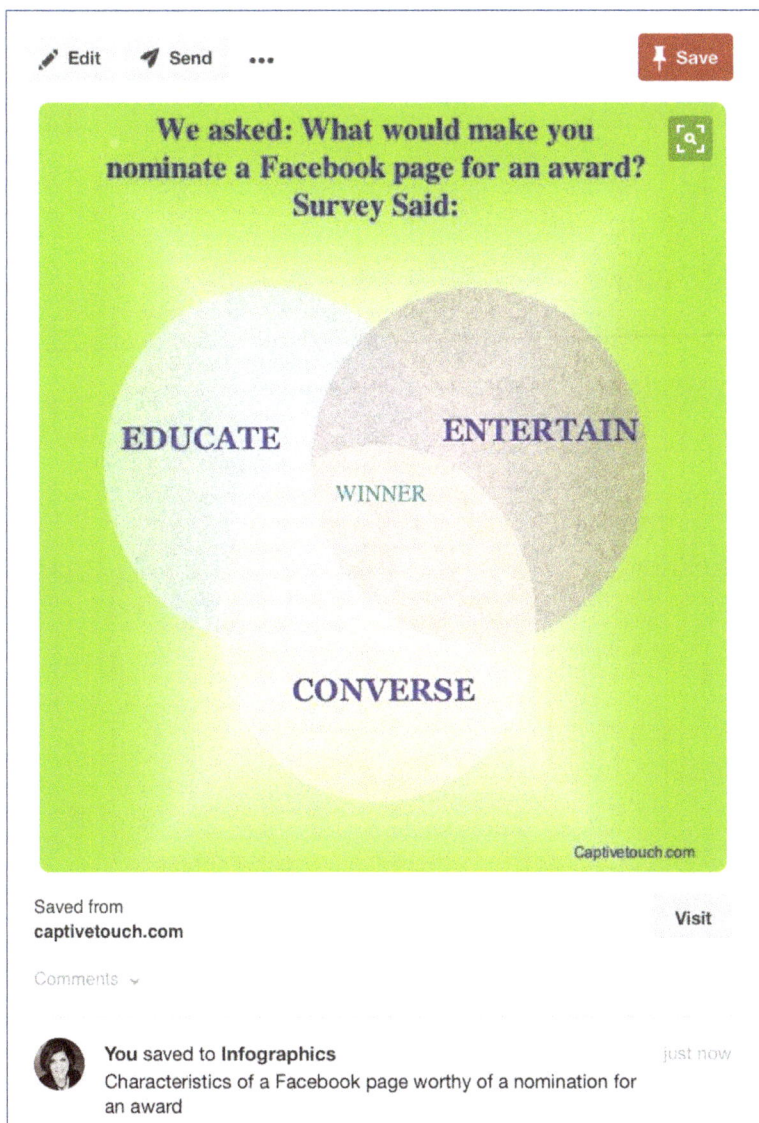

Figure 7.18: Pinning from a website populates the description and URL

Rich Pins

"Rich Pins" are a specialized category of content, which are "rich" in information that is not normally included in a regular pin. As mentioned above, when an image is pinned from a website, the URL and the name of the media file is automatically populated into Pinterest. Rich Pins, however, render additional information into Pinterest, and the type of information depends on which Rich Pin is being used.

There are six types of Rich Pins (at the time of this writing) summarized below:

1. *App pins* include an "Install" button when an image of the app is saved (or pinned) onto Pinterest. This way, when Pinterest users view an app pin on their mobile device, they can install the app without leaving the platform.

2. *Place pins* include a map, address, and phone number when an image is pinned from a website onto Pinterest.

3. *Article pins* include a headline, author, and story description when an image from an article is pinned onto Pinterest.

4. *Product pins* include real-time pricing, and availability of merchandise when an image from a product website is pinned onto Pinterest.

5. *Recipe pins* include ingredients, cooking times, and serving information when images from a cooking website are pinned onto Pinterest.

6. *Movie pins* includes ratings, cast members, and reviews of a movie when content from a website featuring or promoting a movie is pinned onto Pinterest.

To view examples of each type of Rich Pin, please visit this URL https://business.pinterest.com/en/rich-pins. The goal of using Rich Pins is to entice a Pinterest user to click and visit the website that hosts the media that was pinned onto the platform. In order to implement Rich Pins for your website, you will have to add some code to your website and have your Rich Pin verified (7.4). The steps required to set up Rich Pins require advanced knowledge of Web development, which is why they are not covered here.

Pin engagement

Once a pin has been shared on Pinterest, other users can pin the image on their own boards ("Repin" or "Save"), "Like" the pin, share it privately to their contacts, share it publicly in social media channels, comment on the pin, embed the pin into their own websites outside of Pinterest, or report it if they find the pin offensive (Figure 7.19).

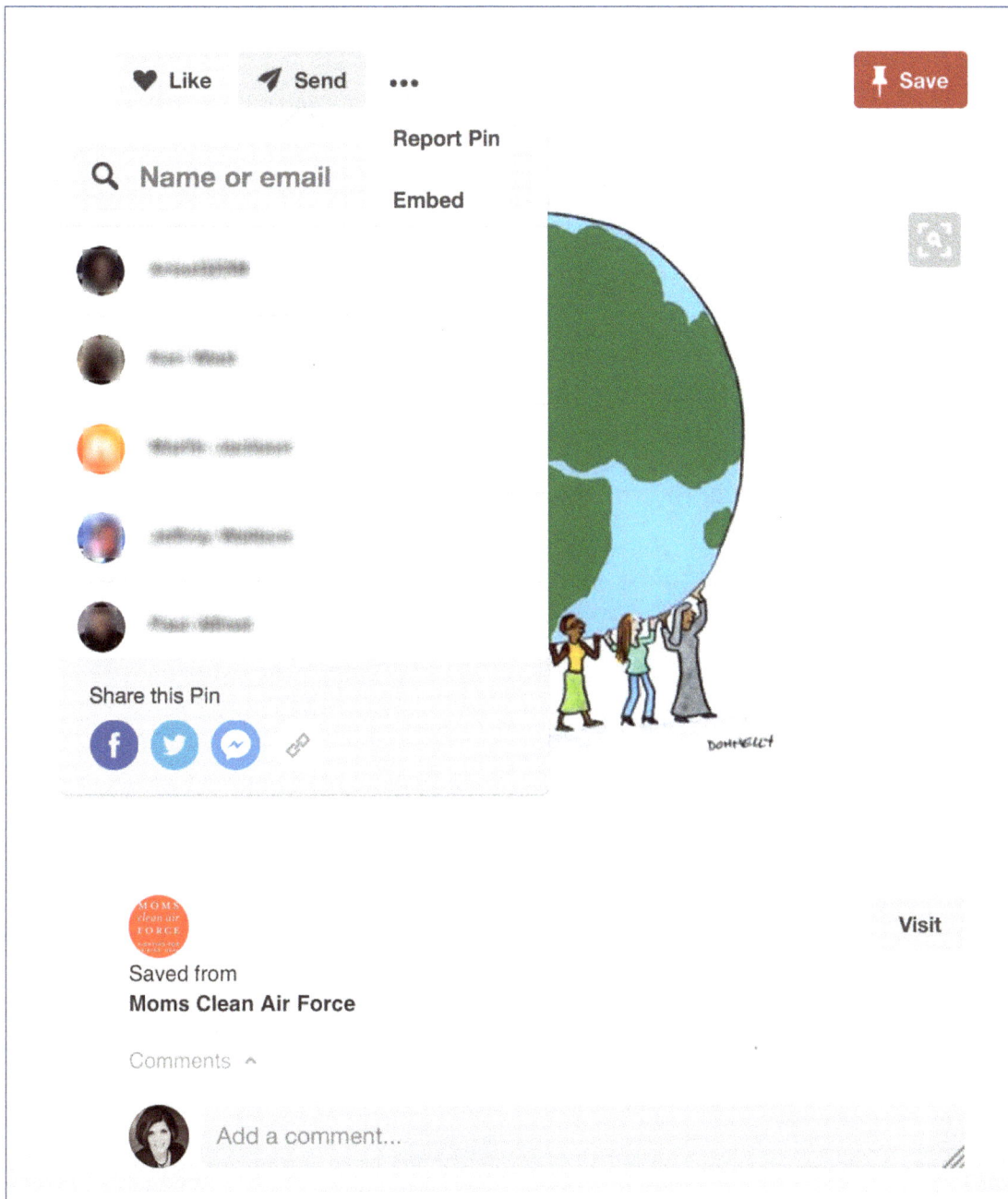

Figure 7.19: Different ways of engaging with a pin

Community

Pinterest users create communities by inviting others to pin to common boards. There is no easy way to search for communities on Pinterest, as there is on Twitter (using hashtags) and Facebook (searching for groups). If you were to find a community of pinners, you would have more success finding pinners who have a large, actively engaged following.

Analytics

Before covering different sections of Pinterest analytics, let's review some terminology:

- Impressions. The number of times a pin showed up in the home feed, search results, and category feeds—all the different ways your content can be discovered on Pinterest. A category feed is content revealed when one searches Pinterest based on categories (see the Search section of this chapter).

- Repins. The number of times someone saved your pin to one of their boards. Repins are how pins are shared across Pinterest.

- Clicks. The number of clicks to your website from pins on your profile.

Analytics on Pinterest are presented in three categories: "Your Pinterest Profile," "Your Audience," and "Activity from Website." Each section of Pinterest analytics also allows you to choose the type of device that was used to view and interact with your content (Figure 7.20).

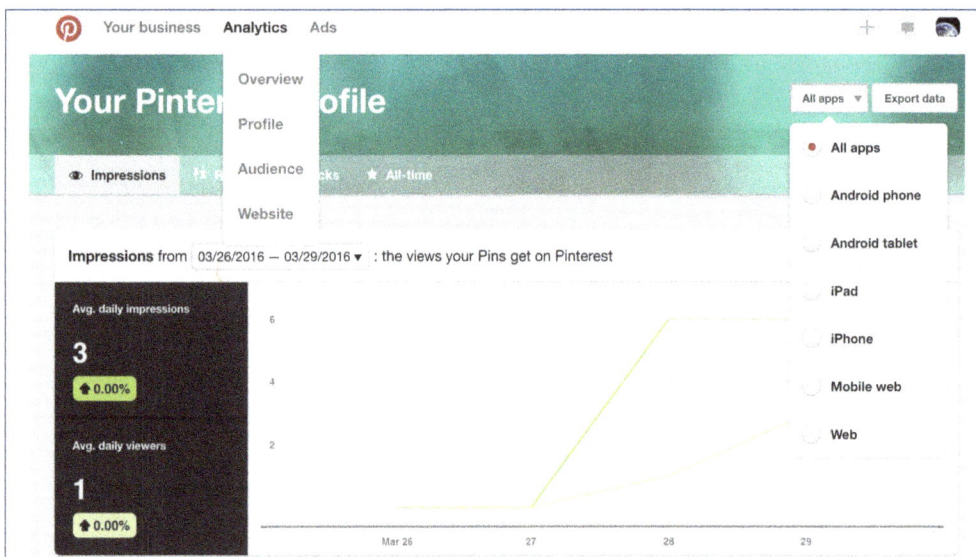

Figure 7.20: Overview of Pinterest analytics

Your Pinterest profile

This section of Pinterest analytics reports on Impressions, Repins and Clicks received by content you have pinned to your boards. This data will help you judge the level of activity on your content over time, and you can also set the time period for these activities (Figure 7.21)

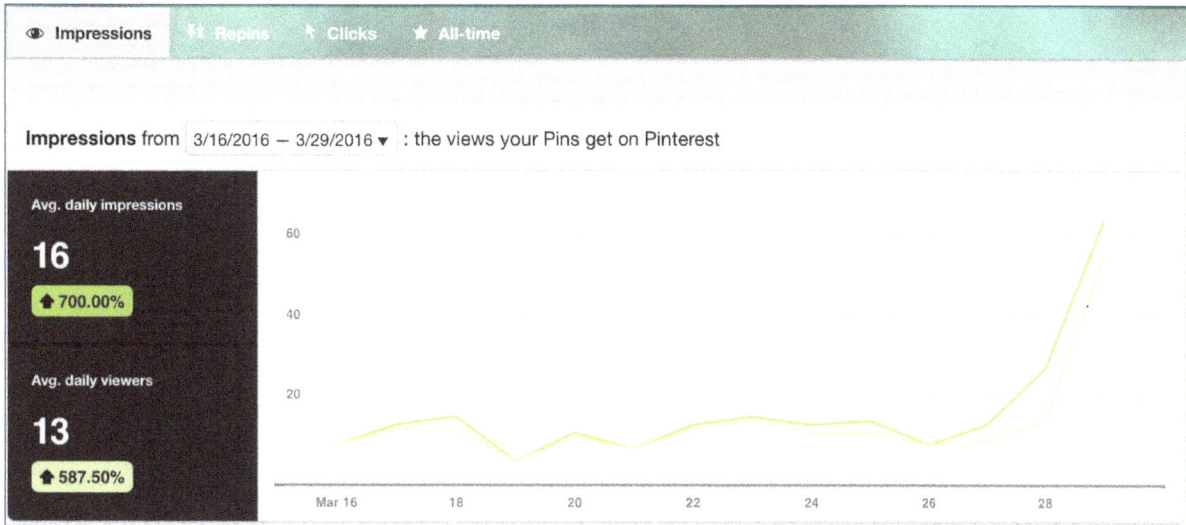

Figure 7.21: Pinterest analytics—"Your Profile: Impressions"

Pinterest analytics also reports on the pins and boards on your profile that have the greatest number of Impressions, Repins and Clicks within the most recent thirty days (not shown). This data will help you determine what type of content resonates most with your followers.

Your Audience

"Your Audience" section of Pinterest analytics provides data about the total number of viewers of your content, as well as the number of viewers who actually engage with your content; the latter is the more important metric (Figure 7.22). "Your Audience" will also provide a distribution of the location, gender, and language of your audience (not shown).

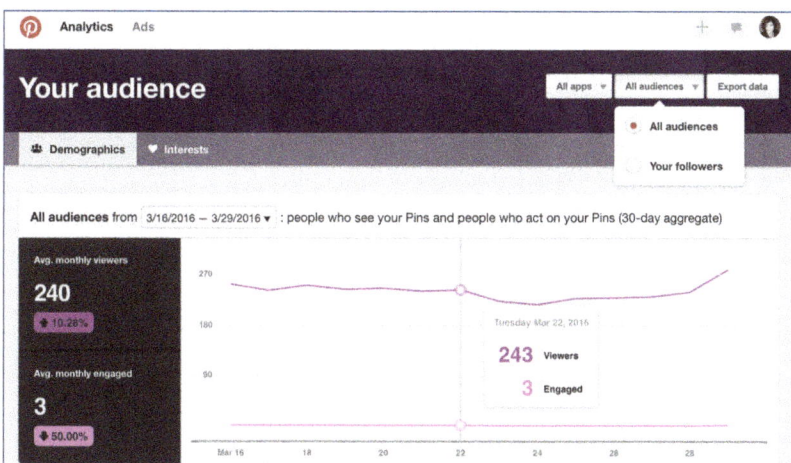

Figure 7.22: Pinterest analytics—Your audience

Another powerful section of "Your Audience" analytics is data about the interests of your audience. This data will reveal the top categories of content with which your audience interacts on Pinterest (Figure 7.23). This data will allow you to fine-tune your content to better match the interests of Pinterest users who engage with your content. How can you fashion your climate action message in a way that overlaps these interests?

Data presented in the "Interests" section also show Boards into which your content tends to be saved and brand pages with which your audience tends to engage (not shown). The latter metric is useful for knowing what Pinterest pages are competing with you for attention of your audience, and the former helps you determine if there is a match between the way you categorize your content into boards, and the way your audience does: the closer they match, the better.

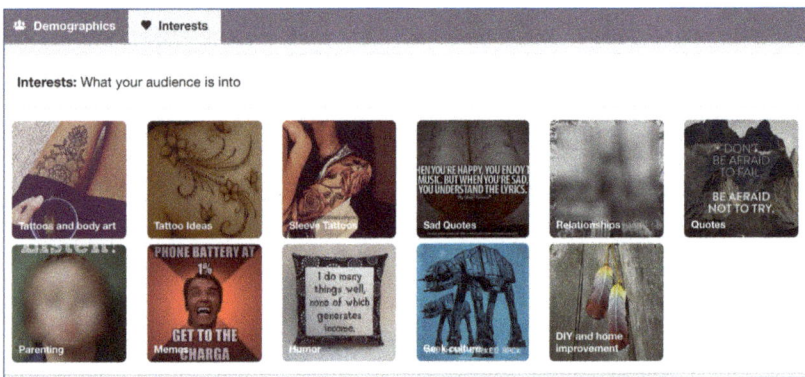

Figure 7.23: Pinterest analytics—Your audience-Interests

Website

The "Your Website" section is where you'll be thankful that you took the time to verify your website. The analytics here help you understand the level of engagement with the pins that originated from your website (Impression, Repins, and Clicks) (Figure 7.24).

Figure 7.24: Pinterest analytics—Your website

Pinterest analytics show you this performance over time, in addition to revealing the best-performing pins and boards into which the content from your website was pinned within the past thirty days (not shown). This data is important because it will help you discover Pinterest boards and pinners who identify with your content enough to pin it to their Pinterest profiles. In other words, Pinterest analytics serves your most engaged website visitors to you on a silver platter, an audience that you may not have otherwise captured.

The "Original Pins" section of Pinterest website analytics shows how many unique pins were created from your website on a daily basis, as well as the most recent unique pins. This data will help you understand the latest trends for pinning original content from your website. Note that you can view the boards onto which your content was pinned, and by extension the users who pinned that content. Next, you should follow each user who pinned content from your website.

The "Pin it" button section of Pinterest website analytics shows data about the level of engagement with this button on your website on a particular day. It will show the number of times the button was viewed, clicked on, and used to pin an image from your website to Pinterest (not shown).

If you notice, there is an "All-time" section in the website and profile analytics (Figure 7.21 and 7.24 top menu), which reveals the best-performing pins from your website or those you posted on your profile. This data is presented as pins that have been shared the most, and ranked the highest in Pinterest search. It also shows "Power Pins," which are pins with the highest performance in all engagement categories (repins, clicks, comments, search). Use this data to learn from your best-performing media so you can replicate their success when creating new ones.

In addition to reporting such events, Pinterest analytics also follows what happens to these pins over a seven-day period in terms of engagement. It will show a second graphic that reveals the number of Impressions and Repins, as well as Clicks back to your website (not shown). The best kinds of images on your website are the ones that entice your audience to pin (or save) them to Pinterest, and makes *their* audience curious enough about the content of your website to click out of Pinterest to view it. Use these metrics to judge the effectiveness of the images on your website or their "stickiness."

Outreach on Pinterest

Similar to Twitter and Facebook, outreach on Pinterest can be done through advertising or by building relationships organically (no ads), or a combination of both. When it comes to advertising, you can either pay Pinterest to promote (show) your pins to a particular demographic on Pinterest, or you can create buyable pins, which allows Pinterest users to shop directly on Pinterest. Buyable pins are particularly useful for e-commerce websites, but promoted pins have broad appeal. Similar to chapters for other social networking sites, Pinterest advertising is not covered in any more detail in this book.

Just like Facebook and Twitter, your outreach workflow starts with audience research and discovery. Whether or not you choose to use advertising on Pinterest, your outreach will involve essentially the same steps, which are shown in Figure 7.25. The ultimate goal of your efforts on Pinterest is to discover and connect with Pinterest users who engage with your content, and you can discover them in two ways. On the one hand, you can begin connecting and engaging pinners whom you discover via Pinterest search. On the other hand, you publish content on your blog, pin images associated with that content onto Pinterest, and monitor Pinterest analytics to identify and connect with your website visitors who pin your content on their boards. Once you connect with your target audience on Pinterest, take the time to build a relationship with them and hold discussions around climate change.

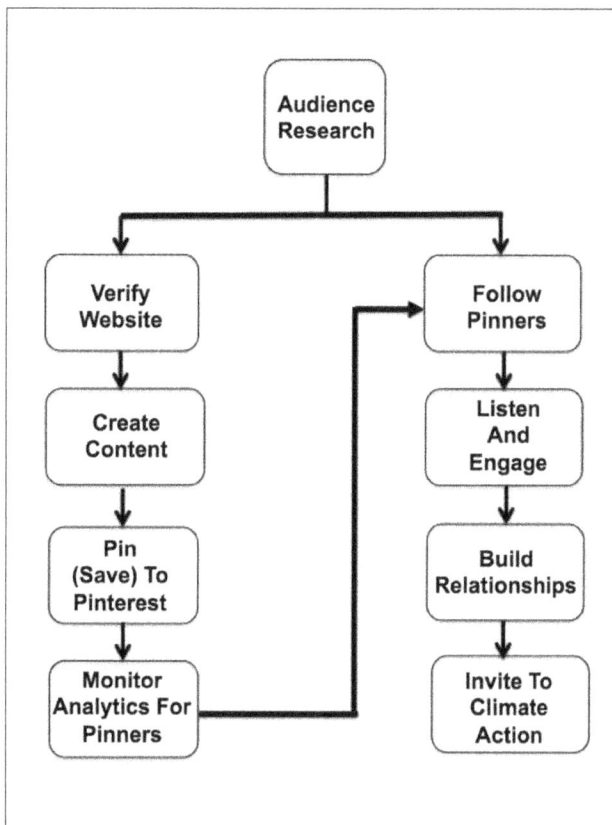

Figure 7.25: Workflow for outreach on Pinterest

Here is an example of a pin that I created on Pinterest to encourage conversation about climate change with Cthulhu fans (Figure 7.26). Before creating this pin I performed a search on Pinterest to browse through content commonly posted by Cthulhu fans. I learned that one popular type of content on Pinterest pinned by Cthulhu fans is "Cthulhu Art." To increase my chances of attracting the attention of Cthulhu enthusiasts, when creating the description for the pin that I uploaded, I used "Cthulhu Art" in the description of the pin. Also, note that the description invites a conversation about climate change and its connection to the image that I posted.

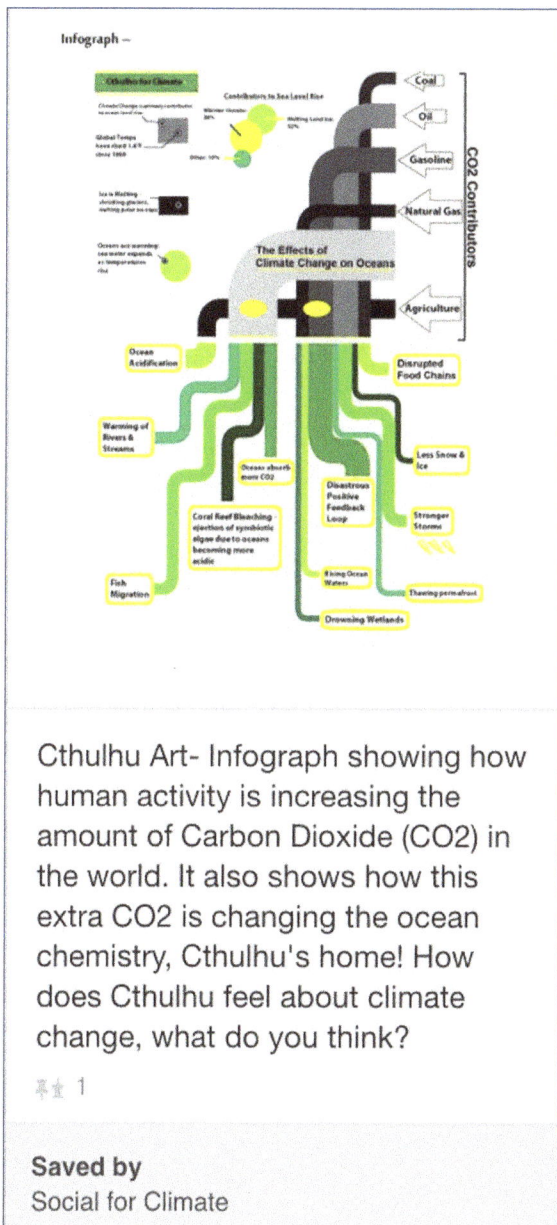

Cthulhu Art- Infograph showing how human activity is increasing the amount of Carbon Dioxide (CO2) in the world. It also shows how this extra CO2 is changing the ocean chemistry, Cthulhu's home! How does Cthulhu feel about climate change, what do you think?

Saved by
Social for Climate

Figure 7.26: Pinned image of Cthulhu Art

Practice what you learned

In this chapter you learned how to set up your own Pinterest profile and verify your website. You also learned how to search for and find your target audience, using the sample profile Linda Goldberg as an example. You learned about "Rich Pins" and how they help enrich your pins with information, and also became familiar with Pinterest analytics. You learned how to use Pinterest analytics to monitor your success and identify your website visitors who save your content onto their Pinterest profiles. Finally, you were presented with a workflow for climate outreach on Pinterest. Now practice what you've learned; use the other two sample profiles, or your own audience profile, and put them to a test. You may think Pinterest is not a good place to find Cthulhu fans or CEOs; try it and you may be pleasantly surprised. Join our Facebook group (https://www.facebook.com/groups/social4climate/) to share what you've learned and to ask questions.

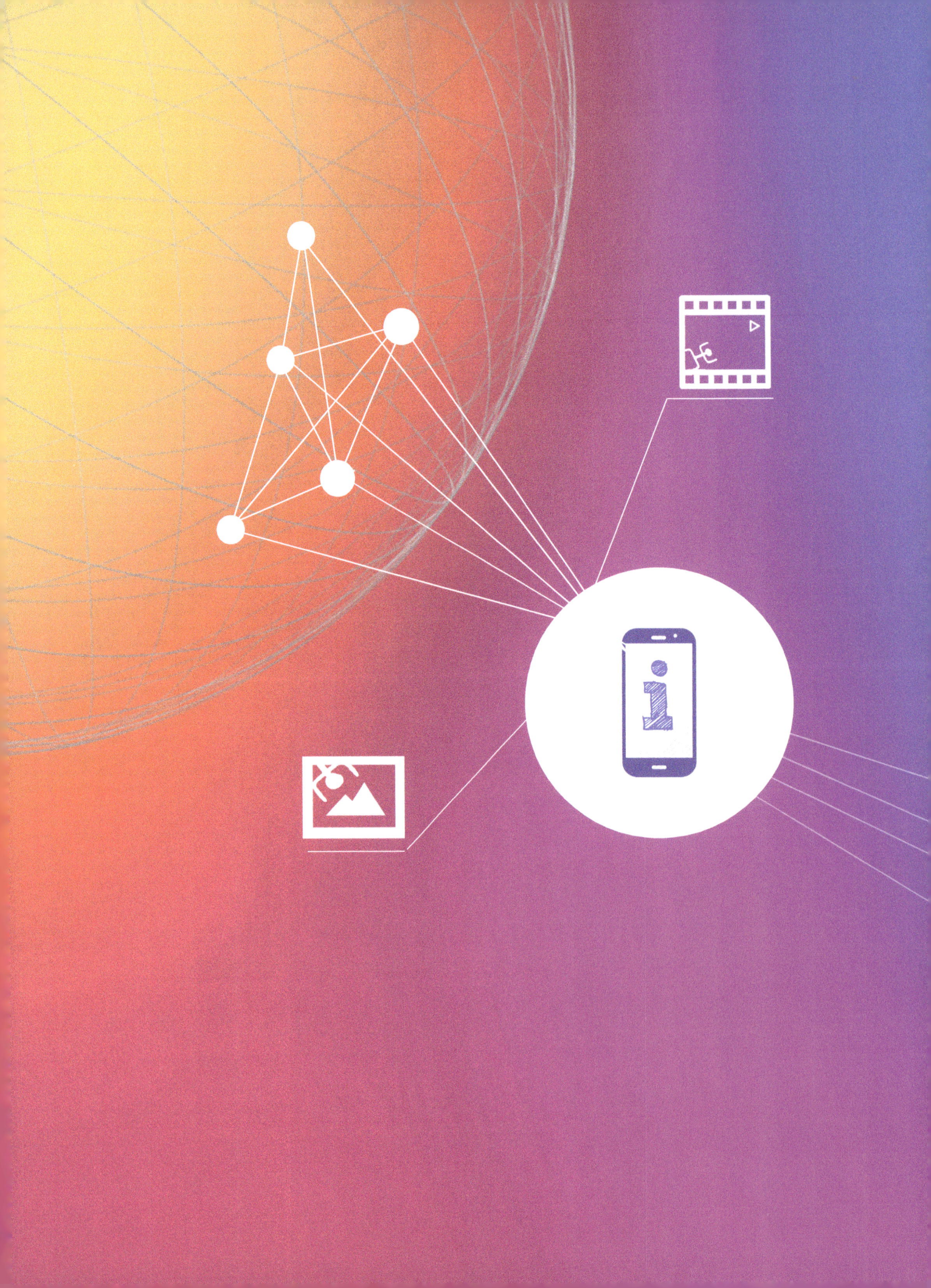

Chapter 8

Instagram

Instagram was launched in 2010 as a mobile photo-sharing application. As mobile devices penetrated every aspect of our lives, Instagram's popularity skyrocketed and Facebook acquired it in 2012. Ever since the acquisition, Instagram has been adding new features, which have fueled its explosive growth. A majority of Instagram users are between the ages of 18 to 29 (7.2), suggesting this platform may be a good place to find Cthulhu fans like Paul Draper (see sample audiences in Chapter 2). Therefore, I will be using this sample profile to showcase Instagram's various features.

As of the writing of this book, Instagram offers only personal profiles. However, by the time this book gets into your hands, business profiles will likely be available. A recent article on the Instagram blog announced the imminent launch of business profiles, loaded with many features, including

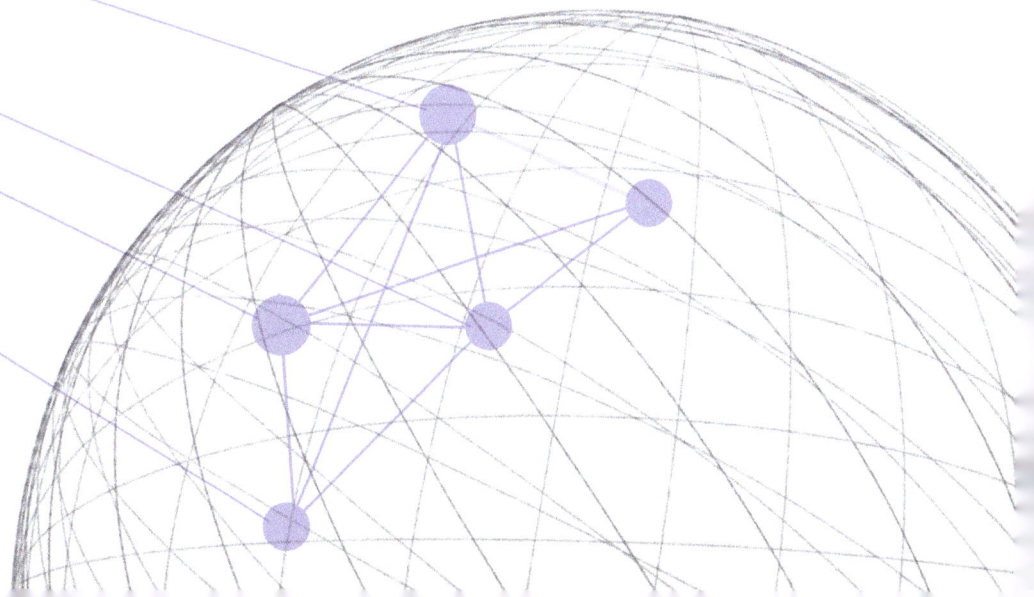

a "Live Contact" button, directions to the business, the ability to promote status updates (ads), and access to rich analytics (8.1). Further details about the new business profiles are not yet available at the time of this writing, so my focus for this chapter is on the personal Instagram profile that everyone has been using. Although I am only covering personal Instagram profiles, your outreach efforts will largely remain the same as how you would be using a personal profile. The Instagram mobile app is comparatively sparse in its number of available features as compared to other social networking platforms covered in this book. Instagram also offers a Web application, which has fewer features than the mobile application. More features may become available soon for the Web application; however, as an alternative, you can use a desktop application called Grids. Grids has been developed by ThinkTime Creations LLC and better resembles the mobile app. However, one caveat is that one cannot post content from the Web application to Instagram, or from any other third-party application for that matter. In an effort to encourage mobile uploading, Instagram does not make its Application Programming Interface (API) available to third-party providers. An API is a window through which third-party applications can access social networking sites to post content and capture data on behalf of the users of these platforms. There are, however, tools that have developed a work-around (8.2), but I am not going to cover them in this book. If you prefer to post content to Instagram from your desktop or want to create and preschedule your content for posting on your account, you may want to give these tools a try. However, sharing real-time moments with your followers is more authentic and stays true to the initial purpose of Instagram: to share your life on the go. The images I present in this chapter will be either from the Instagram mobile or the Grid desktop app.

Profile

An Instagram account, which you need to create by using an app on your phone, contains a profile photo, a description which cannot be longer than 150 characters, and one live link. The parts of an Instagram profile are labeled in Figure 8.1:

1 and 2. This shows two alternative forms of viewing your feed (what other users post).

3. A map of all the different locations from where you shared content. This only shows photos in which you geo-tagged the location.

4. These are media in which you have been tagged.

Feed: Content posted by users you follow.

Search: Where you can search content by keywords and hashtags.

Camera: Where you capture photos or videos to share.

Activity: This shows engagement of the Instagram community with your content, in addition to the activities of those you follow (the content they liked, and commented on, and accounts they followed).

Profile: Your profile as shown in Figure 8.1.

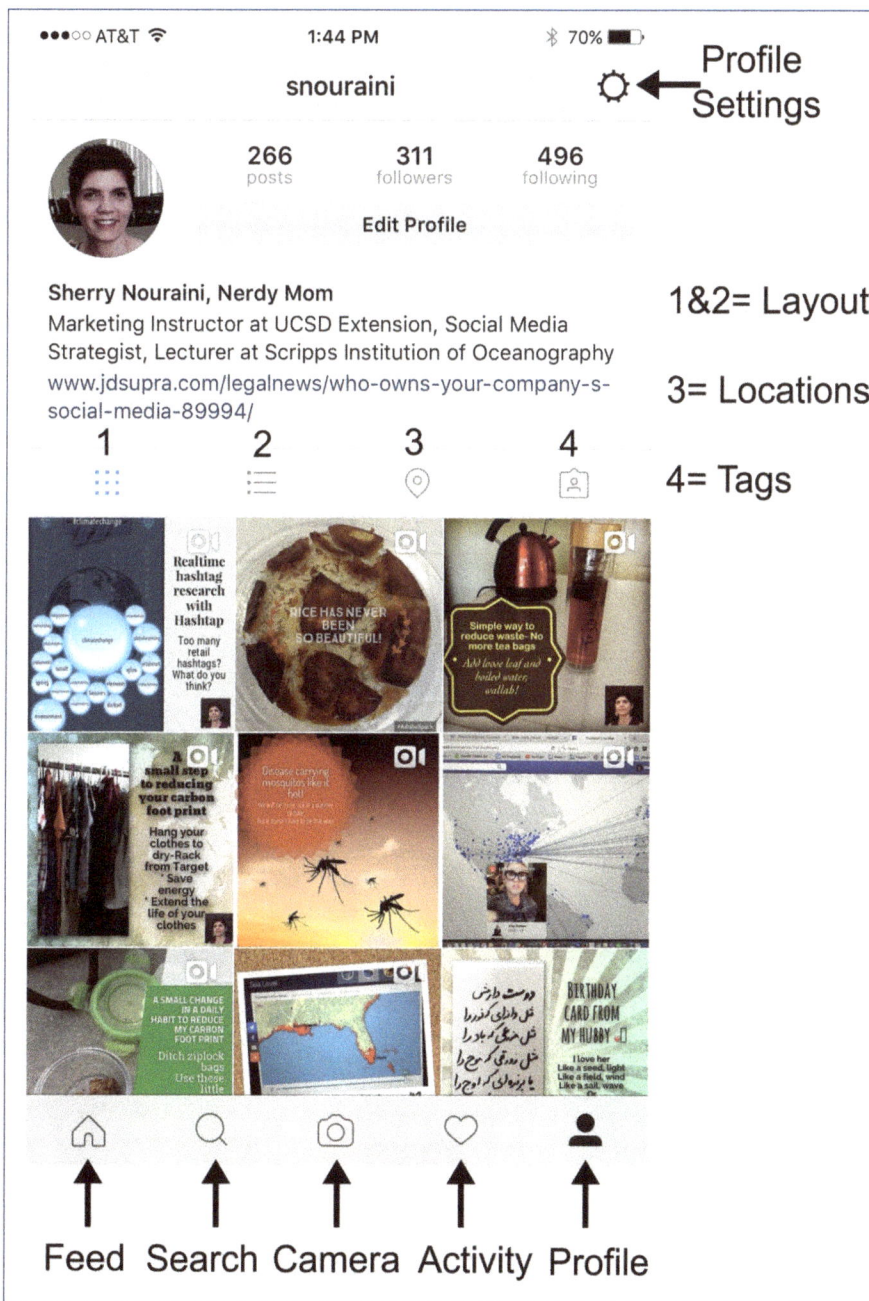

Figure 8.1: Instagram profile

Your profile is the only place on Instagram where a live link is allowed (a clickable URL). When sharing a blog article or a link to an event on Instagram, users temporarily change the URL on their profile and post a photo or a video related to the link. They then include a phrase, such as "link in my profile," in the caption of the image or the video. However, on some tools, such as Grids, links are also live in your status updates. So when you share a link, include it both in your profile and in your media captions to allow your followers who use tools, such as Grids, to have access to live links right in their feed.

Organization

There are no organization features available on Instagram, so serious users usually resort to third-party applications to manage their communities. There are plenty of Instagram management apps, most of which require a paid subscription. The most affordable tool is Iconosquare, which not only allows organization of your community, it also provides analytics (see the analytics section of this chapter).

Iconosquare is a Web application that allows a user to interact with their community, but uploading content into Instagram is not available.

Once you create an Iconosquare account, you can organize your Instagram following by creating a "Feed." There are two premade feeds: "Your Media" (the media you've posted), and "Your Likes" (the media you've liked). Iconosquare allows you to create four more (as of the writing of this book). With each custom feed, you have the ability to follow a maximum of fifty profiles and twenty hashtags. To create feeds, look under the "Manage" menu item and choose "My Feeds." Create a feed, give your feed a name, and assign either hashtags or users to the feed. You can see an example of the feeds I've created in Figure 8.2. You can assign the users that you follow to feeds directly from their profiles. To do so, visit the "Community" section of your Iconosquare profile, find the list of your following, and assign them to feeds (Figure 8.3).

Search

Like any other search engine, you can identify content on Instagram using keywords. Based on the profile we have constructed for Paul Draper, "Cthulhu" seems to be the best keyword. An Instagram search will reveal users, hashtags, and places that contain this keyword. There are reports of Instagram testing a "Search by Categories" feature. As of the writing of this book, this feature is not yet available, but that may change by the time this book gets to your hands.

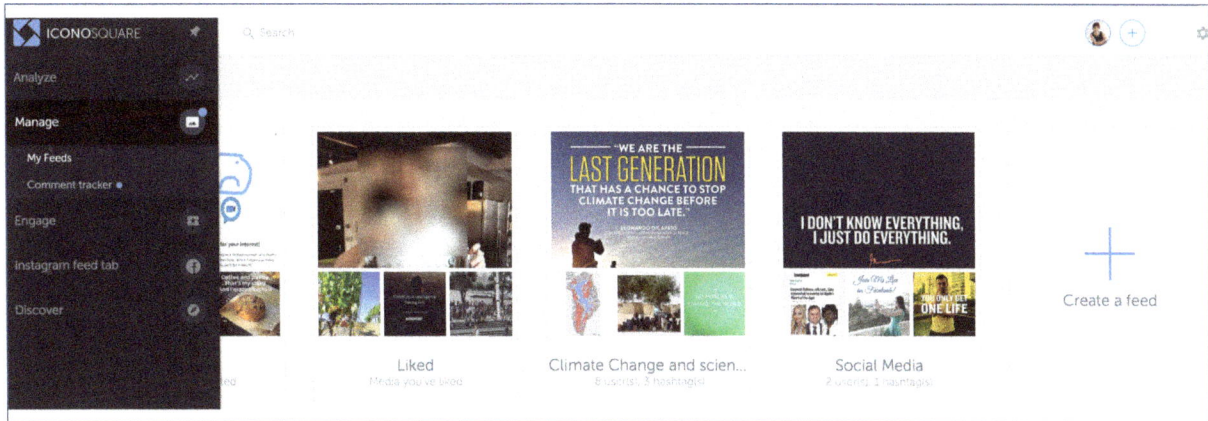

Figure 8.2: Creating content feeds in Iconosquare

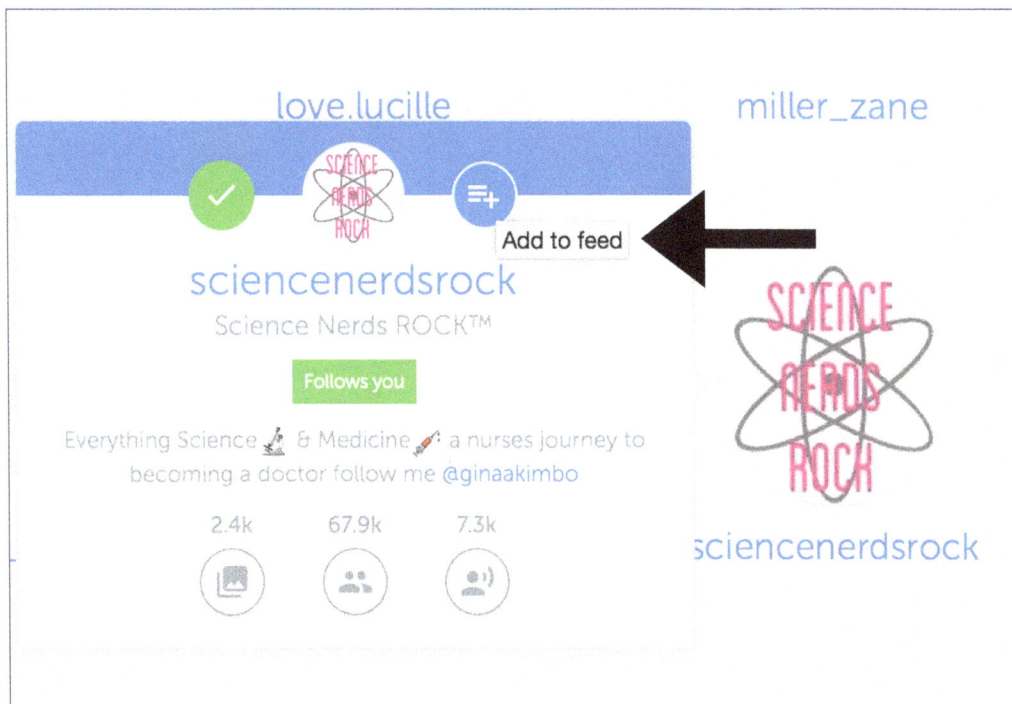

Figure 8.3: Organizing your following in Iconosquare

Upon searching for #Cthulhu (Figure 8.4), what is revealed are top hashtags that include the number of posts containing them, and profiles containing the word #Cthulhu.

Note that since I searched with a hashtag, the "Top" and "Tags" results are identical. Had I used a plain keyword ("Cthulhu" as opposed to "#Cthulhu"), the "Top" tab would reveal a combination of top profiles and hashtags. The "Places" tab showed no results as no location includes the keyword.

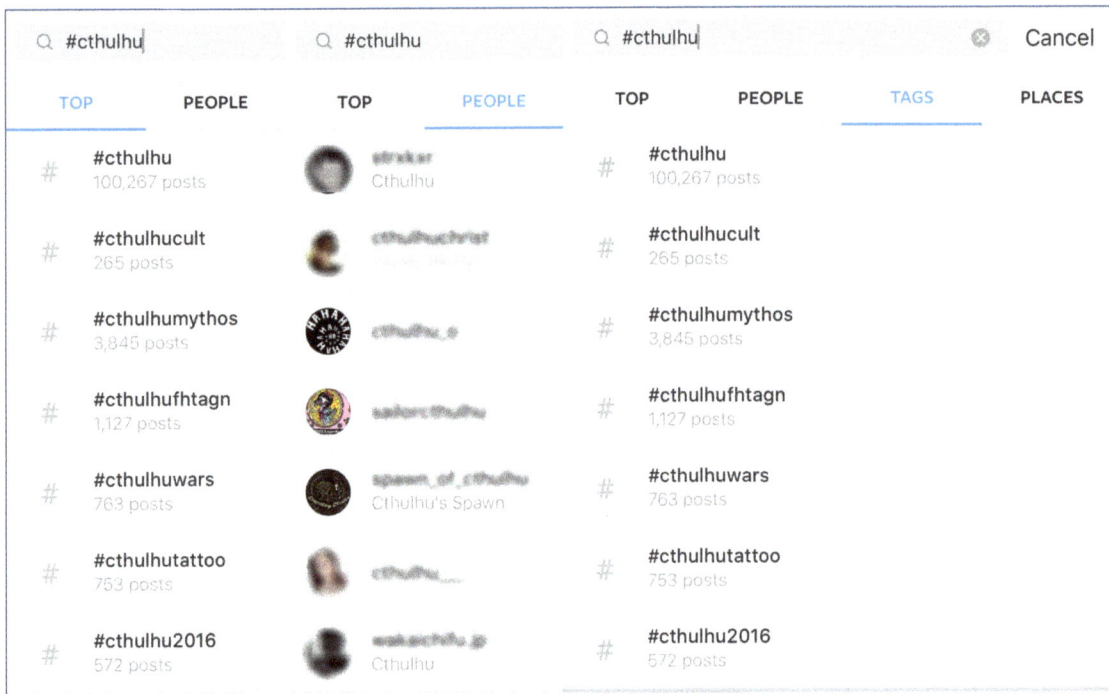

Figure 8.4: Instagram search results with keyword #Cthulhu

The aim is to find Instagram users who are passionate about Cthulhu, the monster. To increase the chances of discovering them, one can browse through the hashtags discovered for the ones that express this devotion. Three hashtags seem to fit this criteria: #Cthulhucult, #cthulhutattoo and #cthulhuforpresident. Clicking on each of these will reveal "Top" and the most recent Instagram content containing these hashtags (Figure 8.5). Clicking on any of these posts will reveal profiles of users who shared them (not shown). If you are looking to connect with Instagram users who fit the profile of Paul Draper, you would probably follow these accounts, and assign them to a feed in Iconosquare.

Status updates and terminology

Instagram allows uploading photos and videos with a maximum length of 60 seconds. Users can also create a video from multiple photo clips on their camera rolls. Once a piece of media has been uploaded into Instagram, a user can edit it and change its appearance by using filters. Figure 8.6 shows the steps involved in creating an image post on Instagram.

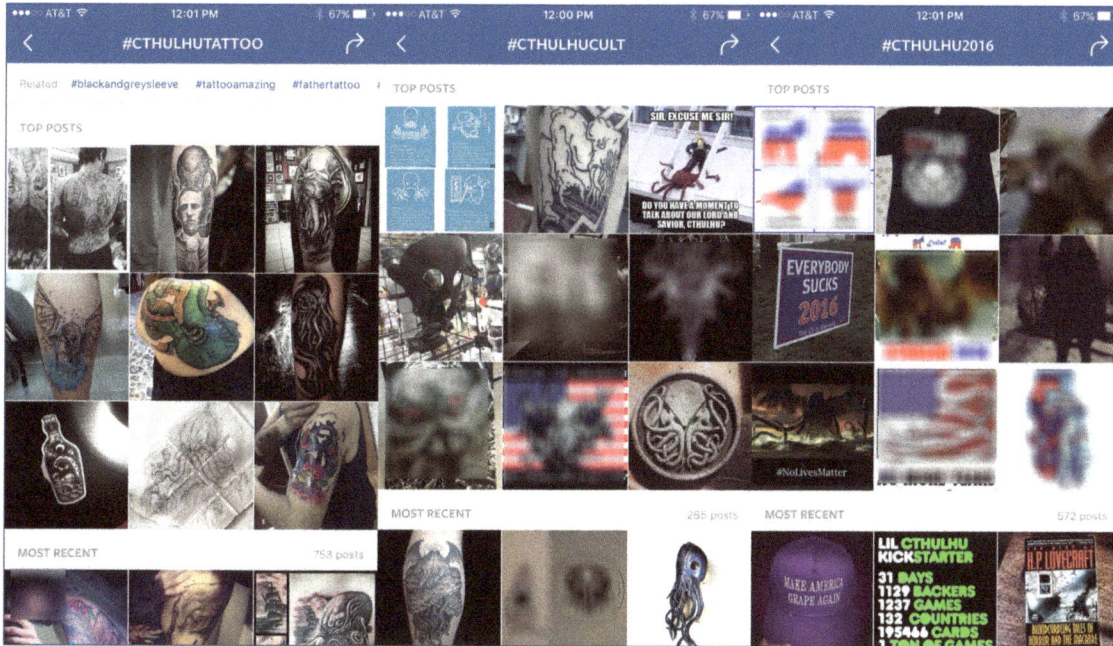

Figure 8.5: Exploring #Cthulhucult and #CthulhuTattoo and #chtulhu2016 search results

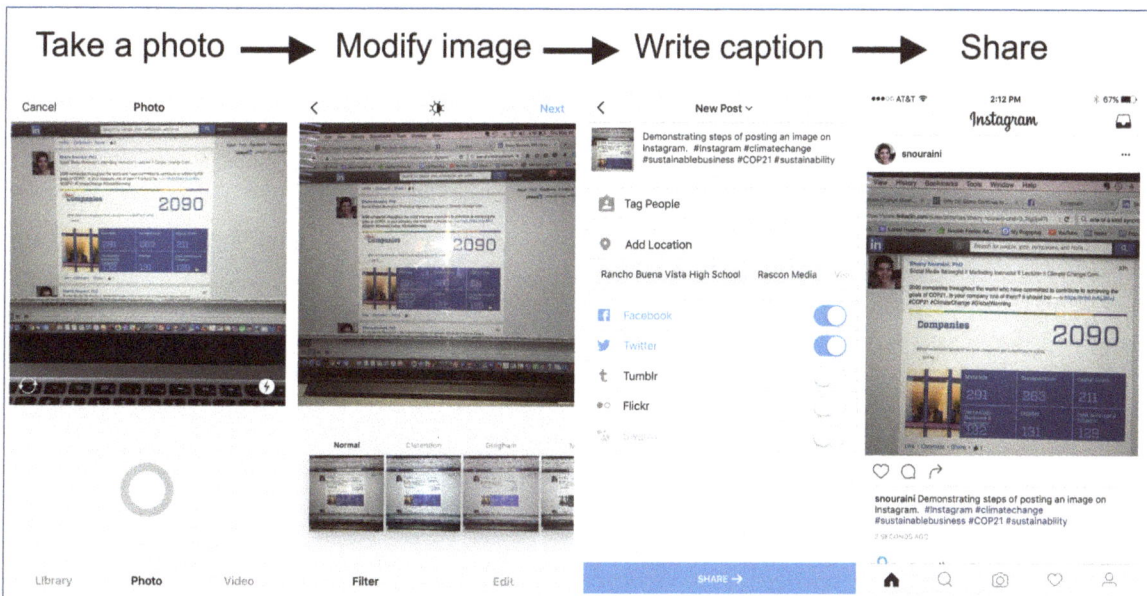

Figure 8.6: Steps in posting an image on Instagram

When writing captions for your posts, don't forget to include as many relevant hashtags as you can (there is a limit of thirty hashtags per caption and comment).

Until recently, status updates on Instagram appeared in a chronological basis. However, Instagram has started using an algorithm, similar to what Facebook uses, to show users more of the content that is most relevant to *them*. At the time of writing this book, inclusion of an algorithm is a brand-new development on Instagram, so there is not sufficient information as to the criteria used in deciding which content that users see and when. Suffice it to say that if you create valuable relevant content, your target audience and followers will likely see them.

Instagram Stories

In an effort to help users create more spontaneous and engaging content, Instagram recently introduced a new feature called Stories. Through this feature, Instagram allows users to post photos and/or videos that disappear after 24 hours. Stories can be produced either by choosing recent photos or videos from the phone camera roll, or by taking new pictures or videos as one normally does on Instagram. Users also have the option of adding text, colorful doodles, as well as emojies (via the phone's keyboard) to their media in order to further enhance their message.

When a user publishes a new story, a colorful ring appears around his/her profile photo. Stories can only be viewed by clicking on the user's profile photo; however, users have the option of featuring photos or videos from their story on their profile in an effort to entice their followers to view the complete story.

Given that story telling is a time honored and powerful method of communication, Stories are a positive added feature that will allow you to connect with your audience via creative story telling. A Story can be a combination of photos and videos, and your story becomes longer as you add new media to it. Just like you would for videos, follow the SUCCES rules of creating engaging content, keep your stories short, and make the beginning of your Story super engaging to entice your viewers to watch to the end.

The steps to creating a story in iOS are described below:

1. Visit the Instagram home page and click on the plus sign on the top left (red arrow in Figure 8.7) or swipe to the right. Notice also the row of profile photos on the top of the screen when you visit the home page, these are users you follow who have published a story in the past 24 hours.

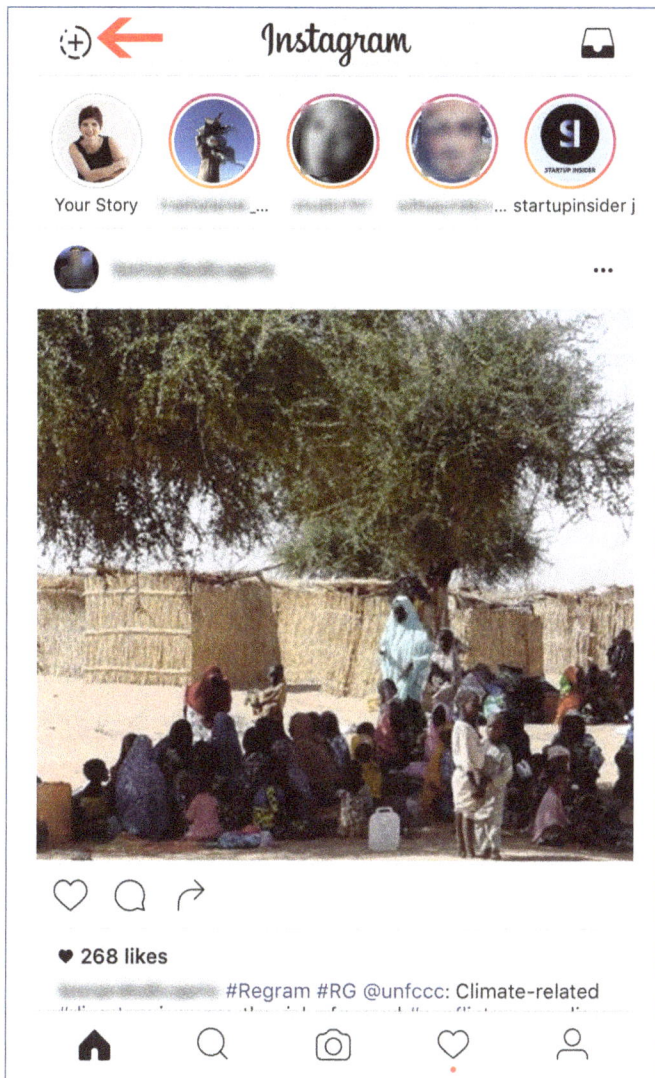

Figure 8.7: How to start creating an Instagram Story

2. To create a story from media in your camera roll swipe down to view and choose pictures or videos you want to add to your story. To create a story from brand new media just take a picture or record a video (not shown)

3. View the top right corner of your photo or video and you will notice icons for either adding doodles (pen symbol) or text and emoji (Letter A symbol). Click on any of these buttons to choose an option. Figure 8.8 shows a story being made from a photo I captured from my desktop modified with a pink pen (doodle) and Figure 8.9 shows addition of emojies from the keyboard of my phone.

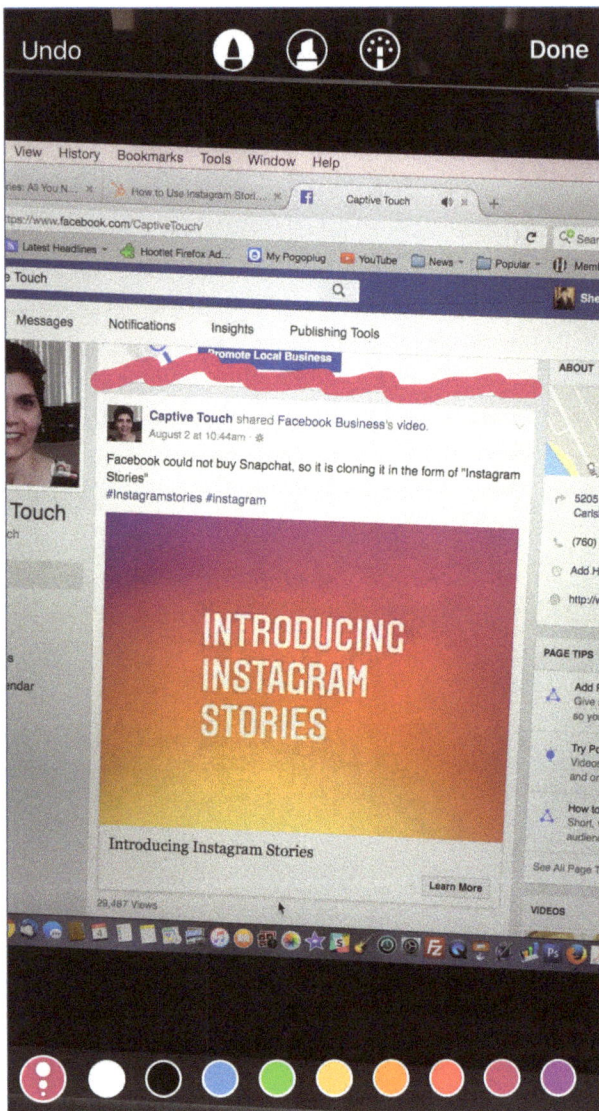

Figure 8.8: Adding doodles to an image

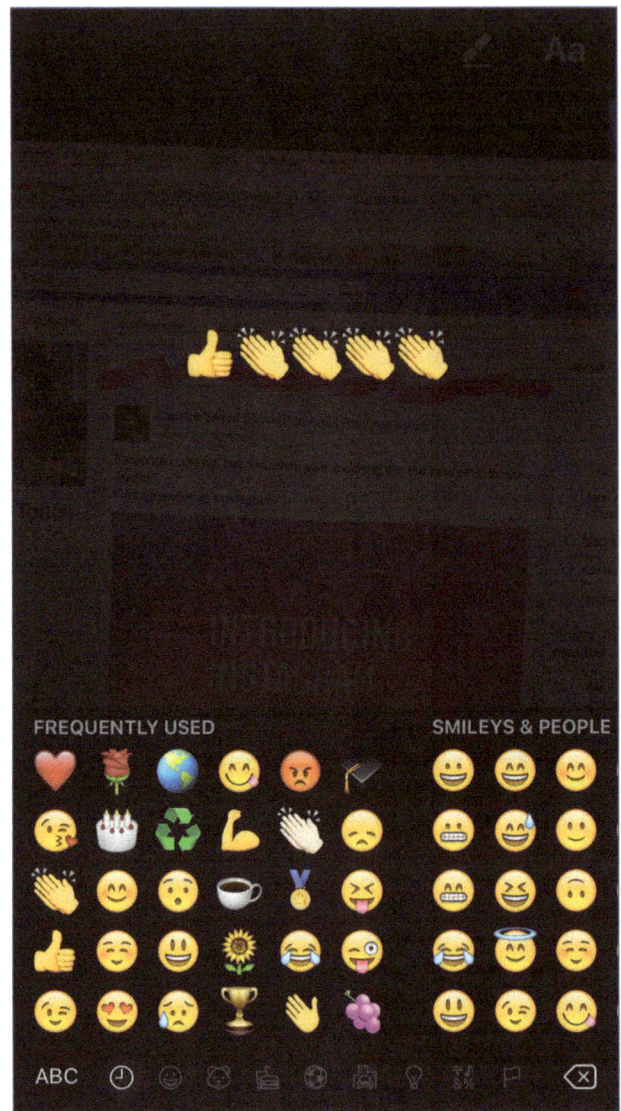

Figure 8.9: Adding emojies to an image

4. Once you are done modifying your media, click the checkbox on the bottom center of the screen to create a story or add media to a story already created (Figure 8.10).

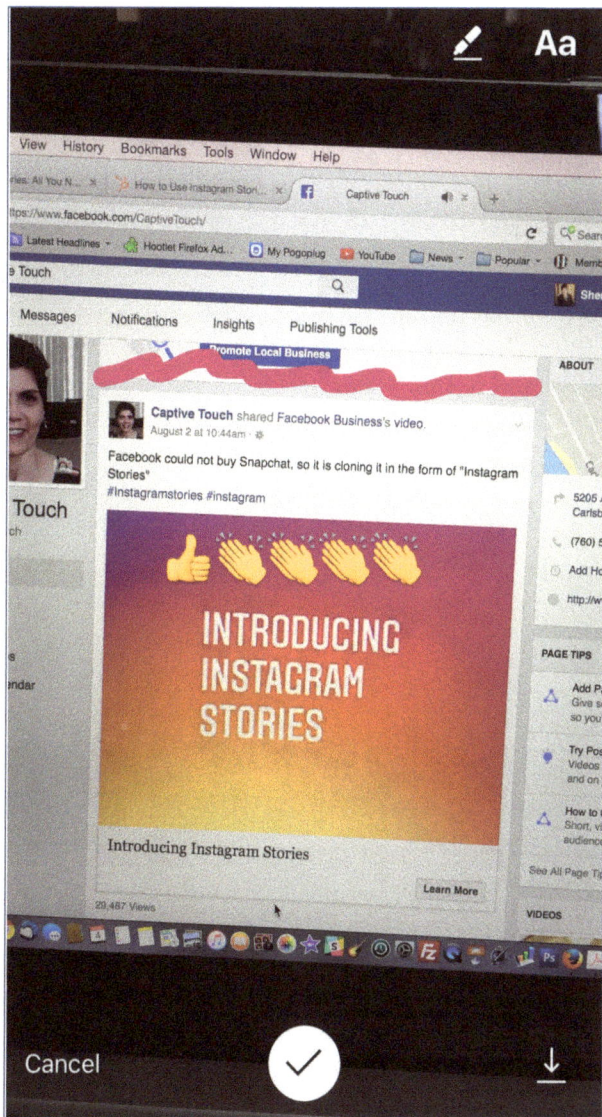

Figure 8.10: Modified media before creation of a story

5. The arrow on the bottom right corner of the screen (See Figure 8.10) will allow you to delete the media used to create a story, save it as a photo on you phone, or share it as a post on your Instagram profile (not shown). You can get to Story Setting from the same menu, which will allow you to exclude specific users from viewing your story, and control who can send you a private message about the story you just published. You can allow everyone to message you, limit messaging to people you follow on Instagram or just turn the messaging option off (not shown).

Another great feature of Instagram Stories is that you can see who has viewed your story. To identify your viewers, visit your story and swipe up on the screen. This action will reveal the names and profile photos for Instagram users who have viewed your story (not shown).

Community

There is really no "Community" feature on Instagram; rather users organically form communities around particular topics. Although, Instagram accounts with a large, engaged following could be considered a community, with the owner of the account being an influencer. So depending on your communication strategy on Instagram (see Figure 4.27), you can either connect with individual users who tend to use a particular hashtag (for example, #Cthulhuforpresident), or connect with and approach an Instagram influencer in a particular niche.

Influencers tend to have a large, engaged community, which leads to their content being featured as a "Top post" upon searching Instagram with a keyword. For example, searching for #Cthulhuforpresident reveals a number of "Top posts" associated with accounts that have a large following (not shown).

Analytics

Unlike other social networking sites we have reviewed so far, Instagram does not offer an analytics feature (although this feature will be available for the new business profiles). However, you can view the number of "likes" and comments each of your Instagram uploads are receiving by browsing your profile. Instagram also shows the number of views for each of your uploaded videos. If you click on the "like counter" for your video, Instagram will reveal the identities of those who liked your video and whether or not you are following them. This is valuable information, as it allows you to easily discover and connect with users who are interested in your content.

Other than the few insights provided by Instagram personal profiles, not much else is available in terms of analytics, which is why serious Instagram users resort to using third-party tools, such as Iconosquare. For just $28.00 per year (this price may change) you can get access to a host of useful data as a Pro account holder, some of which I feature below.

Iconosquare analytics in a Pro account comes in four sections:

1. *Overview* provides an overall summary of activities, engagement, following, and followers of an Instagram account.
2. *Community* provides further insights into the following and followers of an Instagram account.

3. *Content* analyzes trends in the activity of an Instagram account.

4. *Engagement* provides data about the level of engagement with the content posted by an account, and offer suggestions for improvements.

I am only going to review the "Community" and "Engagement" sections in Iconosquare, and offer ways in which this data can be used to improve your outreach efforts through Instagram.

Community

The "Community" section of Iconosquare analytics reports on follower growth over time, as well as a breakdown of the number of gained and lost followers for each day. The dates for which this data can be visualized are customizable (Figure 8.11).

Information about the location and size of the communities of followers of an Instagram account can be found in "Community Structure" (Figure 8.12). This information is helpful in understanding where in the world your content can potentially reach.

"Community Structure" also reveals the percentage of your followers who have a private account. The more private accounts that are following you, the less the chances are that your message will be spread far and wide.

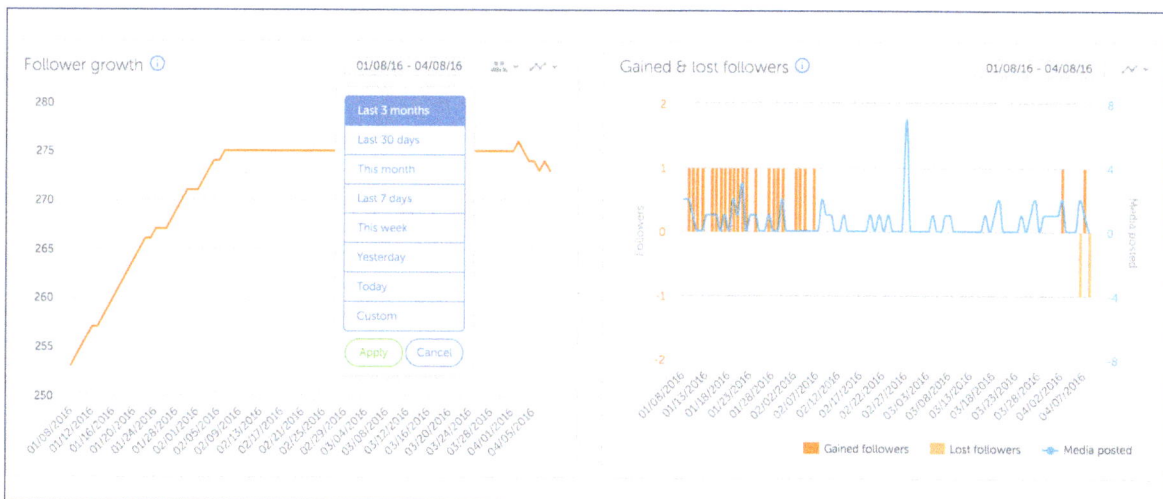

Figure 8.11: Iconosquare Community analytics—Follower growth, and Gained and lost followers

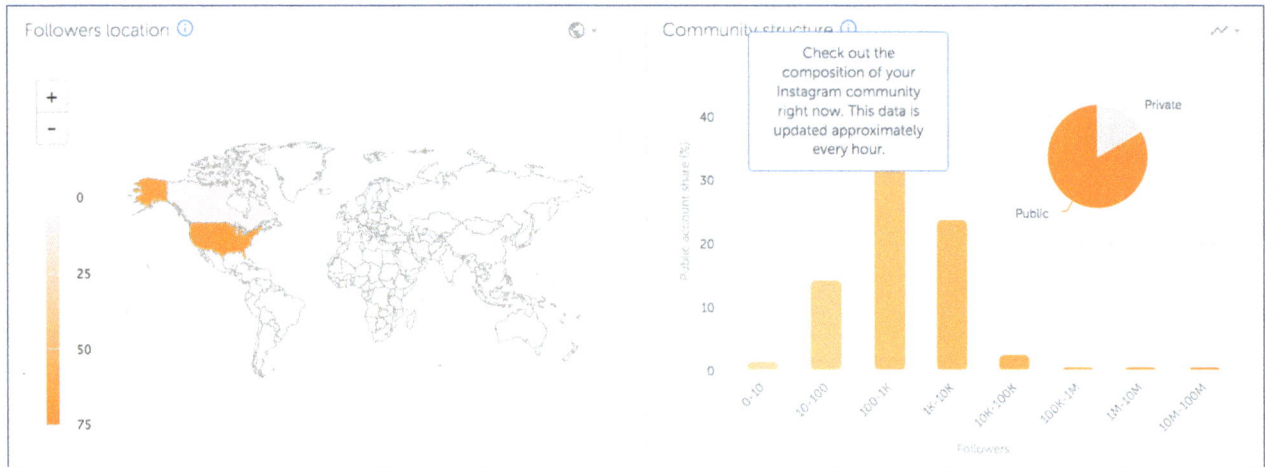

Figure 8.12: Iconosquare Community analytics—Community Structure

Community analytics also break down your community, based on the nature of your relationship: whether or not you are following them and if *they* are following you (not shown). You can also view a snapshot of your followers who are influential on Instagram (Figure 8.13). This information is useful in two respects:

1. You can stop following accounts that have lost interest or are not interested in your content, unless they are influencers with whom you want to build a relationship in the future.

2. You can explore the members of your audience whom you are not following. If they interact with your content, you should follow them back and build relationships.

3. Pay attention to your influential followers, as they can help spread your message far and wide. If they are following you, chances are they are interested in what you have to share.

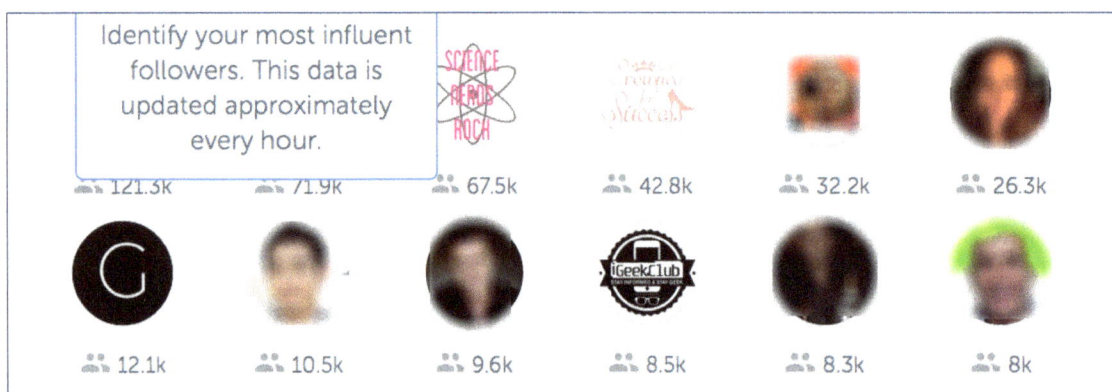

Figure 8.13: Iconosquare Community analytics—Influential followers

Engagement

Here you can find data to help optimize your activities on Instagram, as explained below:

1. Like and Comment History: chronological trends in the average number of likes and comments your media receives (Figure 8.14). Scan through the data and explore the media you posted on these days, and identify the ones that have more than average likes and comments. This can give clues as to what types of media resonate with your audience the most.

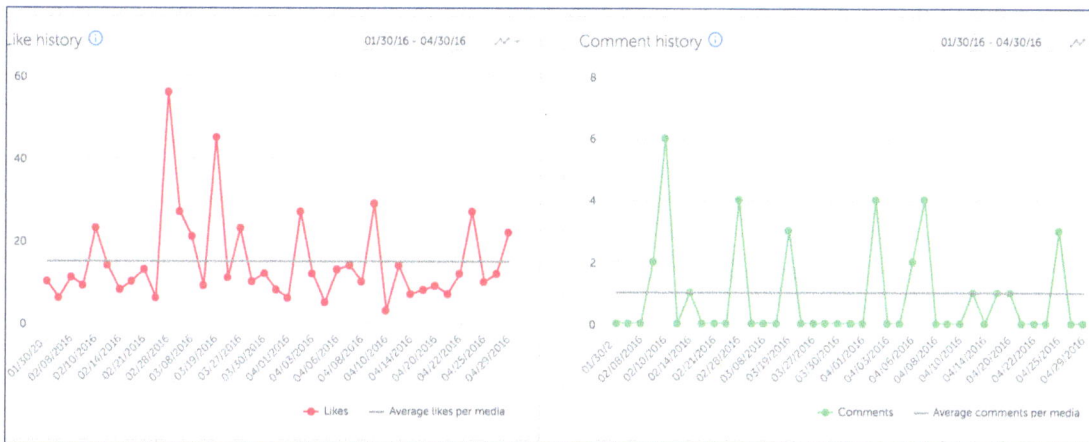

Figure 8.14: Iconosquare analytics—Like and Comment history

2. Best time to post: Iconosquare calculates the best time to post content for your account based on the time of day your content received high numbers of engagement (Figure 8.15). This data should help answer the often-asked inquiry about the best time to post.

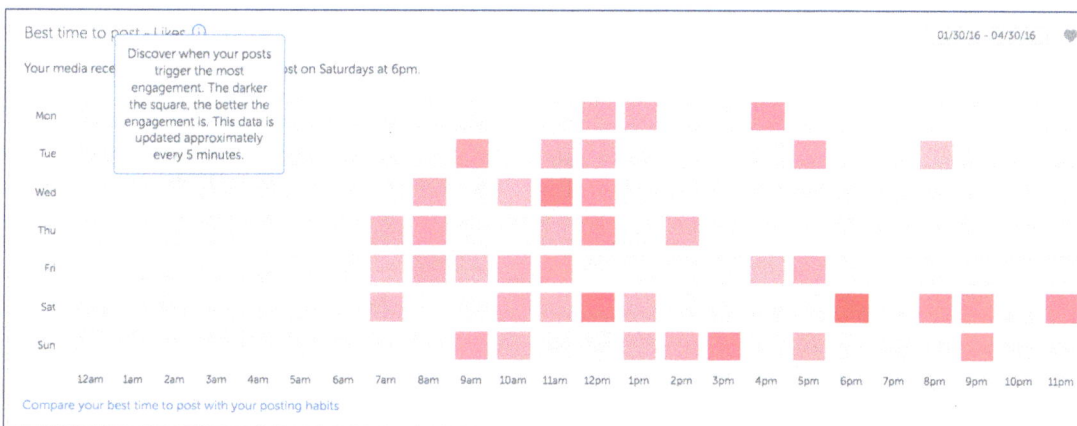

Figure 8.15: Iconosquare analytics—Best time to post

3. Most frequently used tags: Hashtags you tend to use the most (not shown). Use this data to learn about your own hashtag habits and compare to those hashtags that your target audience tends to use the most. There should be an alignment of frequency of use between the two.

4. Love rate: Number of likes received divided by total number of your followers within a thirty-day period, multiplied by 100 (Figure 8.16).

5. Talk rate: Number of comments received divided by total number of your followers within a thirty-day period, multiplied by 100 (Figure 8.16).

6. Engagement rate: Total number of likes and comments received divided by total number of your followers within a thirty-day period, multiplied by 100 (Figure 8.16). The quantities mentioned in items 4 through 6 should always be in a steady or upward trend. If these numbers decrease, you should investigate ways you can reverse the trend.

7. Most liked, commented on, and engaged media: Media that has received the most likes, comments, or likes and comments combined (Figure 8.16). This data can be valuable in guiding the design of the media you post: do more of what works the best.

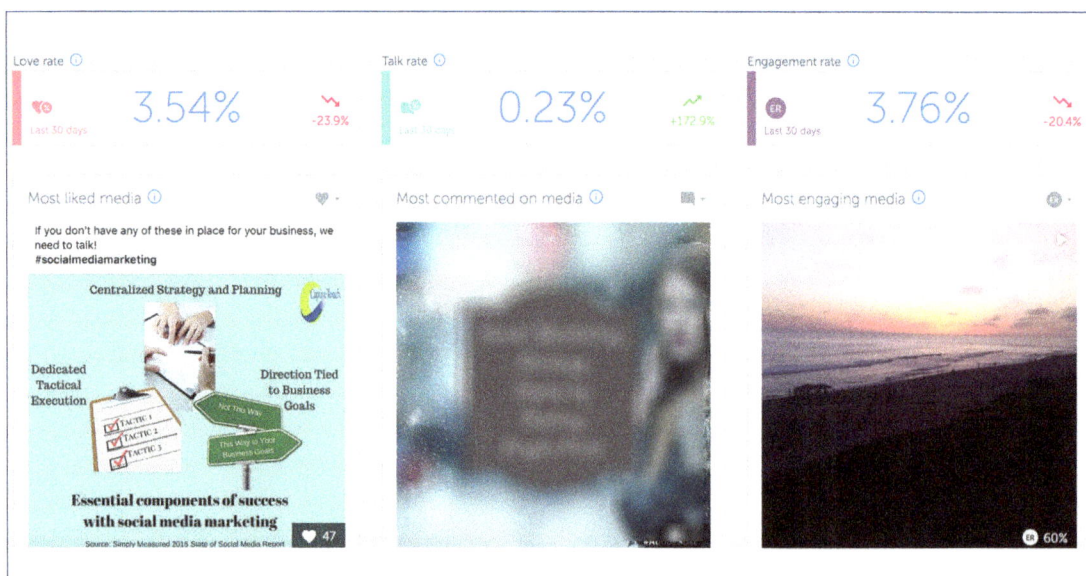

Figure 8.16: Iconosquare analytics—Love rate; Talk rate; Engagement rate; Most liked, Most commented on, and Most engaged media

Outreach on Instagram

The workhorse of engagement on Instagram is primarily the hashtag, just as it is on Twitter. Finding the most effective hashtags on Instagram is key to success, and it is the means by which communities are built on this platform, just like it is on Twitter. For this reason, your workflow for outreach through Instagram is pretty much identical to that of Twitter. Therefore, if you follow the Twitter workflow chart in Figure 5.24, you would be on the right track. However, I want to add some additional information below, which pertains more specifically to Instagram.

Engagement on Instagram is more effective if you provide value and are creative with your content. Here are some tools that can help you create more engaging content right on your phone:

I. Hyperlapse: a free mobile app by Instagram that allows you to create and post Hyperlapse videos.

II. Layout: a free mobile app by Instagram that allows you to create and post photo collages.

III. SparkPost: a free mobile app by Adobe that helps create and post artistic and animated multimedia. This is a great app for the artistically challenged people like me.

IV. Instatag: a free mobile app used for hashtag research, and creating tag lists that are frequently used.

V. Ripl: a free mobile app with a paid version that allows creation of artistic images and animated posts. The paid version ($10.00 per month) has an expanded creative suite that can be white-labeled for your brand, and also provides engagement analytics. I love this app because it provides animations that are visually optimized, removing anxiety about my lack of formal training in graphic design. At the same time, even though the design elements are already set, Ripl allows for customizations of text and graphics so I can get my content to look exactly the way I want it. Additional benefits include sharing my content to other social networking sites as well as Instagram, and viewing engagement analytics for my content on all these platforms. You can also schedule posting to Instagram of the content you create on the app.

Since hashtags are the main driving force of discovery on Instagram, be sure to include as many relevant and trending hashtags as possible in your image captions. To increase the total number of relevant hashtags included in a post, some power Instagram users also include them in a comment as soon as they publish an update. They do this because a hashtag research on Instagram discovers them even if they are included in a comment. There are at least two ways you can find relevant and trending hashtags:

1. Perform a search on Instagram with the main hashtag you are planning to use. Instagram will show you Top posts (largest amount of engagement), the most recent posts that use your hashtag, and a number of trending and relevant hashtags (Figure 8.17). You can get ideas for additional hashtags to include in your post either from those suggested by Instagram, or browse Top posts and see what additional hashtags *they* are using. Once you post your content, be sure to do a search again and see if your post is ranking in the search results. Your goal is to always try and rank in the search results.

2. Use an app called HASHME and do a search with your main hashtag. The app will show you a series of hashtags based on relevance and popularity, and will allow you to copy the top thirty tags (maximum allowed by Instagram on a post), or select individual ones so you can post them onto an Instagram post you are crafting (Figure 8.18).

Finally, remember that Instagram is a visual medium, so include as much as your message in the image or video you are posting. Do not rely too much on the caption of your image for communicating a point, as my anecdotal observations indicate that people tend to use Instagram on the go and will not invest too much time reading captions. This is an especially important point if you are trying to catch the attention of an audience who is disengaged or uninterested in climate change.

Let's look at an example of an Instagram post I put together in Ripl, which is targeted to Cthulhu enthusiast Paul Draper. As shown earlier, #Cthulhuforpresident is currently a popular topic of conversation in the community of Cthulhu enthusiasts on Instagram. A member of this community who is a climate change activist can raise awareness about sea level rise (or other consequences of climate change) by crafting a post similar to Figure 8.19. This post not only communicates enthusiasm about Cthulhu, but also incorporates a climate change message with a humorous twist. In addition, the caption of the image poses a question to entice engagement and includes relevant hashtags discovered via the HASHME App using #Chtulhuforpresident as a seed.

Figure 8.17: Search results for #Cthulhu on Instagram

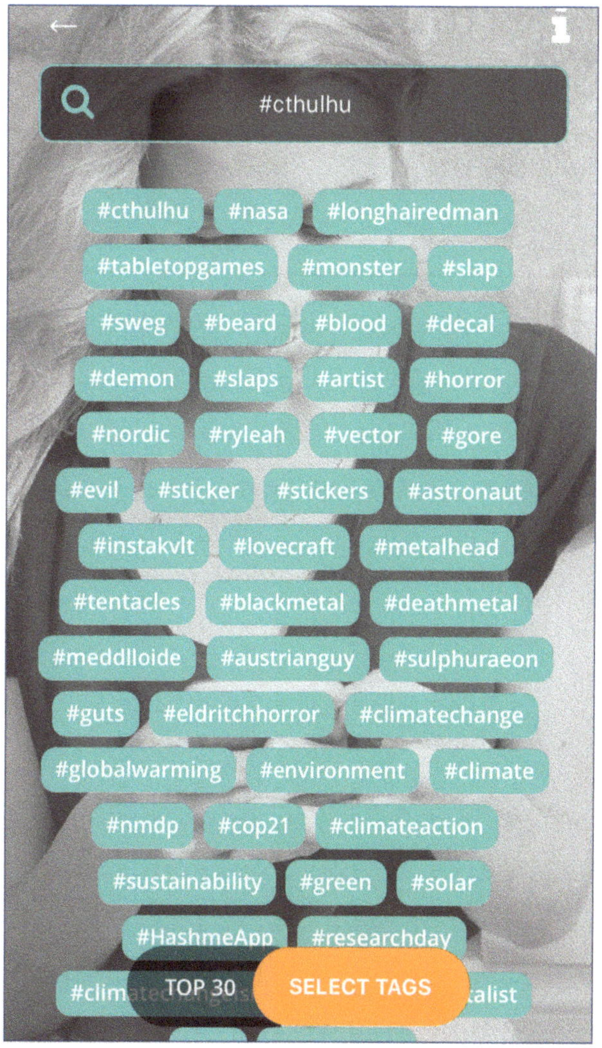

Figure 8.18: Search results for #Cthulhu on HASHME

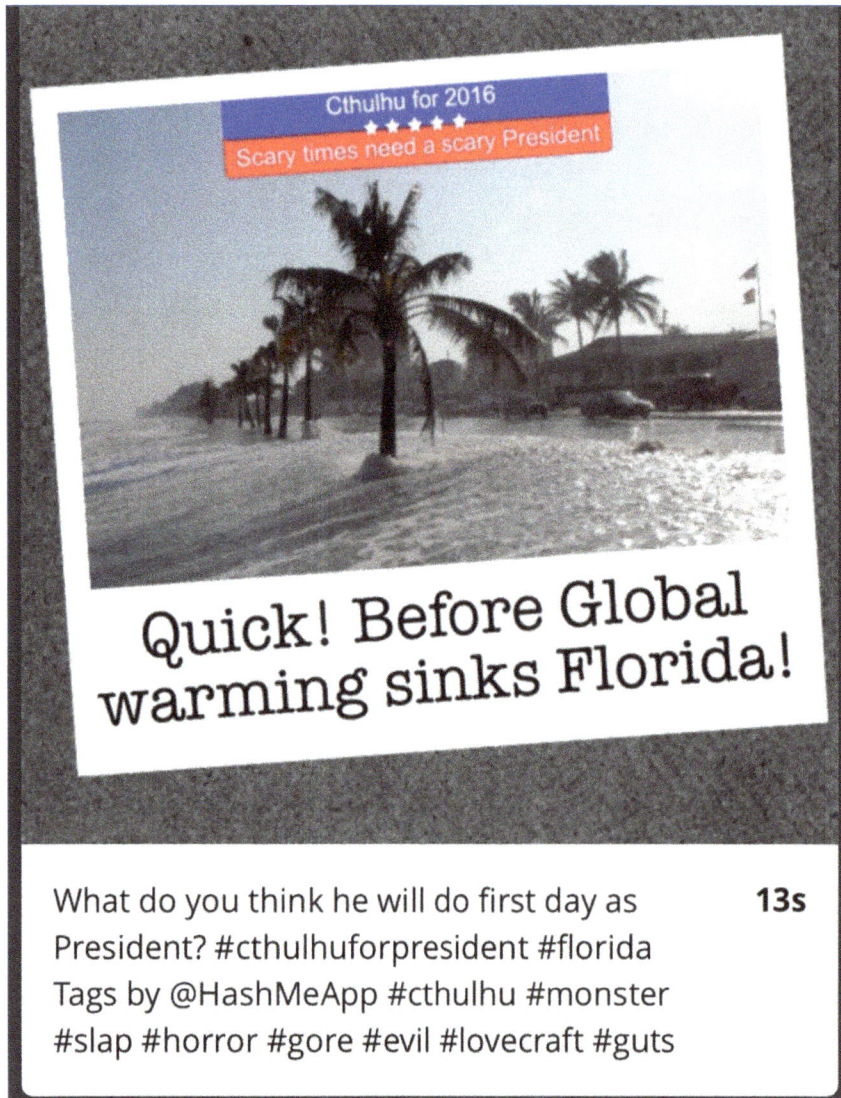

Figure 8.19: Example of an Instagram post targeted to Cthulhu enthusiasts;
Image of Florida Credits: Dave/Flickr Creative Commons/CC BY 2.0

Practice what you learned

In this chapter you were introduced to Instagram and its various features. You also learned about Iconosquare, an Instagram management and analytics tool, and other mobile apps, which offer features that enhance your outreach efforts. Starting with the sample profile Paul Draper, you learned how to search for and connect with this niche audience based on their passion for Cthulhu.

Now, it is your turn to set up your own Instagram account, and use the other sample profiles, or your own audience, to put to practice what you've learned. Be sure to join our Facebook group (http://facebook.com/groups/social4climate) to show us what you've done and get answers to your questions.

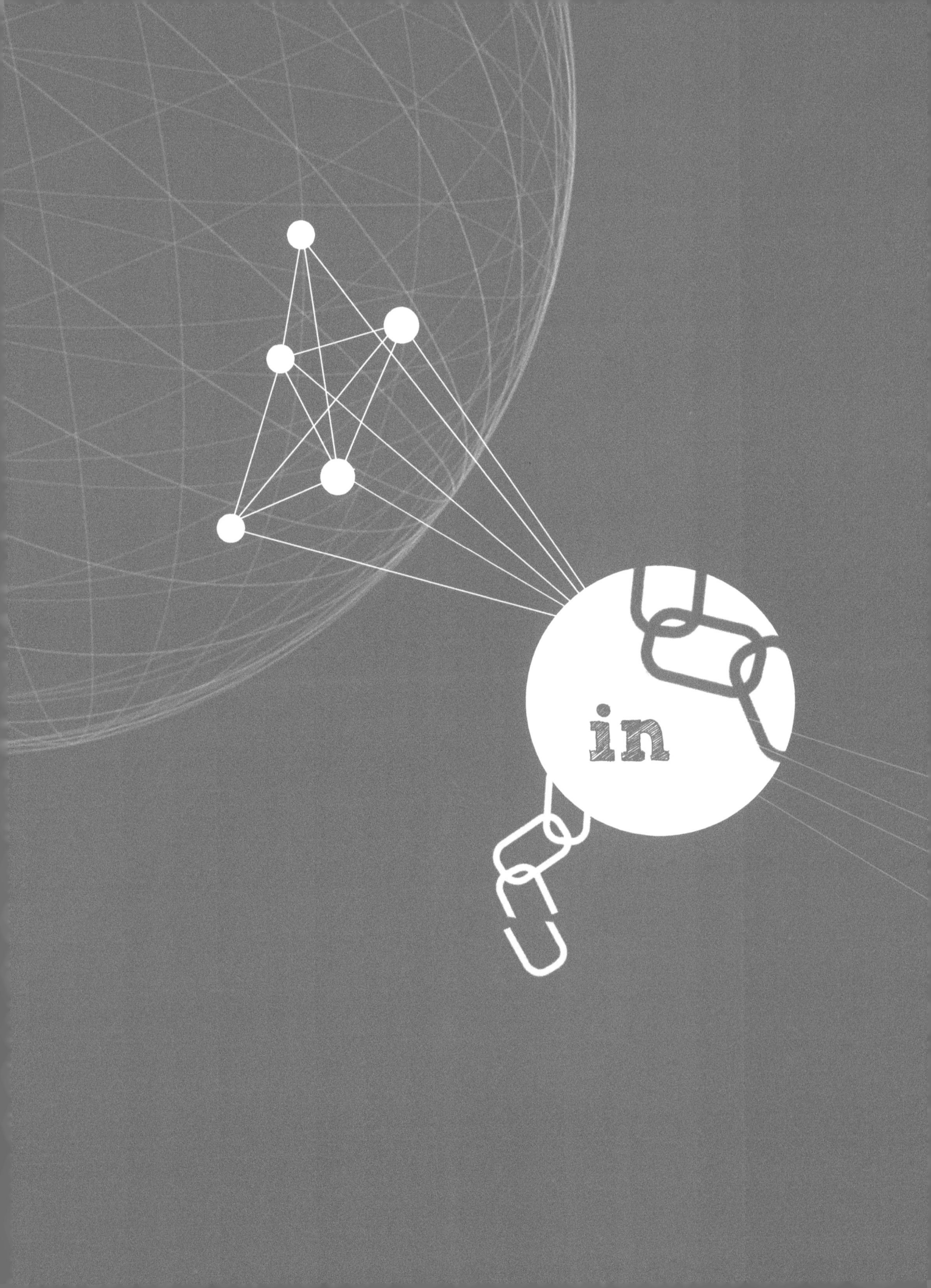

Chapter 9

LinkedIn

LinkedIn is a professional networking site and publishing platform with more than 400 million users worldwide (9.1). I don't know too many scientists who have a complete LinkedIn profile and who leverage this professional platform for networking and building a personal brand. This lack of presence on LinkedIn is simply a crime against climate change (and science) communication efforts, as the business and professional community can play a major role in building a sustainable future. A professional network, LinkedIn is well positioned as a medium for communication with business leaders about the dire effects of—and solutions for—unsustainable business practices. Let's review the different features of LinkedIn, which can help facilitate this communication.

Profile

LinkedIn offers both personal and company profiles. For the purposes of this book I will focus on the former, as the personal profile provides more opportunities for forming connections with individual business leaders. LinkedIn is also a great tool for building a personal brand by telling your professional story and communicating by publishing articles. A strong personal brand combined with the ability to communicate via publishing on a reputable platform enhances your credibility and allows your audience to develop trust in your expertise and points of view. Before diving into the different features of a LinkedIn personal profile, let's review important considerations about the visibility and privacy of your profile.

Controlling what the public sees on your profile

Once you create your LinkedIn profile, do a Google search with your name and you will see that one of the top search results will be a link to this profile. You can decide which sections of your profile are available to the public (those who are not your LinkedIn connections) by following these steps (Figure 9.1):

1. From the "Profile" top menu, choose "Edit Profile."
2. Hover your mouse over the little arrow next to the "View profile as" button and choose "Manage Public Profile Settings."
3. Look to the right and uncheck boxes for sections of your profile that you don't want to be visible to the public.
4, While you are there, also create a custom URL for your LinkedIn profile. This will make it easier to share a link to your LinkedIn profile as it is more easy to read and remember.

Important consideration for making changes to your profile

Landing a new job or earning a new certificate is something you may be excited to share with your connections, but when you want to make small changes to your profile, or keep some of the changes you make private, then you need to make use of the feature I am about to describe. Before editing your profile, take note of the "Notify your network" button, and choose either "Yes" or "No" for notifications. Choosing "Yes" will lead to showing an activity for you in your connections' news feeds, whereas a "No" will hide these activities. Avoid cluttering your connections' news feeds with unnecessary activity updates by choosing "No." You can find this button on the right-hand side of the "Edit Profile" page (Figure 9.2).

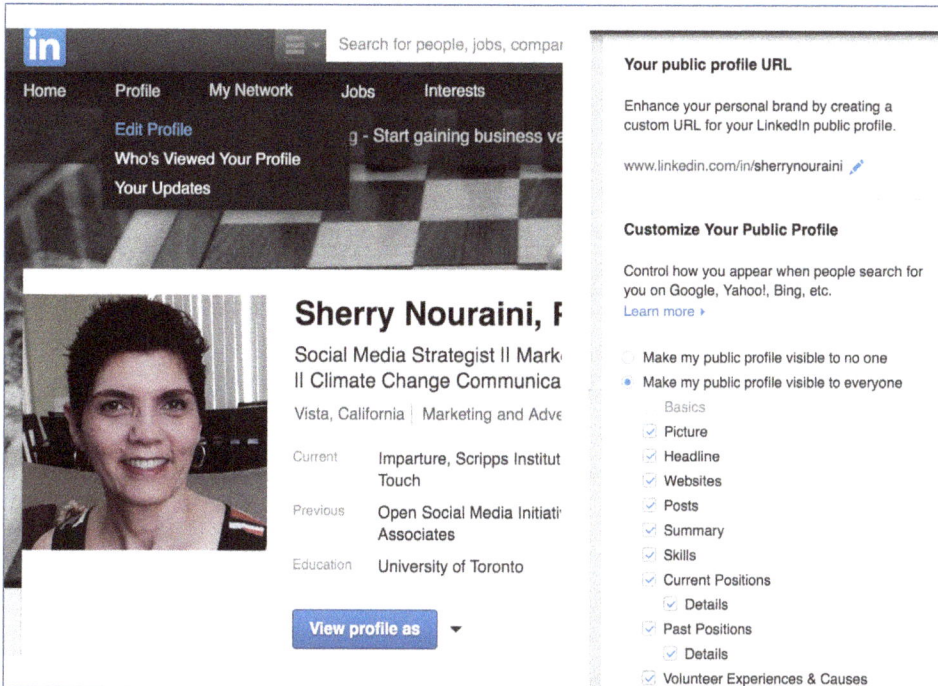

Figure 9.1: LinkedIn profile—controlling what the public sees

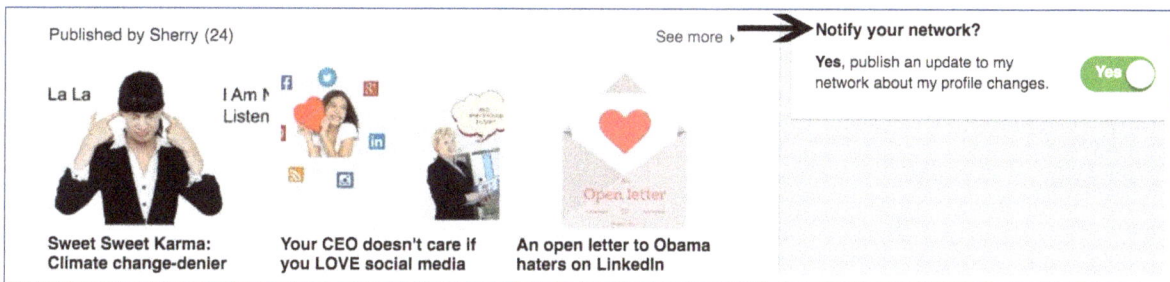

Figure 9.2 LinkedIn profile—Network notification settings

LinkedIn personal profile sections

A LinkedIn personal profile contains a snapshot of a user's identity represented by a profile photo as well as the following, most of which are self-explanatory (as labeled in Figure 9.3):

1. Name
2. Headline
3. Location
4. Industry
5. Current and past job titles
6. Education
7. Contact info
8. Profile URL

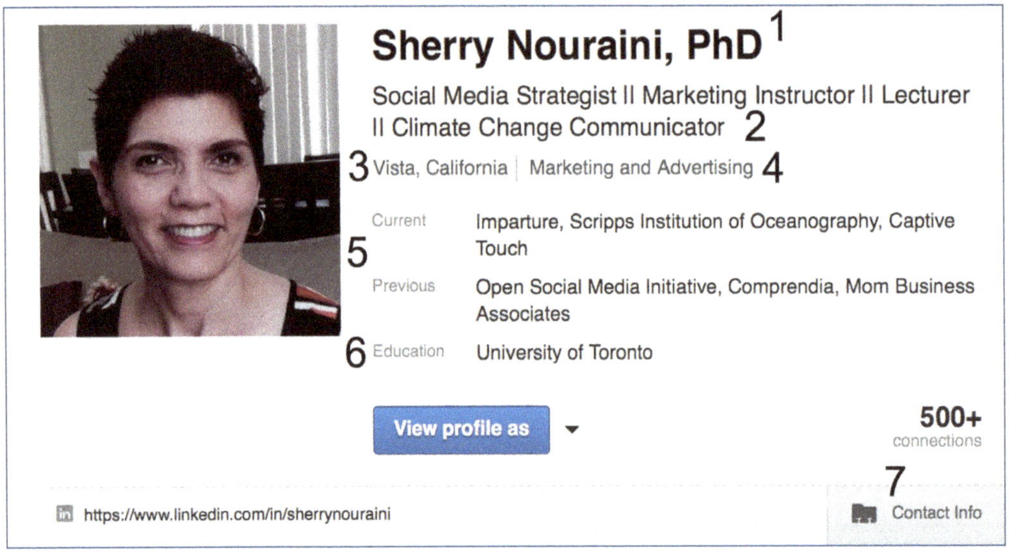

Figure 9.3: LinkedIn profile snapshot

When crafting your headline (labeled 2 in Figure 9.3), describe your professional identity and your specialty. When writing your headline, include keywords or phrases, stated in four words or less, and separate each key phrase with a symbol. The goal is to have each key phrase stand on its own and not connected to the next with commas, periods, or other connectors. Note that there are character limits to the various sections of a LinkedIn profile. For a summary of the latest information about these limitations, please review the link provided in Reference 9.2.

To complete your contact information, click on the "Contact Info" icon to view the fields for entering your data (Figure 9.4). For the "Websites" section, a drop-down list provides a number of options for any links you may want to include. If you do not want to use any of the options, just use "other," then enter a URL and the associated description.

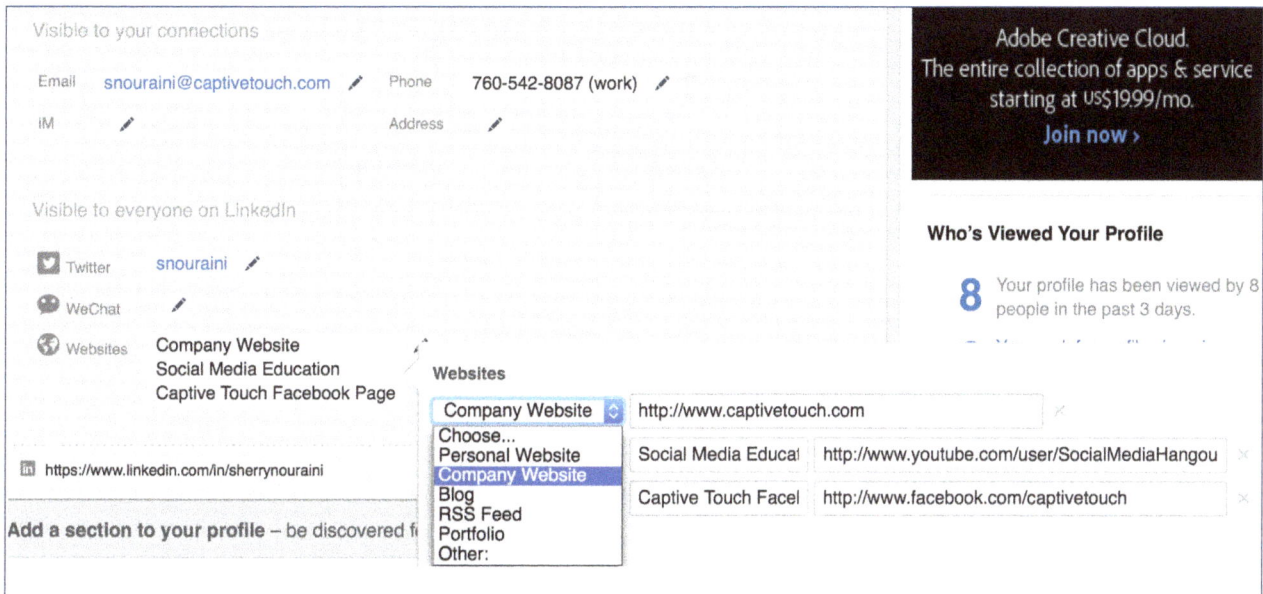

Figure 9.4: LinkedIn profile—Contact info section

In addition to the profile snapshot mentioned above, LinkedIn offers the following features for users to showcase their accomplishments, expertise, and interests:

1. Summary. You can think of this as an equivalent of the Abstract section of a grant application or publication, or an executive summary. Here, you have a maximum of 2,000 characters to provide a summary of your accomplishments, experiences, and strengths (Figure 9.5). In addition to text, a LinkedIn Summary can include multimedia or files. The ability to embed media offers a great opportunity to include a video introduction or samples of slide presentations to give viewers of your profile a more detailed sense of your expertise.

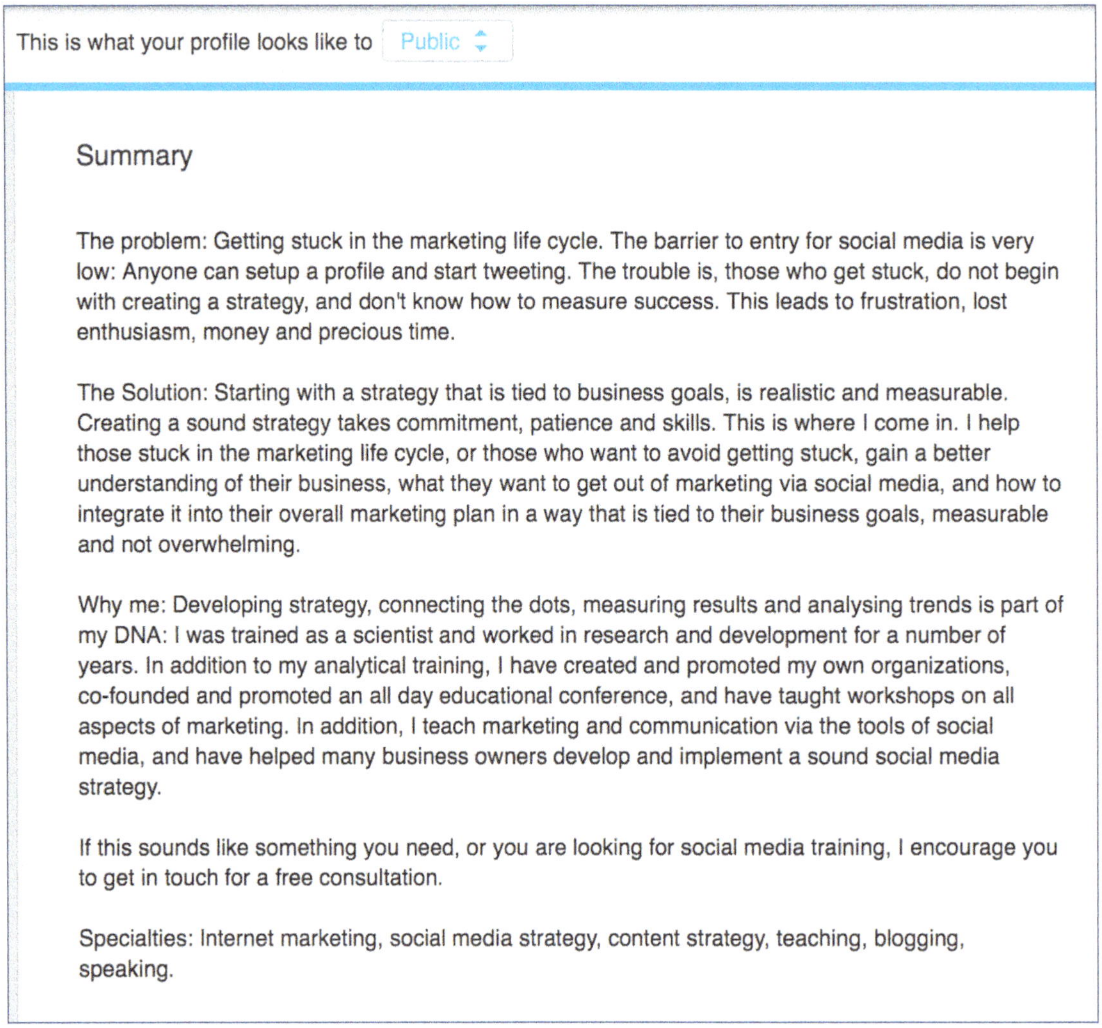

This is what your profile looks like to Public ⬍

Summary

The problem: Getting stuck in the marketing life cycle. The barrier to entry for social media is very low: Anyone can setup a profile and start tweeting. The trouble is, those who get stuck, do not begin with creating a strategy, and don't know how to measure success. This leads to frustration, lost enthusiasm, money and precious time.

The Solution: Starting with a strategy that is tied to business goals, is realistic and measurable. Creating a sound strategy takes commitment, patience and skills. This is where I come in. I help those stuck in the marketing life cycle, or those who want to avoid getting stuck, gain a better understanding of their business, what they want to get out of marketing via social media, and how to integrate it into their overall marketing plan in a way that is tied to their business goals, measurable and not overwhelming.

Why me: Developing strategy, connecting the dots, measuring results and analysing trends is part of my DNA: I was trained as a scientist and worked in research and development for a number of years. In addition to my analytical training, I have created and promoted my own organizations, co-founded and promoted an all day educational conference, and have taught workshops on all aspects of marketing. In addition, I teach marketing and communication via the tools of social media, and have helped many business owners develop and implement a sound social media strategy.

If this sounds like something you need, or you are looking for social media training, I encourage you to get in touch for a free consultation.

Specialties: Internet marketing, social media strategy, content strategy, teaching, blogging, speaking.

Figure 9.5: LinkedIn personal profile Summary

2. Posts. This is the LinkedIn publishing platform where users can publish original blog articles. I will further cover posts in a later section of this chapter.

3. Skills and Endorsements. Create a list of your skills in this section. This will allow your connections who are familiar with your work, to vouch for your skills (Figure 9.6).

Such is the intention behind the skills section; however, this feature is subject to abuse by some users who give and receive fake endorsements. Many genuine LinkedIn users dislike the Endorsement feature, but employ it nevertheless as a necessary evil. They balance this by finding comfort in the fact that they can remove fake endorsements from their profiles, and choose to only give genuine endorsements.

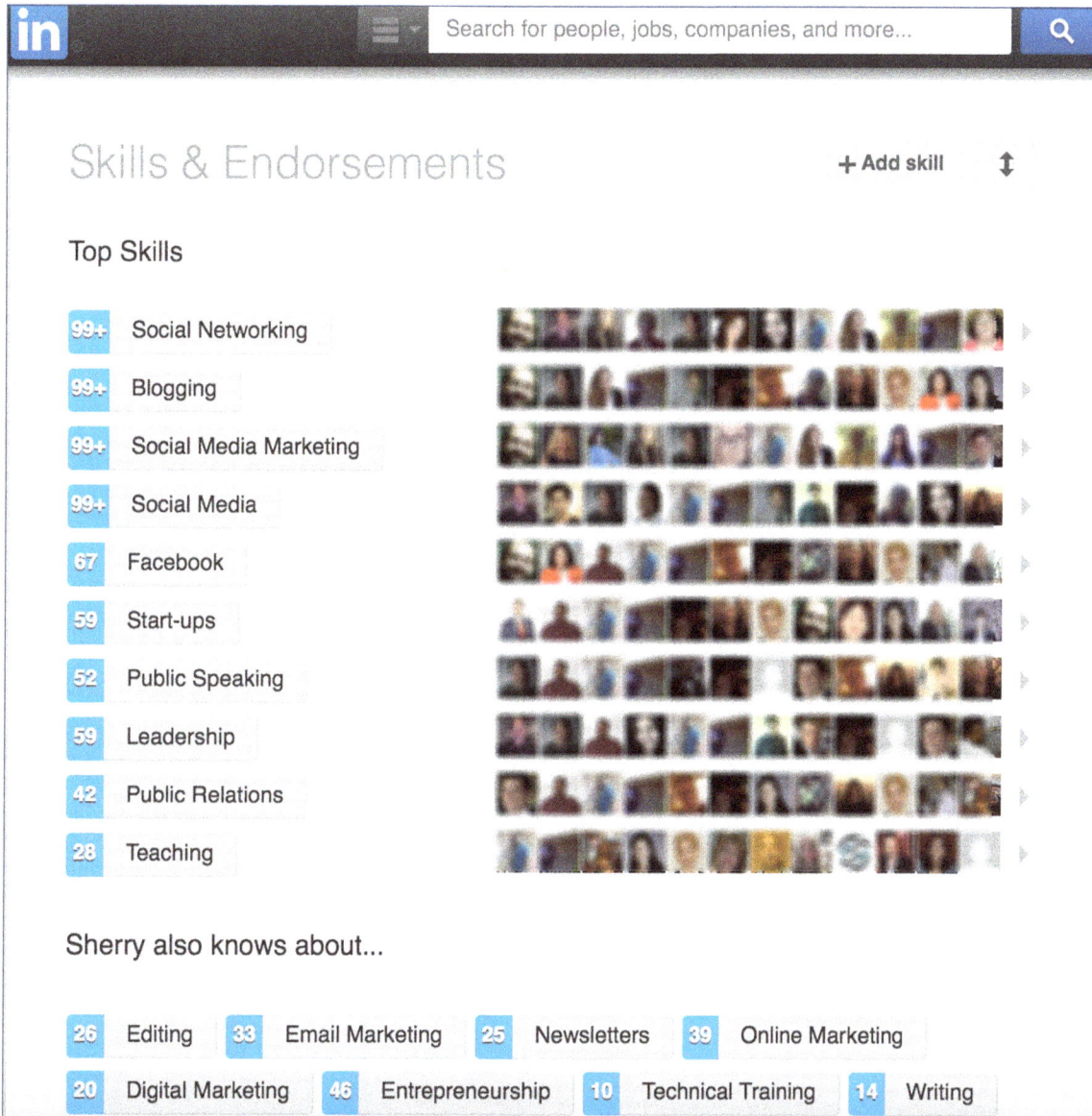

Figure 9.6: LinkedIn profile Endorsements

4. Experience. This is part of a LinkedIn profile that resembles a résumé, as it is a list of positions held along with associated accomplishments for each position. Note that in addition to text content, you can include all sorts of multimedia for each of the positions you've held in the past or currently hold (Figure 9.7).

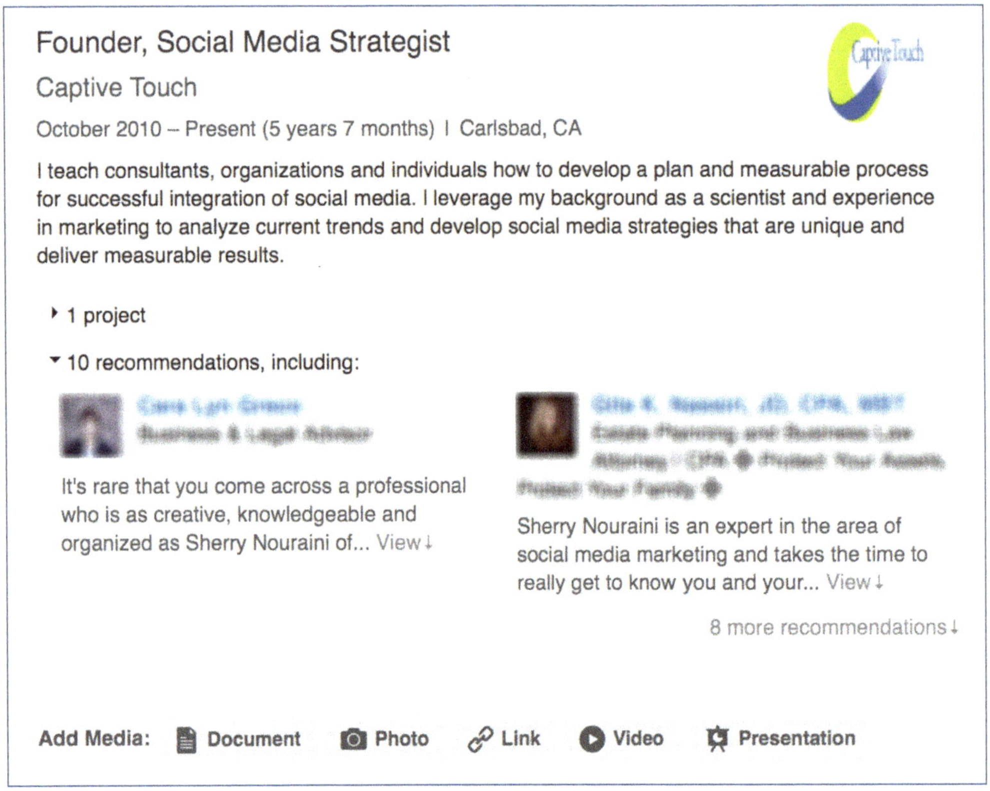

Figure 9.7: LinkedIn profile Experience

5. Recommendations. This is an important part of a LinkedIn profile, as it is a written recommendation from your connections. The fact that it requires someone connected to you to sit down and describe why you are worthy of a recommendation provides a lot more credibility than a one-click endorsement. Therefore, make an effort to connect with people who are most familiar with your work and actively seek recommendations. To ask for recommendations, find the menu to the right of "View profile as" (see the beginning of this chapter), choose "Ask to be recommended," and follow the steps to send one of your connections a message asking to be recommended. When you ask for recommendations, specify a particular "Experience" section of your profile. When a connection writes a recommendation for you, you will receive a message from LinkedIn containing the recommendation. You are given the choice to accept the recommendation and add it to your profile, ignore it, or ask the recommender for some editing. When a recommendation is accepted, it will appear in the associated "Experience" section (See bottom of Figure 9.7), in addition to a separate "Recommendations" section of your profile (not shown).

6. Publications: Enter a list of your publications (not shown).

7. Groups: Automatically generated by LinkedIn, this is a list of groups you have joined (not shown).

8. Following: Automatically generated by LinkedIn, this is a list of LinkedIn users you are following. Note that this is different from "connecting." When you connect with a user, you are automatically following their updates and can direct message them. However, when you "follow" a user, you can only see their activities in your LinkedIn news feed (not shown).

9. Projects: Displays a list of the projects with which you have been involved (not shown).

10. Certifications: List of any certifications you've obtained (not shown).

11. Honors and Awards: List your honors and awards, which would be a good place for student grants or fellowships or any other type of award (not shown).

12. Test Scores: Enter your test scores, especially if they're high (not shown).

13. Volunteer experience and causes: List any volunteer experience you've had, organizations you support, and causes about which you care. In this section, you can also list types of volunteer opportunities in which you want to participate. This will help nonprofits discover you on LinkedIn (not shown). In fact, there is a specific page on LinkedIn where such opportunities are posted (http://volunteer.linkedin.com).

14. Personal details: Enter your personal interests, birthday, and marital status. I normally discourage listing a birthday, as it is such a frequently used method to verify one's identity (not shown).

15. Patents: Enter any patents you've been granted, or those that are pending (not shown).

Note that there is no need to fill out all of these sections of a LinkedIn profile, only ones that are relevant to you and the goals you want to accomplish. However, the more complete a profile, the higher chances are of establishing a reputation and identity on LinkedIn.

Organization

The news feed on LinkedIn is a combination of status updates posted by your connections (or companies and users you follow), as well as their activities, which include people to whom they connected, articles they published, groups they joined, group discussions in which they participated, and comments or likes they gave to content on LinkedIn. The news feed also contains articles shared by LinkedIn Pulse, which is the publishing/content curation feature of LinkedIn (more on this later). Unlike Twitter or Facebook, there is no option to filter your news feed for updates posted by specific groups of your connections. However, you can choose to view the most recent activities or top activities, which are status updates with the most comments and likes. To toggle between these two choices, visit your home page on LinkedIn and hover your mouse on the three dots on the top right corner of the news feed. This will reveal menu items that will help you choose between "Top Updates" or "Recent Updates" (Figure 9.8).

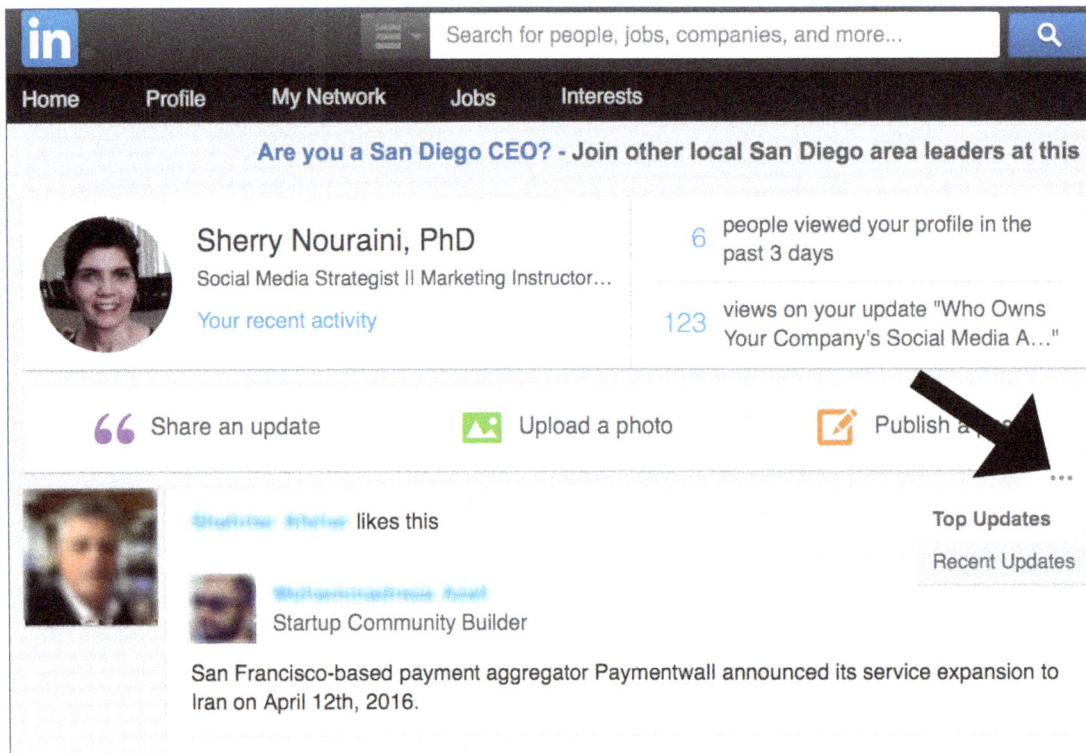

Figure 9.8: LinkedIn news feed—choosing Top versus Recent updates

Other features of LinkedIn include ability to create tags for each of your connections. Tags will help you sort through the list of your connections and find the ones that fall under a particular category—for example, friends, family, coworkers, etc. To create tags, visit the "Connections" section of LinkedIn to view a list of your contacts. Choose the "Tag" option on the lower left of each listing and tag the contact by using one of the premade options or by creating a new one (not shown). You can then use the search feature to filter your contacts based on a particular tag. For example, I have used "science" to tag and organize all of my connections who work in some capacity in the scientific field.

LinkedIn also allows you to create private notes for each of your contacts, and set reminders for contacting them. Private notes are only seen by you and not your contacts. Examples of private notes may include reminders of where you first met a contact or anything else to help you remember your history with them. To create notes and reminders, just visit the profile of your contacts. You will see a box underneath their profile photo in the "Relationship" section (Figure 9.9).

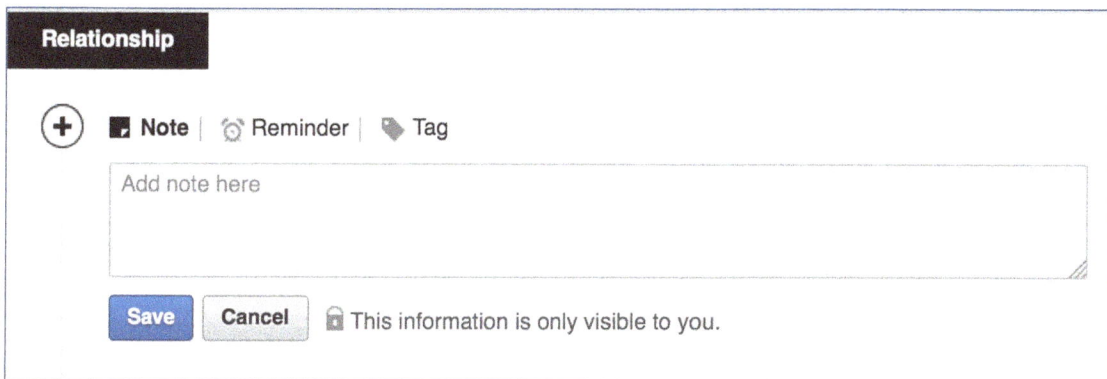

Figure 9.9: LinkedIn–Creating notes and reminders for your contacts

Search

LinkedIn search is a gold mine of information, and the size of the prize depends on whether or not you have a free or a paid account. I am going to focus on search features available with a free account, which will give you plenty of opportunities to find and connect with your target audience.

Basic LinkedIn search can be simple or advanced. Simple search is focused on keywords, and advanced search adds filters for obtaining more focused search results. LinkedIn allows users to search the entire database ("All" in Figure 9.10) or use seven different categories to narrow down the source of information (Figure 9.10):

1. People: LinkedIn users
2. Jobs: LinkedIn job listings
3. Companies: LinkedIn company profiles

4. Groups: LinkedIn groups

5. Universities: LinkedIn university pages

6. Posts: LinkedIn publishing platform

7. Inbox: LinkedIn messages

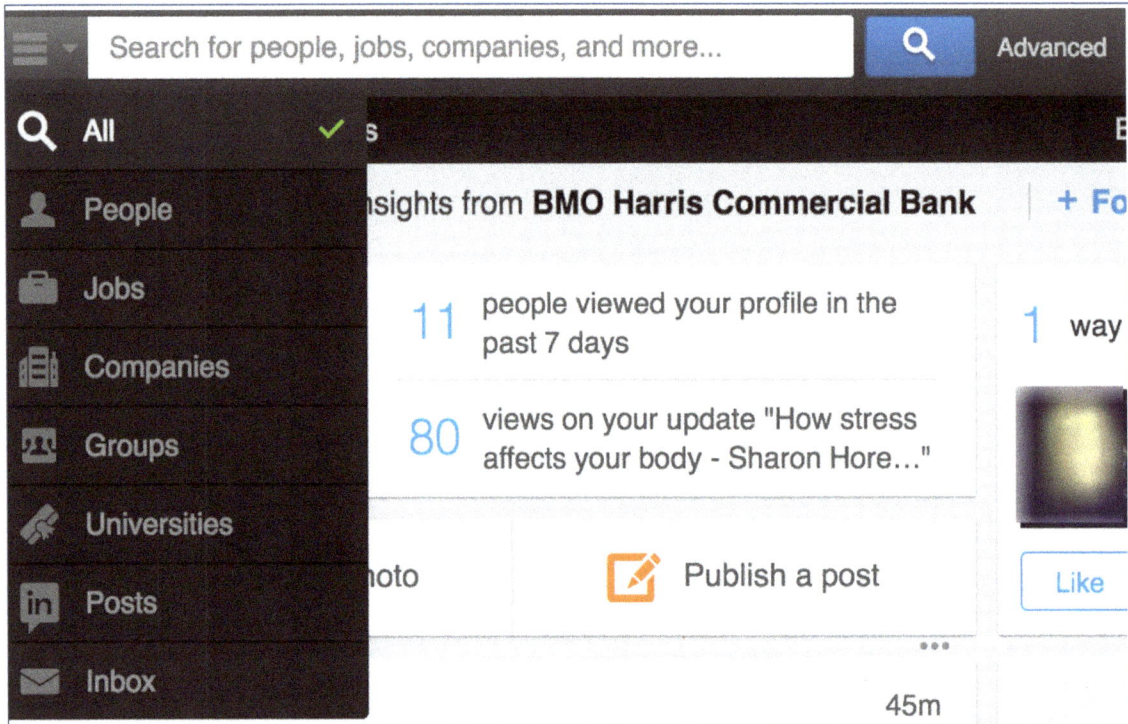

Figure 9.10: LinkedIn Search

Let's try an example search on LinkedIn to better understand the search feature. Since LinkedIn is a professional network, it seems like a good place to search for and find the Alex Donovan example profile, the CEO who is interested in running a sustainable business but is not sure how (see Chapter 2). Since we are looking for a particular profile, we'll start with the "People" category. Unlike search engines from other social networks—which allow finding anyone in their universe—"People search" on LinkedIn limits the results to profiles of users only in your first-, second-, and third-degree connections, as well as members with whom you share groups. With this in mind, our strategy for connecting with Alex would be something like what is depicted in Figure 9.11.

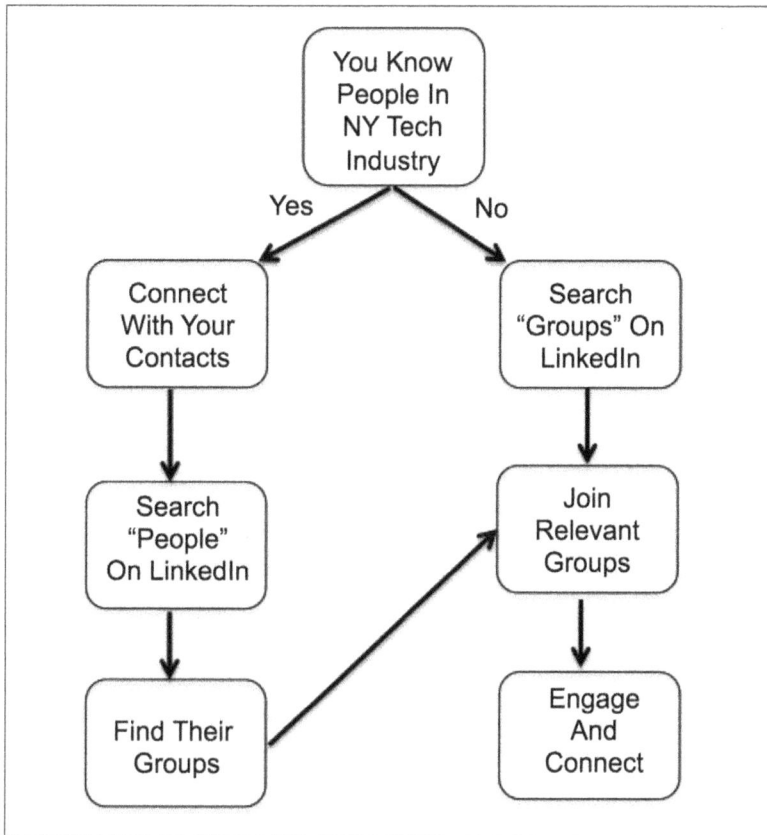

Figure 9.11: Searching for Alex Donovan—a strategic approach

I want to emphasize again the importance of connecting with your target audience based on shared values and interests. For example, if you aim to engage Alex Donovan in developing a sustainability plan for his business, the conversation has to start from a place of common ground. Therefore the strategy outlined in Figure 9.11 is best suited for—although not exclusive to—those already involved in the technology sector with knowledge or expertise in corporate sustainability, or those who have connections with technology business professionals who are advocates and practitioners of corporate sustainability. It will be a mistake to think that you can just connect to random CEOs and start lecturing them about climate change. This is a common mistake made by many sales and marketing professionals who send a sales pitch email as soon as a user accepts their invitation to connect. They *think* they are practicing "social selling," without truly understanding its meaning. You are learning something equivalent to social selling in this book, except you are not selling products or services, but communicating ideas!

Searching for LinkedIn profiles matching Alex

Let's assume you already have LinkedIn connections in New York. To find Alex, use LinkedIn advanced search to find people with the title CEO who are located in New York, working in the Technology industry (Figure 9.12). This search has close to 4,000 results (when searched using my profile), and LinkedIn shows how closely you are related to each profile, in addition to a link to any connections you may have in common (Figure 9.13). Once these users have been discovered, the hard work of narrowing down the search results to the ones who belong to groups that overlap your interests begins. One important reality of connecting on LinkedIn—or on any other social network—is that it takes time and effort to build meaningful relationships (This applies to real life, too, doesn't it?). Be patient, and make connections with people based on genuine overlap of values and interests. Here are some groups joined by LinkedIn users that we discovered by our example search: Executive suite; Software and Technology; Cloud Computing; Chief Information officers; and On Startups, the community for entrepreneurs.

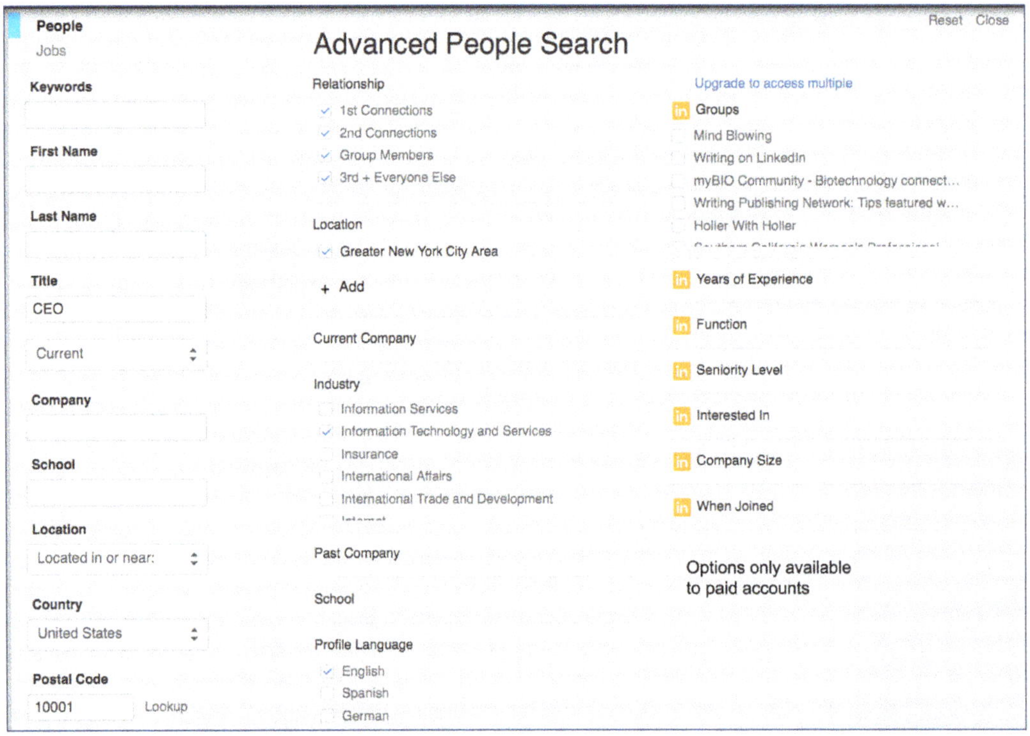

Figure 9.12: Using LinkedIn advanced search to find profiles matching Alex Donovan

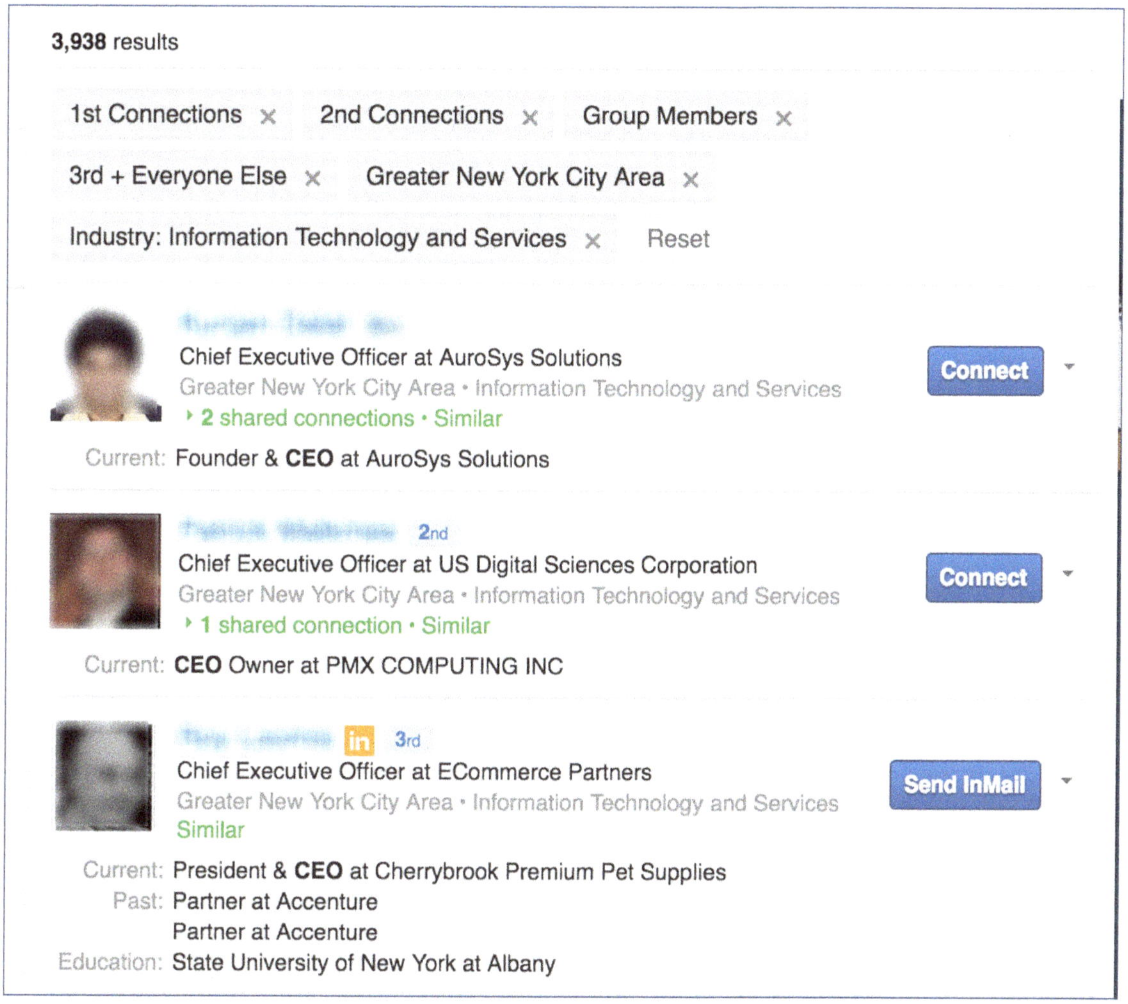

3,938 results

1st Connections ✕   2nd Connections ✕   Group Members ✕

3rd + Everyone Else ✕   Greater New York City Area ✕

Industry: Information Technology and Services ✕   Reset

Chief Executive Officer at AuroSys Solutions
Greater New York City Area • Information Technology and Services
▸ **2 shared connections** • Similar

Current: Founder & **CEO** at AuroSys Solutions

[Connect]

2nd
Chief Executive Officer at US Digital Sciences Corporation
Greater New York City Area • Information Technology and Services
▸ **1 shared connection** • Similar

Current: **CEO** Owner at PMX COMPUTING INC

[Connect]

in 3rd
Chief Executive Officer at ECommerce Partners
Greater New York City Area • Information Technology and Services
Similar

Current: President & **CEO** at Cherrybrook Premium Pet Supplies
Past: Partner at Accenture
       Partner at Accenture
Education: State University of New York at Albany

[Send InMail]

Figure 9.13: Profiles matching Alex Donovan discovered through LinkedIn advanced search

Sending invitations to connect

When it comes to sending an invitation to connect on LinkedIn, find out if you have any connections in common and ask for an introduction. If that is not an option, then sending a direct invitation to connect is the only choice. When you send a connection request, LinkedIn may ask you to share the email address of the person with whom you are trying to connect. This happens due to the following reasons (directly quoted from the LinkedIn help page):

1. The recipient's email preferences are set to only receive invitations from members who know their email address.

2. A number of recipients have clicked "I don't know [name]" after getting your invitations.

3. An invitation has already been sent to the member.

4. You've reached the limit of invitations you can send without email addresses to people you've identified as a "Friend" during the invitation process.

You may also notice that some profiles have an "InMail" option for connecting. This is LinkedIn's messaging option where you have to pay a fee in order to send someone an email (message).

If you cannot get introductions, the best thing to do is to build relationships with your target audience in their LinkedIn groups and ask for a connection in that context. If the target audience you discover on LinkedIn has a presence in other social networking sites—such as Twitter, for example—you can build your relationships within Twitter first, and then ask to be connected on LinkedIn. Where there's a will, there's a way.

Status updates and terminology

You can create content on LinkedIn either by sharing a status update or publishing an article. There are important differences between these two types of content:

1. Character limit: users are limited to 600 characters for a status update, but can include up to 40,000 in an article (9.2).

2. Visibility: The visibility of a user's status updates and articles depends on their privacy settings. Privacy is set by visiting https://www.linkedin.com/settings/activity-visibility and selecting among "Everyone," "Your network," "Your Connections," and "Only Me." When set to "Everyone," a user's status updates can be seen by anyone who has logged in to LinkedIn. However, articles are visible regardless of whether or not someone is logged in. Published articles are also indexed by Google, which means that a LinkedIn user's articles can be discovered as a result of search queries on Google. This is valuable because of the way Google ranks websites for authority on the World Wide Web; LinkedIn is considered a high-authority website and gives its content a higher ranking in search. Most blogs do not enjoy a ranking as high as LinkedIn, so publishing on this platform can exponentially increase the chances of your articles being discovered!

3. Content curation in Pulse: Only published articles have the opportunity to be featured in Pulse. Pulse is a content curation and publishing platform on LinkedIn. This is where LinkedIn users subscribe to and follow content published on major publications, as well as articles written by industry leaders on LinkedIn. In addition, some articles published by LinkedIn users are curated into various channels within Pulse. This is where the opportunity lies for additional outreach and exposure, because some of these channels have thousands of followers, most of which are not connected with the publishing user. When an article is featured in a Pulse channel, all of these subscribers will have the opportunity to view it in their news feed! The criteria used by LinkedIn for choosing which article to feature in a channel is not entirely known; however, the following seems to be factors that increase the chances of this happening:

    i. Articles need to be more than three paragraphs.

    ii. Article topic needs to benefit a wider readership rather than a user's immediate connections.

    iii. Articles needs to be timely, and should take on the news in a particular industry or a larger societal and world issue.

To get ideas as to structuring your article to increase the chances of being featured, visit Pulse (https://www.linkedin.com/pulse/discover) and review the various channels. Try to fashion your articles to fit channels that are most relevant to your topic. Also, browse through some of the articles that are already featured in these channels to get a better idea as to what LinkedIn is looking for in a featured article. In the end, the choice of articles featured on Pulse is up to LinkedIn's editors. Follow the guidelines outlined above, focus on delivering quality content, and you'll drastically increase—although not guarantee—the chances of your articles obtaining a feature spot on a Pulse channel.

Native video on LinkedIn

As I mentioned earlier in Chapter 6, native video is a rising trend within social networking platforms. With the introduction of the "LinkedIn Record" mobile app, LinkedIn has officially joined this trend. For the time being recording and publishing native video on LinkedIn is available to a select number of industry leaders and VIPs, i.e. LinkedIn Influencers. These influencers can choose among various topics and record a 30 second video to answer a particular question about that topic. Once the video is recorded, influencers share the video with their connections and followers who can view, like and comment on this content.

As regular users, you and I can download the LinkedIn Record app and sign in with our LinkedIn credentials, but we will not be able to access the content of the app, except for viewing the following message:

> *"Thank you for your interest! Record is available only to a limited number of industry leaders and VIPs at this time. When it opens up more broadly, we'll be in touch!"*

Although it is disappointing that the rest of us cannot access this feature for now, we can be relatively certain that we will be able to do so soon. So download the app and wait for a notification from LinkedIn about opening up Record to everyone else, I did.

Community

Community building on LinkedIn happens in groups. There are two different types of groups on LinkedIn: Standard and Unlisted. Similarities and differences between the two types of groups are listed in Table 9.1. Each LinkedIn user can join a maximum of 100 groups, and each group can contain up to 20,000 members (9.2).

When you join a group, be sure to read "Group rules" and follow them. Some groups have strict guidelines for frequency and types of content that can be posted. For more on group etiquette in general please review the section on "A few words of clarification and caution" in Chapter 6.

Analytics

By using analytics on most social networking sites, we can learn about which actions have been taken on our status updates or profiles, and who has taken those actions. It is safe to assume that those who take action on our content are interested in what we have to say. But what if we could capture data about those who *may* be interested in our message, or *might* be interested in connecting with us? What if we could come across profiles that we had not considered as an audience suitable for our message? LinkedIn analytics offers these insights, in addition to a host of other data about the performance of your efforts on this platform. Below I describe how relevant LinkedIn profiles can be discovered *before* they take any action on your content or profile.

Table 9.1: Group types on LinkedIn

Feature	Group Type	
	Standard	Unlisted
Search visibility	All search engines	No search engine, but group profile can be viewed via group URL
Group post visibility	Only group members	Only group members
Invitation to join	Members can invite their first-degree connections	Only group manager or owner can send invites [1]
Approval of join requests	Members can accept requests from first-degree connections	Only group manager or owner can accept join requests
Logo of group on members profile	Group logo will show on members' profile	Group logo will be visible on members' profile only by other group members
[1] However, current members can send someone the URL for the group, through which a request to join can be submitted.		

Profile views

An important section of LinkedIn analytics is the "Profile views" section, where users gain intelligence about those who have visited their profile pages. Your LinkedIn profile may attract views due to a number of reasons including being discovered in search, or because of your activities attracting the attention of other users. Some of these profile views may be from your connections, but if you are actively publishing articles or sharing content, many of them won't be. Here is an opportunity to form new connections with people who may have escaped your attention!

To find analytics on your profile visitors, hover your mouse over "Profile" on the top menu in the home page of your LinkedIn account and click on "Who's viewed your profile" (Figure 9.14).

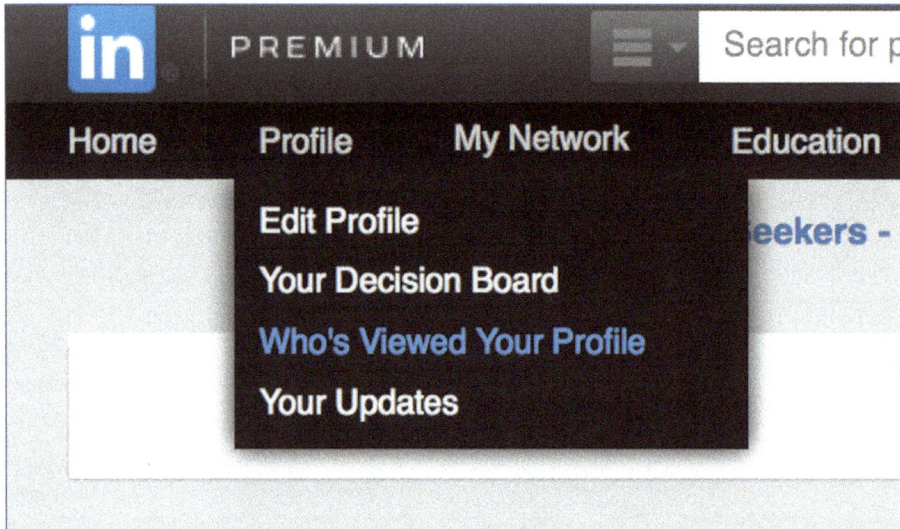

Figure 9.14: Locating data about your profile views

Data in "Who's viewed your profile" is divided into four sections:

1. A ninety-day trend in your profile views and your activities on LinkedIn (Figure 9.15). Here you can attempt to correlate the frequency and types of activities you have undertaken on LinkedIn to the number of views your profile is receiving. This data may help you identify types of activities that tend to increase profile views. LinkedIn also provides a projected number of profile views for the current week you are analyzing the data. If LinkedIn anticipated a downward trend for your profile views, you may want to consider being a bit more active on that week to reverse the trend.

2. "Viewers who worked at" reveals companies in which your profile viewers work (Figure 9.16). Many of your profile visitors may not show associations with any company, perhaps because they found your profile through a Google search, or they chose to view LinkedIn in private mode. Knowing what companies your profile views come from may alert you to organizations that you had not considered for your outreach efforts!

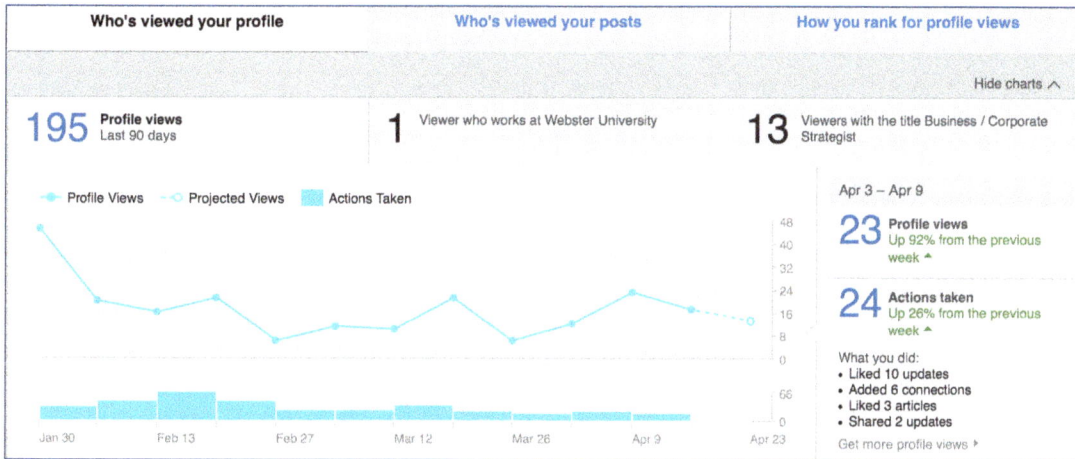

Figure 9.15: LinkedIn analytics—Ninety-day trends of your profile views

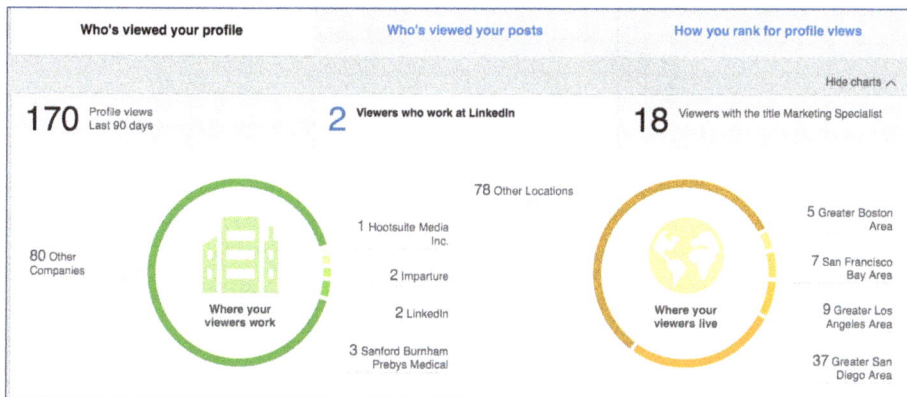

Figure 9.16: LinkedIn analytics—Where your profile viewers work

3. Industries and job titles of your profile viewers can be found in the far right section of "Who's viewed your profile" (Figure 9.17). This is important demographic data about the types of individuals that your profile attracts.

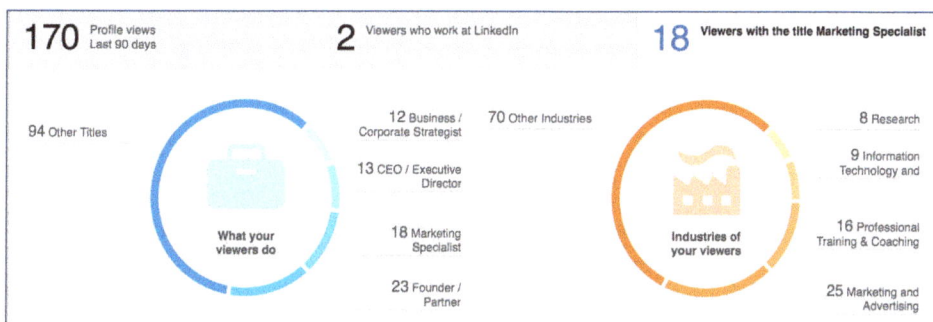

Figure 9.17: LinkedIn analytics—Where your profile viewers work

4. If you scroll down on the page, you can view details about each individual who has recently viewed your profile (Figure 9.18). You really can't get this level of specificity about members of a social network who have yet to take action on any of your content! Leverage this opportunity either to study their profiles, send them a message, or connect with them. Note that some users decide to view your profile in private mode, in which case, no details about them are revealed. This kind of secret profile peeping behavior tends to be practiced by recruiters, or your competitors.

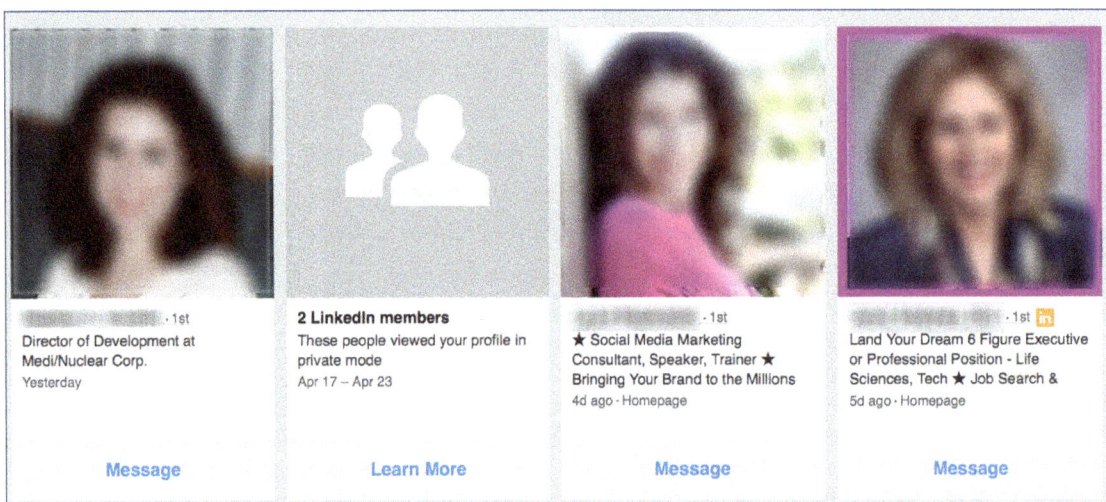

Figure 9.18: LinkedIn analytics—Your profile viewers

Who's viewed your posts

I have already covered the benefits of publishing articles on LinkedIn in terms of increasing the reach of your message outside of your immediate connections on LinkedIn. "Who has viewed your posts" allows you to measure the success of your outreach efforts through publishing, and reveals details about readers of your articles, and those who liked, commented on, and shared your articles. This data is broken down into three sections:

1. Trends of article views, which can be filtered for the last seven, fifteen, thirty days; six months; or one year (Figure 9.19). The analytics for article views reveals one of the strongest arguments for taking the time to publish on LinkedIn: your articles are timeless! An article you wrote a year ago could be discovered on LinkedIn, viewed, and shared a year later. One challenge with online publishing

is the rapid rate at which interest in an article decays, so one always has to be writing new content. But the articles you publish on LinkedIn can become a gift that keeps giving long after its initial date of publication.

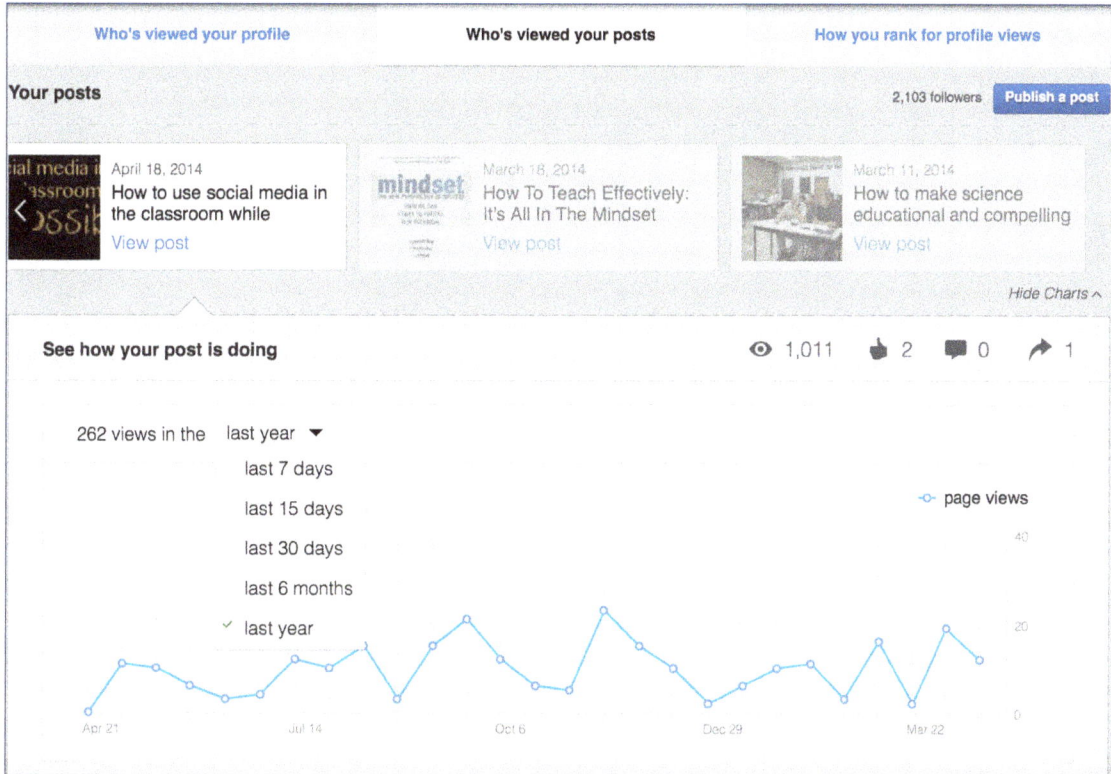

Figure 9.19: Who has viewed your posts–View trends for your articles

2. Demographics of your readers gives a statistical overview of "Top industries," "Top job titles," "Top locations," and "Top traffic sources" for your article (Figure 9.20). When you publish an article on LinkedIn, be sure to share the article in your other social networking profiles because LinkedIn will show you which sources perform the best for attracting readers. The demographics of your readers provide valuable intelligence into how closely you are targeting your intended audience, and help you discover demographics that you may have not considered!

3. "Who's responding to your posts" displays profiles of LinkedIn users who either commented, liked, or shared your article on LinkedIn (not shown). These are readers who identify with your message the most, so be sure to get to know their profiles and the groups to which they belong, and then connect with them.

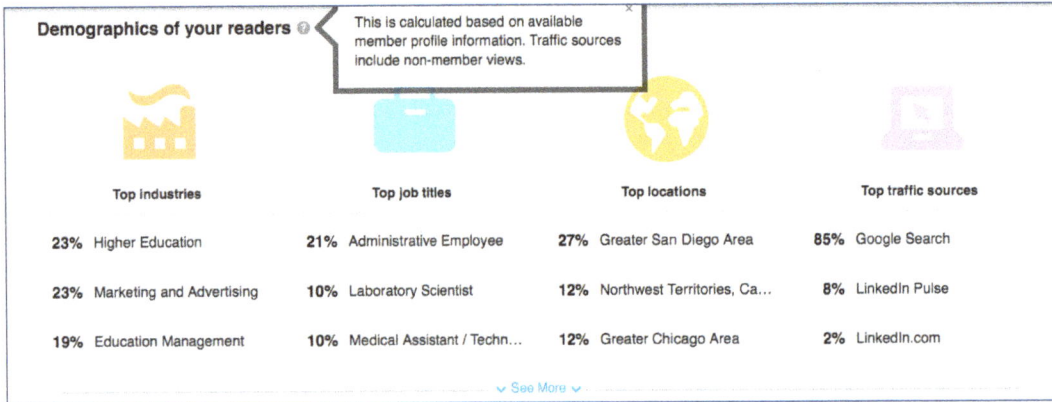

Figure 9.20: Who's viewed your posts—Demographics of your readers

How you rank for profile views

This is an important metric as it allows you to gauge the success of your efforts to attract attention to your profile (Figure 9.21). Your profile views are presented as compared with that of other LinkedIn users, and those who are associated with organizations or companies to which you belong. There's also comparative data between you and professionals working in your industry, but it is only available to paid LinkedIn members. One powerful way of increasing your profile ranking is to regularly publish articles of high impact and value. Publishing on LinkedIn is a golden opportunity for your outreach efforts to reach a large audience.

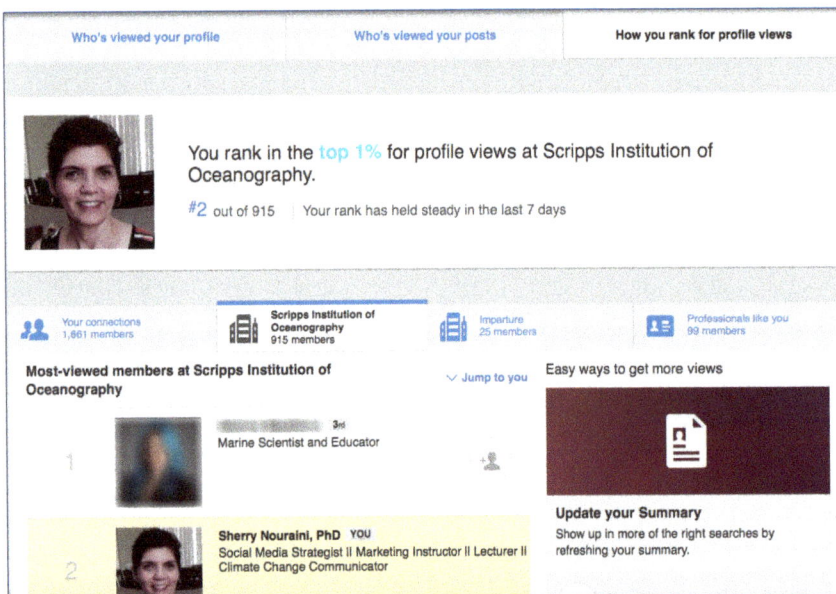

Figure 9.21: LinkedIn analytics—Ranking of your profile

Outreach on LinkedIn

Every outreach workflow you've seen so far begins with Audience Research, and LinkedIn is no exception. However, since this platform offers a robust and popular publishing platform, it provides the opportunity for a two-pronged approach when doing outreach. Either way, the end point will be the same, which is joining relevant groups to make connections (Figure 9.22).

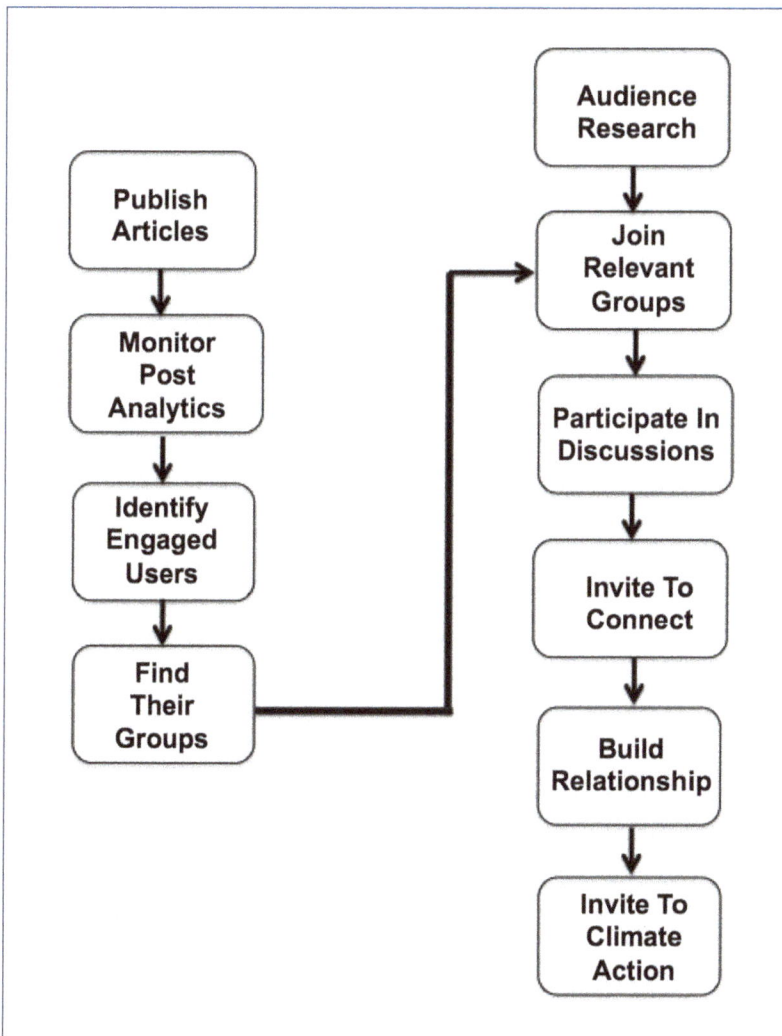

Figure 9.22: Workflow for outreach on LinkedIn

In Chapter 4 of this book, you learned how to set up your own blog and write articles that are search engine–optimized. If you find blogging on two separate platforms to be taxing on your time, you can always repurpose an article you've written for one platform to post on the other. Just change the title and structure of the article when you repurpose the article. For example, compare the article I wrote on a blog for training purposes (9.3) with a repurposed version on LinkedIn (9.4). Resist the temptations to post the exact same article on the two platforms, as Google does not take kindly to duplicate content.

When you publish on LinkedIn, keep a close eye on likes, comments, and shares of your article. Some LinkedIn users defeat the purpose of publishing on LinkedIn by leaving the platform as soon as they post an article. Be sure to publish when you know you will have time to keep track of interactions with your article, reply to comments, send thanks for shares, and review the profiles of your audience. Once you examine the profiles of your engaged audience, be sure to follow the workflow to connect with them and engage them in conversations, as this is the start of any meaningful outreach. Also, be sure to share your article far and wide on other social networking sites, as you don't want to miss an opportunity for your message to be discovered outside of LinkedIn.

Practice what you learned

In this chapter you learned about LinkedIn and its valuable features. You learned how to use LinkedIn search to find users matching our example profile Alex Donovan, and how to use LinkedIn publishing to extend your reach on and beyond the platform. Setting up a profile and forming connections on LinkedIn is not only a powerful way of enhancing your climate action efforts, it is also a great tool in your career-building arsenal. I strongly encourage you to publish on LinkedIn and leverage this great platform to strengthen your personal brand. So build or complete your profile, and then take LinkedIn on a test run for finding communities for your target audience, or the other two example profiles in this book. I would love to hear from the results of your efforts, so please join our Facebook group, share what you find, and ask for help if needed.

Now what?

This chapter ends Section 2 of this book, which focused on teaching specifics about some of the mainstream social networking sites. Social media is a huge topic that is in constant flux. There are also many advanced topics for each social networking site, which I have not covered because they mostly pertain to marketing and advertising. However, rest assured that upon reading this book and applying the concepts to your own individual efforts, you will be well-positioned to get started on a solid, strategic footing. Or if you have already started using social media for science and climate change outreach, you can optimize your approach and be even more strategic and focused. In the next and final section, I will present strategies to help you devise a comprehensive, measurable plan for your outreach efforts. You have learned a great deal of material up to this point. It is time to put your new knowledge together to create a communication action plan.

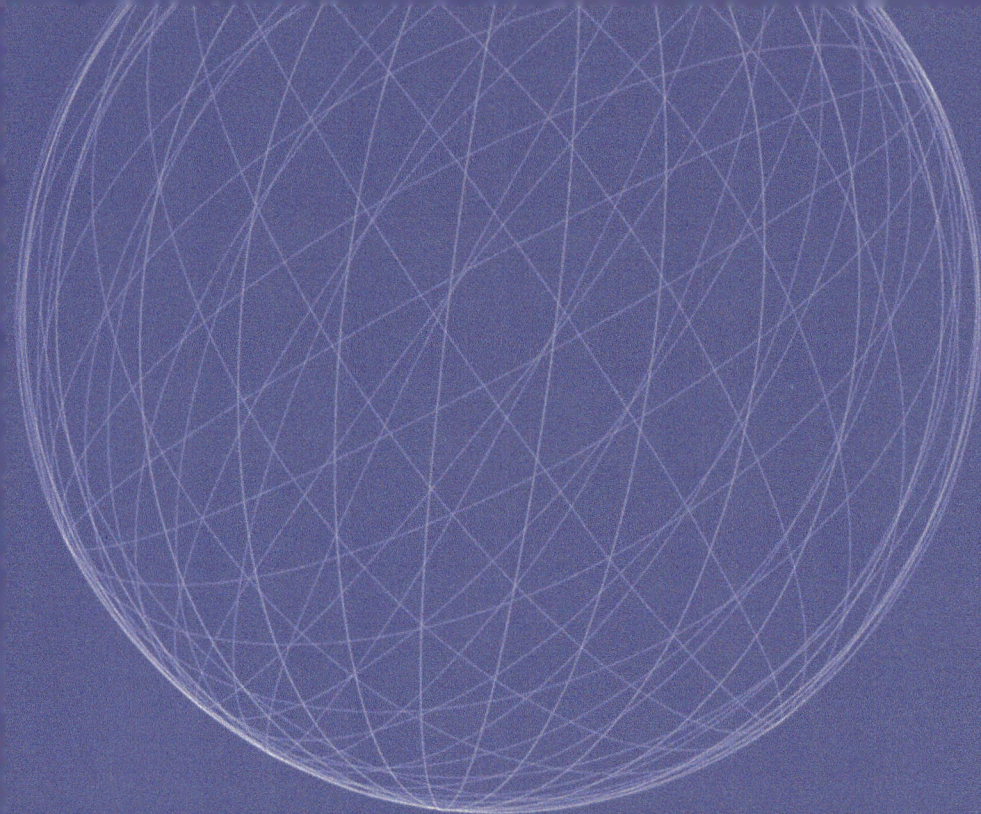

Section 3

Wrap Up

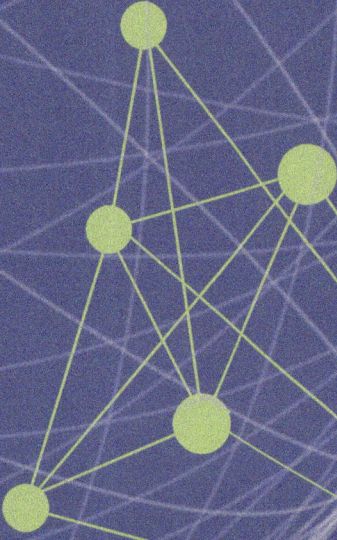

Putting It All Together

Knowledge is power, but this power comes from putting knowledge into practice in a strategic fashion toward achieving a specific and measurable goal. It is the aim of this chapter to help you apply the knowledge you've gained toward a measurable change in awareness and action on climate change. Let's review the concepts you've learned so far. In Chapter 1, you read an overview of the state of climate change communication, and the role that social media has played in this effort. In Chapter 2, you learned strategies to lay the foundation for your climate change outreach efforts by defining who *you* are, and whom you are targeting. This effort helped you develop a mission statement, which serves as a guiding principle for your climate change communication efforts. In Chapter 3, you learned challenges facing climate change and science communication efforts, and the strategies you can employ to overcome them. In Chapter 4, you learned about creating a content strategy, keyword optimization, and setting up your own blog.

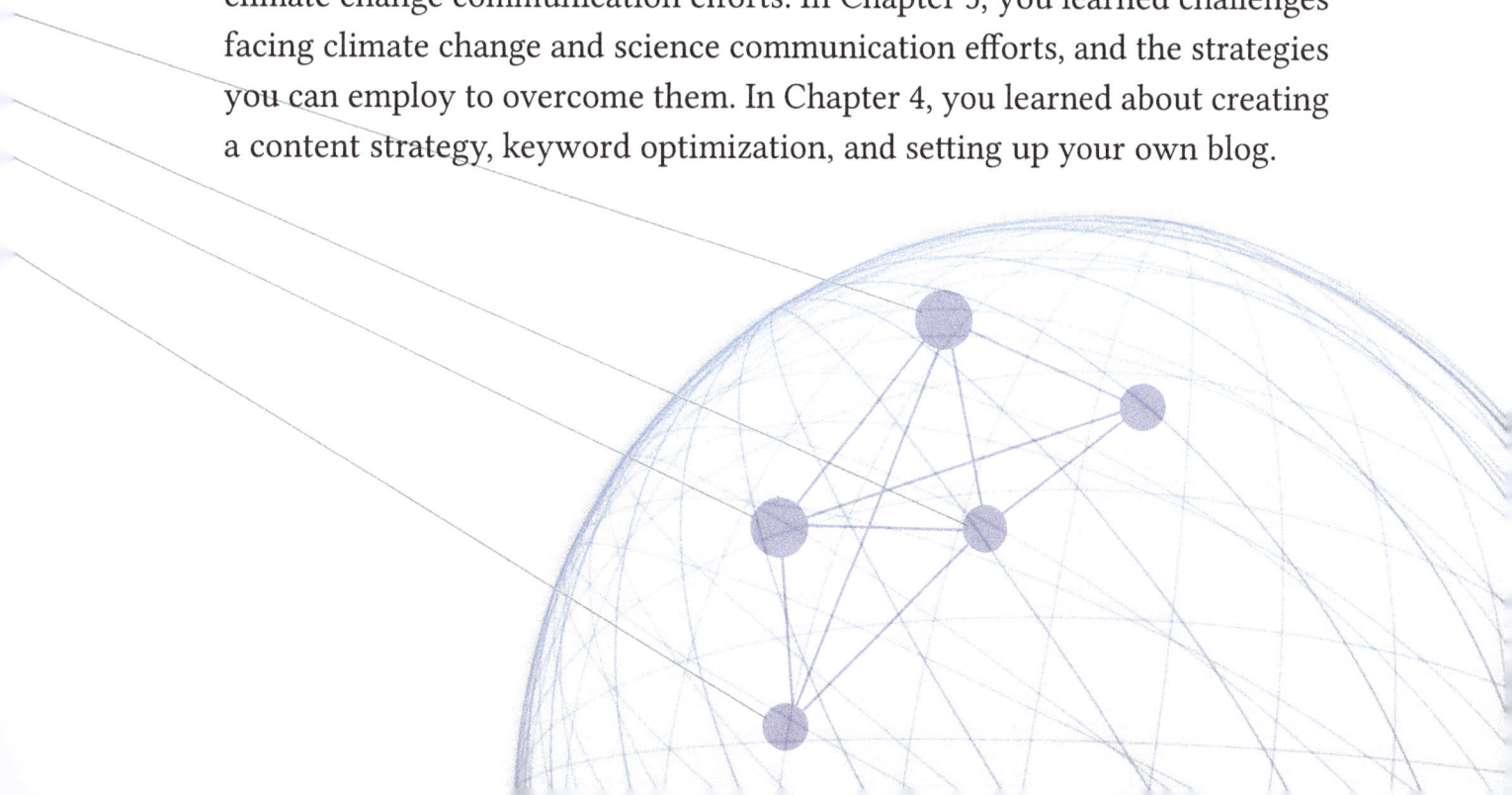

In Chapters 5 through 9, you learned about five mainstream social networking sites, how to search and find communities of your target audience in these social platforms, and how to organize your interactions. You also learned about analytics and metrics offered by each platform so that you can measure your effectiveness, along with sample workflows for your outreach efforts. Now it is time to put into place a plan of action. Here are some points to keep in mind and how to put your plan together.

You don't have to do it all

I have introduced five popular social networking sites, in addition to blogging. Although blogging should form a central part of your outreach efforts, you do not have to be active in all of the social networking sites in order to find success. The choice will depend on one thing: your target audience. Once you solidify your target audience profile, you can search each social network and find the one or two with the most vibrant and active communities of your target audience. Becoming a master of only one or two social networking sites where your target audience is gathering will be enough.

Define specific goal(s)

You have created a solid mission statement, which specifies what you want to accomplish. You have a mission, you know whom you want to target, and why and how you want to target them. But in order to put a plan of action together, you will have to become even more specific. You will need to develop a Specific, Measureable, Assignable (who is going to do it), Realistic, and Time-related (SMART) goal for your outreach efforts (10.1). I will illustrate this with an example.

Say you are fan of H.P. Lovecraft fiction and want to educate your fellow fiction enthusiasts—the Paul Draper sample profile demographic, for example—about climate change using the voice of Cthulhu. Through audience research, you have discovered a Cthulhu fan group on Facebook. Your SMART goal would look something like this:

> *I want to leverage the fact that Cthulhu lives in the ocean to highlight ocean acidification and other damage caused by climate change, and thereby create an engaging conversation. My goal is to create one climate-related discussion every month. I will initiate the discussions by posting a specific topic related to a blog article that I will be writing, along with a link to*

this article. I want to see at least ten unique members of this community participating in the conversation. Using this discussion, my aim is to get at least five community members to acknowledge the need to protect the ocean. I know that since I am new to the group, I will have to spend at least a month getting to know everyone, participate in conversations, and maybe even attend live events in order to build relationships. Once I raise the interest of this otherwise disengaged community in the topic of climate change mitigation, I will enter the next phase of my outreach by suggesting solutions in which the community can participate. It is now May 2016; I want to accomplish my SMART goal by the end of December of 2016, which means I will have to craft nine blog articles, one for each month.

When you state your goals with this level of detail, your tactics and timelines will become clear, and you will know exactly how to measure success.

Measure your effectiveness and success

There is no sense in doing outreach if you cannot measure your level of success or investigate reasons for its failure. Every SMART goal (10.1) has a Key Performance Indicator (KPI), which is a particular event that needs to take place for the goal to be deemed successful. In our example above, the KPI would be the number of Cthulhu fan group members who have acknowledged the necessity for climate change mitigation. If your KPI shows success, you can rejoice, but what if you do not succeed? This is why you need to monitor additional metrics to understand the reasons behind your lack of success. I call these diagnostic metrics. These same metrics can show you whether or not you are on the right track. Note that you don't have to leave measurement until the end.

So, what are some diagnostic metrics you can use in this example? Here is a list:

1. Number of community members who viewed your post. Every time you post something to a group, Facebook shows you how many community members have or have not seen it, and who they are (Figure 10.1).
2. Number of community members who have clicked through the link you've posted. You may be wondering how you can monitor this event. There are many ways to do this, here are few suggestions:

   a. Post a link to your article on your Facebook page, and then share this post to the group. If you remember from Chapter 6, you can view level of engagement on shares in the Facebook page insights (Figure 6.26).

   b. Use a link-shortening tool such as Bit.ly to monitor clicks through the link you shared.

3. Number of community members who have engaged and commented on your content.

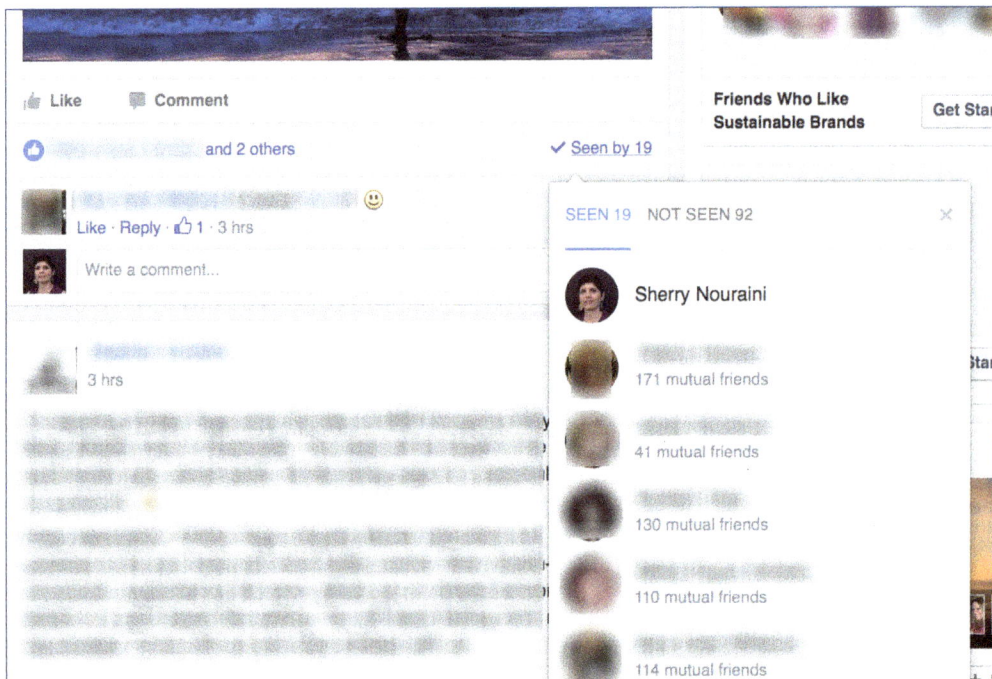

Figure 10.1: Facebook group analytics—who has seen a post

These diagnostic metrics will be invaluable for you to identify bottlenecks that may be preventing you from reaching your goals. For example, if enough members are not seeing your posts, or clicking through the link to your blog post, then your efforts in implementing a change in behavior will not be successful. But do you give up at this point? No! If community members are not inclined to read blog posts, then perhaps share short videos or ask to give a presentation during an upcoming event. When one tactic does not work, pivot and try something else. The future of our planet is depending on it.

As mentioned above, KPI for your SMART goal will be the number of community members who have acknowledged the need to mitigate climate change. Note that acknowledging the need to act is not the same thing as taking action, but if you can't get them to acknowledge the need for a solution, they won't take action! Besides, taking action will come in the next phase of your outreach, which will need its own SMART goal and KPI.

You don't have to create all the content

Creating content, whether it is text or video, can be time-consuming. The good news is you don't have to create all the content that you share in the social media space. Climate change is a subject of many blogs, videos, and infographs—content that you did not create and what is referred to as "third-party content." What you need to do is find a few sources of *relevant* quality climate change–related content, and subscribe to their feeds. Then, every one to three days, review these articles and share the ones that are most suitable for your audience, and start a discussion around them. Never share a link in a community and then run; it is your responsibility to show why the article behind the link is relevant and worth the attention of your community members.

A caveat of this approach is that you'd be sending traffic to other blogs, instead of yours. Here is where a great tool called Sniply (http://www.snip.ly) will come in handy. Sniply is like a link shortener, with many valuable additional features, one of which is adding a "call to action" to third-party content (see Figure 10.2).

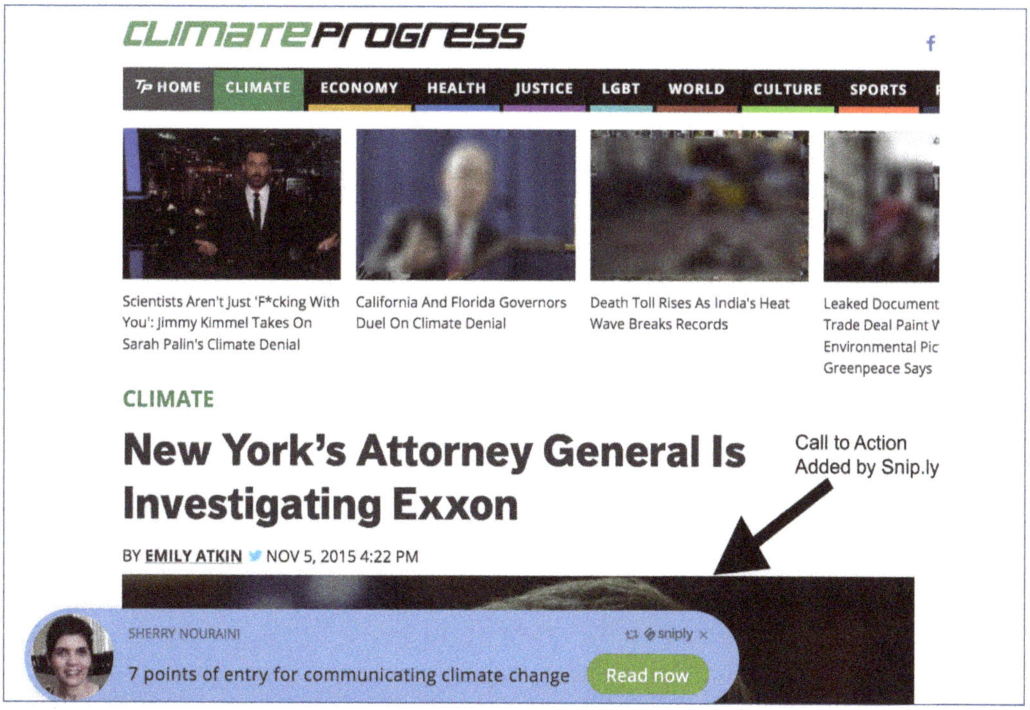

Figure 10.2: Adding a "call to action" for visiting your blog to third-party content

When a third-party content is "snipped" by Sniply, you will get a shortened URL, which not only reveals this content, but also renders an invitation to view contents on your blog. Sniply not only allows you to include a "call to action" for visiting your blog within a third-party content, it will also provide analytics as to the performance of this call to action. Note that if you want your call to action to be successful, you will have to be strategic by matching the topic of the third-party content to that of your own article.

Be cognizant about knowledge gaps

When reaching out to your audience, it is important to be cognizant about where they are in the spectrum of knowledge about climate change. I referred to this spectrum as related to the range of knowledge gaps in the American public in Chapter 3 (see Figure 3.1). For example, a disengaged audience, such as our example profile Paul Draper, is probably at the "Awareness of the Problem" stage, where they are largely uninformed of the negative consequence of climate change to their personal lives and the planet. Engagement with Paul Drapers of the world needs to start making them aware of the consequences of climate change to their lives before moving to a stage of encouraging climate action. This is because if they don't understand why they need to take action, they won't! For example, articles attempting to alarm a disengaged audience about bleaching of coral reefs (10.2) will probably fall into deaf ears if we fail to teach the audience how these ecosystems contribute to their personal lives (10.3). You have to help your audience understand before asking them to act.

This is again where a tool like Sniply can be used strategically to build on the knowledge you share with your audience. For example, your communication with a disengaged audience could begin with an article about economic benefits of coral reefs. Using Sniply, you can include a call to action to read your next article about how climate change is affecting coral reefs (Figure 10.3). Build a sequential map of knowledge for your audience and help them get where you want them to be in a stepwise strategic fashion.

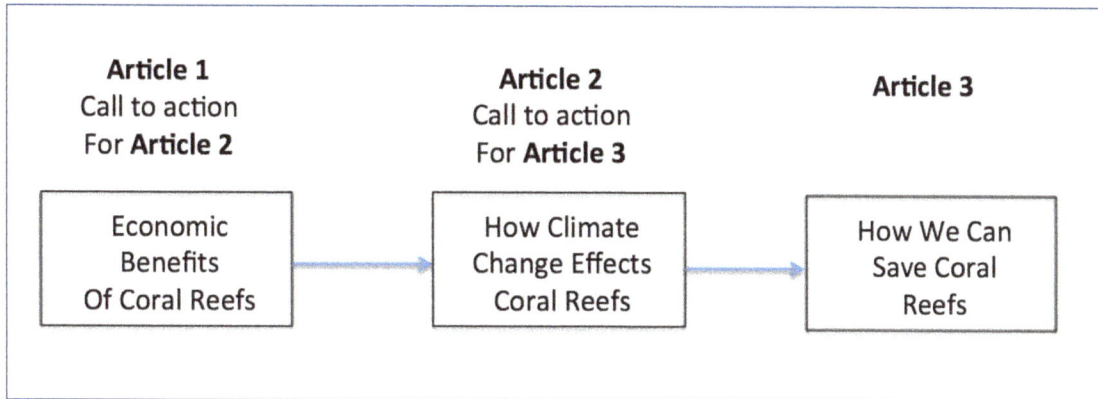

Article 1 Call to action For **Article 2**	Article 2 Call to action For **Article 3**	Article 3
Economic Benefits Of Coral Reefs	How Climate Change Effects Coral Reefs	How We Can Save Coral Reefs

Figure 10.3: Building a sequential strategic content plan

Leverage the mobile revolution

The world is becoming increasingly mobile, and our phones are getting increasingly smarter, allowing us to accomplish tasks on the go. For most of your social media content creation and interactions, you really don't have to be sitting at your desk behind a personal computer. Take advantage of the mobile revolution and capture content from the real world as you go about your day to send a message. Take, for example, the Instagram status update I shared (Figure 10.4) while I was using a water filtration station at the college where I teach. I use this water filter often to replenish the water in my steel water bottle. Why not use this opportunity to send an environmentally friendly message?

In addition to capturing photos with your phone, you can also leverage the mobile versions of social networking sites of Facebook and Twitter to post live video broadcasts. You can use live video streaming to host interviews, report on live events, and capture real-time reports about environmental effects of climate change. The powerful tools in the palm of your hands afford you to be a citizen journalist; take advantage of this power to enhance your climate change mitigation efforts.

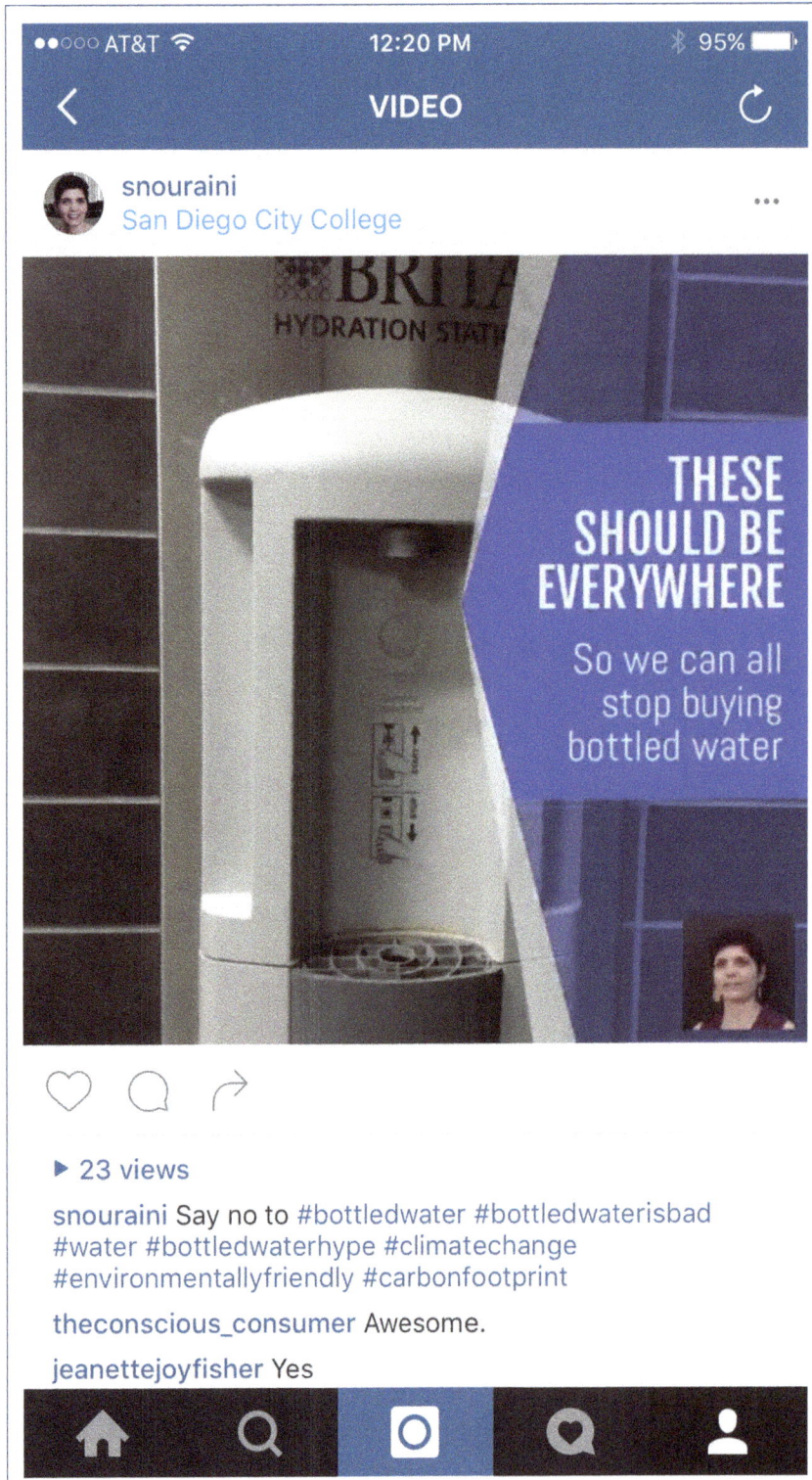

Figure 10.4: Sharing my climate-friendly habits via Instagram

Consider copyright and other legal issues

Pictures speak a thousand words, but they can also be a source of trouble if you do not have proper permissions to use them. This is why leveraging the powerful camera in your smart phone to capture your own media should be your first resort. Obviously, you cannot expect to create all the media you need for every topic you want to discuss, which is why occasionally you will have to use photos created by others. If you use stock images, try finding royalty-free sources, and be sure to read and understand the terms of service and copyright permissions associated with each source. For most royalty-free sources, at the least you are asked is to cite and give credit to your source. I have listed a number of royalty-free sources for images in the Appendix.

Aside from copyright issues, you should also be aware of how you approach debunking myths. As a general rule, focus on correcting the misinformation rather than attacking the source, especially if the source is an individual. Unfortunately, many science activists do not follow this rule, and create memes of individual antivaccine and anti-GMO bloggers, such as David Wolfe or Food Babe. Such personal attacks are not likely to be effective in changing minds of those who follow these bloggers, and they also open the possibility of your social media account being suspended if the individual whose claims are being debunked chooses to file a complaint. Abusing Facebook's "community rules" and reporting feature is a tactic that has been used by antivaccine enthusiasts to successfully suspend the accounts of science activists. Simply mentioning the name of an antivaccine individual in the comment section of a Facebook page was enough to file a complaint (10.4). Protect your social media accounts from false accusation by focusing on debunking myths and sharing facts instead of trying to discredit individuals, just like you learned in Chapter 3.

Power of social norms

One simple way to inspire climate action in others is to live the change you want to create and share your life with your offline or online communities. Take an inventory of your own habits. Do you make an effort to walk short distances instead of driving? Do you avoid using disposable items, such as paper plates, cups, and utensils? Do you avoid buying bottled water, and do as much as you can electronically so you don't waste paper? Do you support clean energy startups on Kickstarter and similar sites? If you do any of these things, and more, be sure to share them with your communities, offline and online. If social science and psychological research in climate change attitudes have taught us anything, it is that changing attitudes requires changing behavioral norms (10.5). If you can inspire a few members of your communities to adopt environmentally friendly habits, research has shown that the rest will follow.

Environmentally friendly life habits can take different forms, from the way we consume and commute to how we vote. Transforming our societies to become free of fossil fuels is no small task, as it *must* involve our local, state, and federal lawmakers, who are deeply influenced by the fossil fuel industry. But it is important to realize that members of the public are the ultimate deciding factor, because if they demand a carbon-free economy, lawmakers will follow to gain their votes, and the industry will comply to earn their business. In this digital age, behavioral norms, political activism, and purchase decisions are made largely within the context of social networking sites. We are social creatures, and social networking sites are where we gather, learn about the social norms of our communities, share our stories, and organize around a topic. Human behavior is largely driven by social norms. If we want the attention of the public, we have to meet them in the context in which they develop these norms. I believe that solutions to climate change will ultimately be social. I hope this book has given you the tools to inspire change in others to reach these solutions.

Stay up to date with developments in social media

Social media platforms are always in flux, constantly adding new features to improve functionality and design. This constant improvement is partly fueled by competition between these platforms for loyal users— people like you, me, and our friends and families. The other source of motivation for these innovations is driven by a desire to develop technological advancements in the way we communicate. Regardless of the motivation behind the constant innovation in social media platforms, our communication efforts will be a beneficiary of the availability of new features. This is why it is important to keep up with new innovations in the social media space. I have developed a number of resources to help you keep informed about these developments and their implications for climate change and science communication:

1. Social for Climate Facebook group: Join our Facebook page (www.facebook. com/groups/Social4Climate) where we will answer your questions and keep you up-to-date with developments in the social media space.

2. Social for Climate website: Visit our website (www.SocialForClimate.com) to gain access to our blog and templates to help you better plan for your communication efforts.

3. Social for Climate e-News: If you prefer to receive updates in your email box, you can follow this link (http://eepurl.com/b6WIf1) and subscribe to our monthly news publication. Alternatively, use a QR reader on your smart phone to scan this code and subscribe.

5. Social for Climate Twitter account: To keep up with updates about social media and our occasional Twitter chats, follow @social4climate on Twitter (Twitter.com/social4climate) and keep an eye on the hashtag #social4climate.

6. Other Social for Climate accounts: Follow us on Pinterest (pinterest.com/socialforclimate) to see examples of how we share information in a visual format via images or video.

Finally, I would love to hear your feedback on the contents of this book, so please share your thoughts in an online review.

Acknowledgments

When I started writing this book, I wanted to produce something I could be proud of. Under the meticulous direction of Linda Scott, my editor and owner of eFrog Press, I was able to do so. Linda managed this editing and publishing project from start to finish, and made sure I stayed committed to my goal.

Every author needs a sanctuary, a quiet place for uninterrupted thinking and writing. The quiet writing room at Hera Hub, a co-working space for women, was my sanctuary during the weekdays. Hera Hub is also where I met Linda Scott, and many other amazing women who inspired and supported me every day. I am especially grateful for the kind mentorship, encouragement and guidance I received from Hera Hub member Catherine Mowbray Lorenz, she is truly a one of a kind lady. Thanks to Felena Hanson, an accomplished writer in her own right, and Founder of Hera Hub, for creating this beautiful space, and for her continuous encouragement, advocacy and support.

Thanks to my husband Bahram for taking on additional responsibilities with the kids during weekends so I can have a sanctuary at home to write. I am also grateful for his encouragement and support during the long journey of writing this book. His unwavering love and support made this book possible. I must acknowledge my children, Kimia and Sina, for keeping me balanced.

My sisters, Farah and Firoozeh, deserve special mention. Farah has always been the greatest supporter and cheer-leader of my ambitious projects, and Firoozeh has always graced me with her unconditional love.

Thanks to the Scripps Institution of Oceanography (SIO), Dr. Lynn Russell and Dr. Ellie Farahani, for trusting me in teaching the social media component of this program, and providing the inspiration for writing this book.

A very special thanks to students of the inaugural course in Social Media for Climate Science and Policy, Natalia Zadorkina , Chris Scarsi, and Rebecca Stark whose excellent work provided the foundation for this book.

Last but not least, my parents, Mansour and Tooran, deserve special acknowledgement as they instilled in me the value of integrity, honesty, hard work, empathy and optimism.

About the Author

Sherry Nouraini, PhD, is a Social Media Strategist, Marketing Instructor, Lecturer, and Climate Change Communicator who lives in Southern California. Her rigorous scientific training and experience in marketing makes her uniquely qualified for data-driven and strategic communication, which she offers through her marketing business Captive Touch, and also by teaching Marketing via Social Media within the University of California system.

In preparation for developing a curriculum and teaching social media for climate science and policy at the Scripps Institution of Oceanography, Sherry spent countless research hours reviewing how scientists and activists use social media for outreach. She saw little sign of effectiveness in their efforts. Additional research and observations in conversations around the topic of climate change in the blogosphere and social media confirmed her suspicions, and ignited her desire to be part of the solution in her own way. Toward that end, she combined her new-found knowledge about science and climate change communication with years of experience in scientific research, teaching and social media consulting, to write a step-by-step communication guide. Her aim is to help scientists and climate change activists inspire change in their communities through the tools of blogging and social media. Social Solutions for Climate Change is her contribution to further the cause of climate change communication and mitigation.

Bibliography

Chapter 1

1.1: Cook, John, Dana Nuccitelli, Sarah A. Green, Mark Richardson, Bärbel Winkler, Rob Painting, Robert Way, Peter Jacobs, and Andrew Skuce. Quantifying the Consensus on Anthropogenic Global Warming in the Scientific Literature. *Environ. Res. Lett. Environmental Research Letters* 8, no. 2 (2013): 024024. doi:10.1088/1748-9326/8/2/024024.

1.2: "The 2014 U.S. National Climate Assessment." Accessed June 6, 2016, http://nca2014. globalchange.gov/report.

1.3: Oreskes, N. BEYOND THE IVORY TOWER: The Scientific Consensus on Climate Change. *Science* 306, no. 5702 (2004): 1686. doi:10.1126/science.1103618.

1.4: "National Aeronautics and Space Administration, scientific consensus." Accessed June 6, 2016. http://climate.nasa.gov/scientific-consensus/.

1.5. Morice, Colin P., John J. Kennedy, Nick A. Rayner, and Phil D. Jones. Quantifying Uncertainties in Global and Regional Temperature Change Using an Ensemble of Observational Estimates: The HadCRUT4 Data Set. *Journal of Geophysical Research: Atmospheres J. Geophys. Res.* 117, no. D8 (2012). doi:10.1029/2011jd017187.

1.6: Summary for Policymakers. *Working Group I Contribution to the Fifth Assessment Report of the Intergovernmental Panel on Climate Change Climate Change 2013— The Physical Science Basis*: 1-30. doi:10.1017/cbo9781107415324.004.

1.7: Honisch, B., A. Ridgwell, D. N. Schmidt, E. Thomas, S. J. Gibbs, A. Sluijs, R. Zeebe, L. Kump, R. C. Martindale, S. E. Greene, W. Kiessling, J. Ries, J. C. Zachos, D. L. Royer, S. Barker, T. M. Marchitto, R. Moyer, C. Pelejero, P. Ziveri, G. L. Foster, and B. Williams. The Geological Record of Ocean Acidification. *Science* 335, no. 6072 (2012): 1058-063. doi:10.1126/science.1208277.

1.8: Parmesan, Camille. Ecological and Evolutionary Responses to Recent Climate Change. *Annu. Rev. Ecol. Evol. Syst. Annual Review of Ecology, Evolution, and Systematics*37, no. 1 (2006): 637-69. doi:10.1146/annurev.ecolsys.37.091305.110100.

1.9: Cleland, E., I. Chuine, A. Menzel, H. Mooney, and M. Schwartz. Shifting Plant Phenology in Response to Global Change. *Trends in Ecology & Evolution* 22, no. 7 (2007): 357-65. doi:10.1016/j.tree.2007.04.003.

1.10: Parmesan, Camille, and Gary Yohe. A Globally Coherent Fingerprint of Climate Change Impacts across Natural Systems. *Nature* 421, no. 6918 (2003): 37-42. doi:10.1038/nature01286.

1.11: Langevin, Christian D., and Michael Zygnerski. Effect of Sea-Level Rise on Salt Water Intrusion near a Coastal Well Field in Southeastern Florida. *Groundwater* 51, no. 5 (2012): 781-803. doi:10.1111/j.1745-6584.2012.01008.x.

1.12: "What we know, the reality, risks and responses to climate change." Accessed June 6, 2016. http://whatweknow.aaas.org/get-the-facts/.

1.13: Leiserowitz, A., Maibach, E., Roser-Renouf, C., Feinberg, G., & Rosenthal, S. (2015). *Climate change in the American mind: October, 2015.* Yale University and George Mason University. New Haven, CT: Yale Program on Climate Change Communication. Accessed June 7, 2016. http://environment.yale.edu/climate-communication-OFF/files/Climate-Change-American-Mind-October-2015.pdf

1.14: "Global public opinion about climate change." Accessed June 6, 2016. http://climatecommunication.yale.edu/publications/analysis-of-a-119-country-survey-predicts-global-climate-change-awareness/

1.15: Maibach, E., Leiserowitz, A., Roser-Renouf, C., Myers, T., Rosenthal, S. & Feinberg, G. The Francis Effect: How Pope Francis changed the conversation about global warming (New Haven, CT: Yale Project on Climate Change Communication, 2015).

1.16: "A 50th anniversary few remember: LBJ's warning on carbon dioxide." Accessed June 6, 2016. http://www.dailyclimate.org/tdc-newsroom/2015/02/president-johnson-carbon-climate-warning.

1.17: Center for Research on Environmental Decisions and ecoAmerica. *Connecting on Climate: A Guide to Effective Climate Change Communication.* (New York and Washington, D.C._2014).

1.18: Pisarski, Anne, and Peta Ashworth. The Citizen's Round Table Process: Canvassing Public Opinion on Energy Technologies to Mitigate Climate Change. *Climatic Change* 119, no. 2 (2013): 533-46. doi:10.1007/s10584-013-0709-4.

1.19: "Social networking fact sheet—Pew Research: Why Americans use social media." Accessed June 6, 2016. http://www.pewinternet.org/2011/11/15/why-americans-use-social-media/.

1.20: Schäfer, Mike S. Online Communication on Climate Change and Climate Politics: A Literature Review. *WIREs Clim Change Wiley Interdisciplinary Reviews: Climate Change* 3, no. 6 (2012): 527-43. doi:10.1002/wcc.191.

1.21: Williams, Hywel T.P., James R. McMurray, Tim Kurz, and F. Hugo Lambert. Network Analysis Reveals Open Forums and Echo Chambers in Social Media Discussions of Climate Change. *Global Environmental Change* 32 (2015): 126-38. doi:10.1016/j.gloenvcha.2015.03.006.

1.22: Vicario, Michela Del, Alessandro Bessi, Fabiana Zollo, Fabio Petroni, Antonio Scala, Guido Caldarelli, H. Eugene Stanley, and Walter Quattrociocchi. The Spreading of Misinformation Online. *Proceedings of the National Academy of Sciences Proc Natl Acad Sci USA* 113, no. 3 (2016): 554-59. doi:10.1073/pnas.1517441113.

1.23: "Climate science 50 years later." Accessed June 6, 2016. https://t.co/Cvs36Xw8O4.

Chapter 2

2.1: Dan Ariely, *Predictably Irrational, Revised and Expanded Edition: The Hidden Forces That Shape Our Decisions* (New York: Harper-Collins, 2009).

2.2: "Bridging the gap between actual and reported behavior." Accessed June 6, 2016. http://www.uxbooth.com/articles/bridging-the-gap-between-actual-and-reported-behavior/

2.3: Jennifer Jacquet, Monica Dietrich, and John T. Jost Frontiers. The ideological divide and climate change opinion: "top down" and "bottom up" approaches. *Frontiers in Psychology* 5 (2014): 1-6. doi:10.3389/fpsyg.2014.01458.

2.4: March, James G., and Chip Heath. *A Primer on Decision Making: How Decisions Happen.* New York: Free Press, 1994.

2.5: Leiserowitz, A. & Smith, N. (2010) *Knowledge of Climate Change Across Global Warming's Six Americas.* Yale University. New Haven, CT: Yale Project on Climate Change Communication. Accessed June 07, 2016. http://climatecommunication.yale.edu/publications/knowledge-of-climate-change-across-global-warmings-six-americas/.

2.6: Leiserowitz, A., Maibach, E., Roser-Renouf, C. & Smith, N. (2011) *Global Warming's Six Americas, May 2011.* Yale University and George Mason University. New Haven, CT: Yale Project on Climate Change Communication. Accessed June 7, 2016. http://www.earthtosky.org/content/course-content/2012-mini-course/Knowledege_of_Audience/SixAmericasMay2011.pdf

2.7: Morrison, M., R. Duncan, C. Sherley, and K. Parton. A Comparison between Attitudes to Climate Change in Australia and the United States. *Australasian Journal of Environmental Management* 20, no. 2 (2013): 87-100. doi:10.1080/14486563.2012.762946.

2.8: Leiserowitz, A., Thaker, J., Feinberg, G., & Cooper, D. (2013) *Global Warming's Six Indias.* Yale University. New Haven, CT: Yale Project on Climate Change Communication. Accessed June 07, 2016. http://environment.yale.edu/climate-communication-OFF/files/Global-Warming-Six-Indias.pdf

2.9: "Facebook Newsroom, Company info". Accessed September 23rd, 2016. http://newsroom.fb.com/company-info/

2.10: "Current world population." Accessed June 6, 2016. http://www.worldometers.info/world-population/.

Chapter 3

3.1: Nisbet, M. C., and D. A. Scheufele. What's next for Science Communication? Promising Directions and Lingering Distractions. *American Journal of Botany* 96, no. 10 (2009): 1767-778. doi:10.3732/ajb.0900041.

3.2: Oreskes, Naomi, and Erik M. Conway. *Merchants of Doubt: How a Handful of Scientists Obscured the Truth on Issues from Tobacco Smoke to Global Warming.* New York: Bloomsbury Press, 2010.

3.3: "The facts of the alternative medicine industry." Accessed June 6, 2016. https://www.sciencebasedmedicine.orgthe-facts-of-the-alternative-medicine-industry/.

3.4: "Public and scientists' views on science and society." Accessed June 6, 2016. http://www.pewinternet.org/files/2015/01/PI_ScienceandSociety_Report_012915.pdf.

3.5: "A sneaky new rhetoric is holding back progress on climate change." Accessed June 6, 2016. http://qz.com/563283/a-sneaky-new-rhetoric-is-holding-back-progress-on-climate-change/.

3.6: Haidt, Jonathan. *The Righteous Mind: Why Good People Are Divided by Politics and Religion.* New York: Pantheon Books, 2012.

3.7: Samson, J., D. Berteaux, B. J. Mcgill, and M. M. Humphries. Geographic Disparities and Moral Hazards in the Predicted Impacts of Climate Change on Human Populations. *Global Ecology and Biogeography* 20, no. 4 (2011): 532-44. doi:10.1111/j.1466-8238.2010.00632.x.

3.8: Nyhan, B., J. Reifler, S. Richey, and G. L. Freed. Effective Messages in Vaccine Promotion: A Randomized Trial. *Pediatrics* 133, no. 4 (2014). doi:10.1542/peds.2013-2365.

3.9: Hart, P. S., and E. C. Nisbet. Boomerang Effects in Science Communication: How Motivated Reasoning and Identity Cues Amplify Opinion Polarization About Climate Mitigation Policies. *Communication Research* 39, no. 6 (2011): 701-23. doi:10.1177/0093650211416646.

3.10: Best, Joel. *Social Problems.* New York: W.W. Norton & Company, 2013.

3.11: David J. Hardisty, Eric J. Johnson, and Elke U. Weber, "A Dirty Word or a Dirty World?" Psychological Science 21 (2010): 86-92. doi: 10.1177/0956797609355572.

3.12: "Seeking to save the planet, with a thesaurus." Accessed June 6, 2016. http://www.nytimes.com/2009/05/02/us/politics/02enviro.html.

3.13: Schank, Roger. *Teaching Minds: How cognitive science can save our schools.* New York, NY. Teachers College Press, 2011

3.14: Miller, Anne, and Anne Miller. *The Tall Lady with the Iceberg: The Power of Metaphors to Sell, Persuade & Explain Anything to Anyone.* New York: Chiron Associates, 2012.

3.15: MacEachern, Diane. "LEDs Are An Important Climate Change Solution." *The Huffington Post.* Accessed June 07, 2016. http://www.huffingtonpost.com/diane-maceachern/energy-star-leds-are-an-i_b_5854384.html.

3.16: Heath, Chip, and Dan Heath. *Made to Stick: Why Some Ideas Survive and Others Die.* New York: Random House, 2007.

3.17: Terms That Have Different Meanings for Scientists and the Public. Google Docs. Accessed June 07, 2016. https://docs.google.com/spreadsheets/d/1eEBFGRO1UgA6OYoUF9XgRpwgXdliWIYB4J07qElEv-Y/edit?usp=sharing.

3.18: Andrew David Thaler, PhD. Andrew David Thaler, PhD. Accessed June 07, 2016. http://www.andrewdavidthaler.org/.

3.19: Somerville, Richard C. J., and Susan Joy Hassol. Communicating the Science of Climate Change. *Phys. Today Physics Today* 64, no. 10 (2011): 48. doi:10.1063/pt.3.1296.

3.20: "How to Get Your CFO to Love Your Sustainability Director." Accessed June 7, 2016. https://www.greenbiz.com/blog/2013/02/05/how-cfo-love-sustainability-director.

3.21: Making Sense of Climate Science Denial. EdX. 2015. Accessed June 07, 2016. https://www.edx.org/course/making-sense-climate-science-denial-uqx-denial101x-0.

3.22: Farmer, G. Thomas., and John Cook. *Climate Change Science: A Modern Synthesis.* Dordrecht: Springer, 2013.

3.23: Freedman, Andrew. California Gas Leak Is Emitting 4.5 Million Cars' worth of Pollution Every Day. Mashable. 2016. Accessed June 07, 2016. http://mashable.com/2016/01/13/california-gas-leak-super-emitter/#CgDCA7VDlEqO.

3.24: Wolf, Richard. Supreme Court Blocks Obama's Climate Change Plan. USA Today. 2016. Accessed June 07, 2016. http://www.usatoday.com/story/news/politics/2016/02/09/supreme-court-halts-obamas-emissions-rule/80085182/.

3.25: 10:10 "#itshappening" Climate Change Survey. ComRes. Accessed July 23, 2016. http://files.1010global.org/documents/1010-itshappening-polling-2014-10-22.pdf

3.26: Feldman, L., and P. S. Hart. "Using Political Efficacy Messages to Increase Climate Activism: The Mediating Role of Emotions." *Science Communication* 38, no. 1 (2015): 99-127. doi:10.1177/1075547015617941.

3.27: "Climate Cynicism." Citizens for Global Solutions. 2014. Accessed June 07, 2016. http://globalsolutions.org/blog/2014/11/Climate-Cynicism#.VruqmZMrKjS

Chapter 4

4.1: "Analytics Academy." Analytics Academy. Accessed June 07, 2016. https://analyticsacademy.withgoogle.com/explorer.

4.2: Amerland, Dave. *Google Semantic Search: Search Engine Optimization (SEO) Techniques That Get Your Company More Traffic, Increase Brand Impact and Amplify Your Online Presence.* Pearson Education, 2014.

4.3: "The Future of Digital Marketing and SEO: Art of SEO Book Event." Accessed June 07, 2016. http://www.slideshare.net/ericenge/the-future-of-digital-marketing-and-seo-art-of-seo-book-event.

4.4: "Building for the next Moment." Inside AdWords. Accessed June 07, 2016. http://adwords.blogspot.com/2015/05/building-for-next-moment.html.

4.5: D'Onfro, Jillian. "Facebook Hits an All-time High after Earnings Beat." Business Insider. 2015. Accessed June 07, 2016. http://www.businessinsider.com/facebook-q3-earnings-2015-11.

4.6: "Insights." Case Study: The Evolution of Digital Video Viewership. Accessed June 07, 2016. http://www.nielsen.com/us/en/insights/reports/2015/case-study-the-evolution-of-digital-video-viewership.html.

4.7: "Trends 2016: Rise of Live Streaming." Trends 2016: Rise of Live Streaming. Accessed June 07, 2016. http://www.globalwebindex.net/blog/trends-2016-rise-of-live-streaming.

4.8: "Wistia." Does Video Length Matter? Accessed June 07, 2016. http://wistia.com/blog/does-length-matter-it-does-for-video-2k12-edition.

4.9: "3Play Media." Americans with Disabilities Act (ADA) & Web Accessibility. Accessed June 07, 2016. http://www.3playmedia.com/2016/03/02/americans-with-disabilities-act-ada-and-web-accessibility-requirements-for-video/.

4.10: Ncarucar. "Steroids, Baseball, and Climate Change." YouTube. 2012. Accessed June 07, 2016. https://www.youtube.com/watch?v=MW3b8jSX7ec.

4.11: "Climate Communication." Climate Communication. Accessed June 07, 2016. https://www.climatecommunication.org/.

4.12: "Climate Change and More." YouTube. Accessed June 07, 2016. https://www.youtube.com/playlist?list=PLsmqeqKj7M-p_cC_I81favAvBu4U8-5-2.

4.13: Nisbet, M. C., and J. E. Kotcher. "A Two-Step Flow of Influence?: Opinion-Leader Campaigns on Climate Change." *Science Communication* 30, no. 3 (2009): 328-54. doi:10.1177/1075547008328797.

Chapter 5

5.1: "Coming Soon: Express Even More in 140 Characters | Twitter Blogs." Coming Soon: Express Even More in 140 Characters | Twitter Blogs. Accessed June 07, 2016. https://blog.twitter.com/express-even-more-in-140-characters.

5.2: "Twitter MAU Worldwide 2016 | Statistic." Statista. Accessed June 07, 2016. http://www.statista.com/statistics/282087/number-of-monthly-active-twitter-users/.

5.3: "Company | About." Company | About. Accessed June 07, 2016. https://about.twitter.com/company.

5.4: "Ten Jewish Teachings on Judaism and the Environment." —GreenFaith. Accessed June 07, 2016. http://www.greenfaith.org/religious-teachings jewish-statements-on-the-environment/ten-jewish-teachings-on-judaism-and-the-environment.

Chapter 6

6.1: Constine, Josh. "Facebook Speeds Past 1.55 Billion Users And Q3 Estimates With $4.5B Revenue." TechCrunch. 2015. Accessed June 07, 2016. http://techcrunch.com/2015/11/04/facebook-earnings-q3-2015/.

6.2: Nouraini, Sherry, PhD. "How to Ensure You See All of Your Friends' Updates on Facebook." Linkedin. Accessed June 7, 2016. https://www.linkedin.com/pulse/how-ensure-you-see-trixie-biffs-updates-facebook-sherry-nouraini-phd.

6.3: Oremus, Will. "Who Really Controls What You See in Your Facebook Feed—and Why They Keep Changing It." Slate Magazine. 2016. Accessed June 07, 2016. http://www.slate.com/articles/technology/cover_story/2016/01/how_facebook_s_news_feed_algorithm_works.html.

6.4: "Page Plugin—Social Plugins—Documentation—Facebook for Developers." Facebook Developers. Accessed June 07, 2016. https://developers.facebook.com/docs/plugins/page-plugin.

Chapter 7

7.1: "Pinterest Says It Has 100 Million Monthly Active Users." Marketing Land. 2015. Accessed June 07, 2016. http://marketingland.com/pinterest-says-it-has-100-million-monthly-active-users-143077.

7.2: Duggan, Maeve, Nicole B. Ellison, Cliff Lampe, Amanda Lenhart, and Mary Madden. "Demographics of Key Social Networking Platforms." Pew Research Center Internet Science Tech RSS. 2015. Accessed June 07, 2016. http://www.pewinternet.org/2015/01/09/demographics-of-key-social-networking-platforms-2/.

7.3: "Introducing the Save Button for the World." For Business. Accessed June 07, 2016. https://business.pinterest.com/en/blog/introducing-save-button-world.

7.4: "Pinterest Developers." Accessed June 07, 2016. https://developers.pinterest.com/docs/rich-pins/overview/.

Chapter 8

8.1: Instagram-business. "Coming Soon: New Instagram Business Tools." Instagram for Business. Accessed June 07, 2016. http://blog.business.instagram.com/post/145212269021/new-business-tools.

8.2: "Can You Post to Instagram from Desktop? 7 Options to Try." Louise Myers Visual Social Media. 2015. Accessed June 07, 2016. http://louisem.com/6138/post-to-instagram-from-desktop.

Chapter 9

9.1: "About Us | LinkedIn Newsroom." LinkedIn Newsroom. Accessed June 07, 2016. https://press.linkedin.com/about-linkedin.

9.2: Foote, Andy. "Maximum LinkedIn Character Counts for 2016." LinkedIn. Accessed June 07, 2016. https://www.linkedin.com/pulse/maximum-linkedin-character-counts-2016-andy-foote.

9.3: Nouraini, Sherry, PhD. "Social Media Strategy for Climate Change Communication." Communicating Climate Change With the Six Americas: Where to Begin. Accessed June 07, 2016. http://socialforclimate.blogspot.com/2015/08/communicating-climate-change-with-six.html.

9.4: Nouraini, Sherry, PhD. "Seven Points of Entry for Communicating Climate Change." LinkedIn Pulse. Accessed June 07, 2016. https://www.linkedin.com/pulse/seven-points-entry-communicating-climate-change-sherry-nouraini-phd

Chapter 10

10.1: Doran, George T. "There's a S. M. A. R. T. Way to Write Management's Goals and Objectives." Management Review 70, no. 11 (1981): 35-36.

10.2: Mathiesen, Karl. "Climate Change Warnings for Coral Reef May Have Come to Pass, Scientists Say." The Guardian. 2016. Accessed June 07, 2016. http://www.theguardian.com/environment/2016/mar/22/climate-change-warnings-coral-reef-great-barrier-reef-experts-projections-scientists.

10.3: "Welcome to the USGS Pacific Coral Reefs Web Site." United States Geological Survey Pacific Coral Reefs Web Site. Accessed June 07, 2016. http://coralreefs.wr.usgs.gov/.

10.4: Gorski, David. "Facebook's reporting algorithm abused by antivaccinationists to silence pro-science advocates." Science Based Medicine. Accessed June 22, 2016. https://www.sciencebasedmedicine.org/facebooks-reporting-algorithm-abused-by-antivaccinationists-to-silence-pro-science-advocates/

10.5: Ross, Lee, Kenneth Arrow, Robert Cialdini, Nadia Diamond-Smith, Joan Diamond, Jennifer Dunne, Marcus Feldman, Robert Horn, Donald Kennedy, Craig Murphy, Dennis Pirages, Kirk Smith, Richard York, and Paul Ehrlich. "The Climate Change Challenge and Barriers to the Exercise of Foresight Intelligence." *BioScience* 66, no. 5 (2016): 363-70. doi:10.1093/biosci/biw025.

Appendix

Facebook page types

- Local business or Place: For brick and mortar businesses
- Company, Organization, or Institution: For businesses with no physical location, or those with multiple locations
- Brand or Product: Businesses that sell products through multiple retailers or resellers
- Artist, Band, or Public Figure: For promoting a band or for personal branding
- Entertainment: Businesses in the entertainment industry
- Cause or Community: For grassroots efforts

Graphic design tools

- Easel.ly

  http://www.easel.ly/

  Easel.ly is an online tool for creating infographs. It provides pre-constructed templates that can be customized via drag and drop elements.

- Piktochart

  https://piktochart.com/

  Similar to Easel.ly, Piktochart offers a customizable infographics template. It also allows creation of presentations that can be displayed directly from the website.

◔ Canva

http://Canva.com

Canva not only offers infographics templates, it also offers free or affordable stock images and design elements that can be customized to image size requirements of mainstream social networking platforms.

Video creation software

◔ GoAnimate

https://goanimate.com/

GoAnimate is a powerful video cartoon creation tool. It allows production of video infographs, 2-D animated videos and whiteboard animations. It also allows uploading of the videos you create to other video hosting sites such as YouTube. It allows import of videos and sound files.

◔ Powtoon

https://www.powtoon.com

PowToon offers similar features as GoAnimate, except with a less powerful whiteboard animation. The styles used in PowToon are provided royalty-free.

Royalty-free Sources of stock images

◔ Unsplash.

https://unsplash.com/

Unsplash is a resource put together by Crew, an invite-only social network for freelance designers, developers, and studios. Images in this resource can be downloaded for free.

◔ Pixabay

https://pixabay.com

All images and videos on Pixabay are released free of copyrights and can be used royalty-free for commercial and noncommercial applications. There is no attribution required. However, if you are going to use media from this site, be sure to read the FAQ section of this website to learn about exceptions to these rules.

www.ingramcontent.com/pod-product-compliance
Lightning Source LLC
Chambersburg PA
CBHW042031220326
41598CB00073BA/7447